The Best ACT® Math Books Ever
Book 1: Algebra

Created by Brooke P. Hanson

The Best ACT® Math Books Ever
Book 1: Algebra

Created by Brooke P. Hanson

Additional Contributions by SupertutorTV

Layout by Jennifer Wang

Cover design by Skyler Thiot

First Edition
Published by Supertutor Media, Inc.

ISBN-13: 978-1732232006
ISBN-10: 1732232008

Access Your Free Online Resources*

FREE ACT calculator programs for download, our pdf format ACT math formula "cheat sheet," list of errata, and more. Find them at:

supertutortv.com/BookOwners

*Creation of an account may be required.

Need Private Tutoring?

Brooke offers tutoring to a limited number of one-on-one private students, both online via Skype and in her Los Angeles office for the ACT, SAT, ISEE/SSAT and writing. She also offers college admissions consulting and essay coaching. For more information, visit:

supertutortv.com/tutoring-information

Join our Mailing List!

Join our mailing list for FREE to keep up to date on our YouTube video releases, new products, tips, reminders and exclusive offers! Sign up at:

supertutortv.com/subscribe

Follow us!

YouTube.com/SupertutorTV
Instagram @SupertutorTV
Facebook.com/SupertutorTV
Twitter @SupertutorTV

TABLE OF CONTENTS

FOREWORD

As a private tutor, I always try to ensure my students have total mastery of the content necessary to crush any exam. On the ACT math section, that means exposure to every piece of math knowledge that might be on the test. But in my 15+ years of tutoring, I could never find all that math knowledge in one place. I tried book after book, but while one had nice lessons, it offered a measly 3 practice problems per section. While another seemed to have drills, it left out major sections covered by the exam. As a result, I was constantly assigning my students a mishmash of pdf worksheets, chapters from their sister's Algebra book, and portions of multiple ACT prep books. True, I had some students score perfectly on the ACT math exam. But I always wanted to have a single source for all the formulas, skills, and drills my students needed to master the math section once and for all. Thus this book series was born.

These books are the culmination of 3+ years of research and work. Every problem in the series is inspired by an authentic ACT problem. With the help of my team, I've analyzed dozens of released official ACT exams, sorting through thousands of problems, to ensure we cover what you need to know. Each chapter is organized by topic area. I list the skills you need to know, introduce each topic, and highlight any key terms or formulas you'll need. Then, I show you at least one step-by-step example of how to approach each type of problem. Each lesson is followed by a problem set with full explanatory answers, typically with ~20 or so problems, enough that you still have more to drill with if you need to peek at the answers until you get the hang of things. If you're wanting more explanation of these topics, I walk through every lesson in this book in our online ACT video based course, The Best ACT Prep Course Ever, at Supertutortv.com. When ideas overlap, I notate at the beginning of each chapter what related chapter you may also want to consider reviewing.

To accommodate the 1000+ practice problems in the series, I've split the material into two volumes. This first volume is everything Algebra. That's not to say this book is "easier" than the 2nd. Many of the most challenging topics on the ACT are in this book. If you're looking for help in Geometry, Arithmetic, Probability or Trig, check out the 2nd volume in this series: The Best ACT Math Books Ever, Book 2: Numbers, Stats, Trig & Geometry.

It's taken a village to finish these books. From cutting out and sorting questions by type to proofreading questions, spooling answer choices, drafting explanations, and drawing geometry figures, over a dozen people contributed to the problem sets in this book, sharing their problem solving know how. Thanks so much to all of you!

Contributors:
Estefania Lahera, Elijah Spiegel, Jennifer Cook, Heidi Chiu, Gabriel Schneider, Lianne Huang, Matthew Morawiec, Nathan Wenger, Megan Roudebush, Daniel Lee, Thierno Diallo, Liane Beatriz Capiral, Leslie Kim, Celestine Seo, Arjun Verma.

We've proofread, but there still may be errors. If you find any, email us at info@supertutortv.com. If any explanations confuse you, don't hesitate to reach out!

HOW TO USE THIS BOOK

Method 1: Do EVERYTHING

If you have 6+ months to prep, at least a couple of hours per week, and want to get the most insanely high score possible, you can do this book and the other book in this series start to finish. Feel free to skip over problems that are crazy easy for you or explanations you don't need as you go. On easier chapters, you can skip right to the problem set. On the other hand, if you're not aiming for a top score, feel free to skip the toughest problems so you can get an all around baseline on the easier ones.

Method 2: Target & Drill

For everyone else, I recommend taking an official ACT practice test and grading it. You can download one for free from the link our resources page at **SupertutorTV.com/resources** if you need one. Then figure out what kinds of questions you missed, identify the corresponding chapters, and "assign" these chapters to yourself. If you're an online course subscriber, I typically will indicate which problem type each question is (i.e. suggest what book chapters to review per the question you missed) in my corresponding explanation video. We have explanations for about 30 math questions from the 2017-2018 Official ACT practice test available free on our YouTube Channel as well: **youtube.com/SupertutorTV**. Go to our page and scroll down to look for the playlist.

Armed with your study list, read or skim each relevant chapter before the problem set, depending on your level of comfort with the particular problem type. (If you are a subscriber to our online course, you can also follow along with our lesson video). Then attempt the problem set. If you like, you can first start with the odds on the first pass, grade and review with the explanations, and then return again to the evens to solidify your learning and see if you can get the rest correct without peeking at any more explanations. Once you're done, take another practice test and repeat!

If you're having trouble finishing on time, focus on all Algebra Core problems, working to overcome your "rusty" habits and timing yourself to ensure you don't take more than about 30-45 seconds for any easy questions. The chapters are not entirely in order of difficulty, though they're approximately in sequential learning order: some chapters build on previously presented skills; others are grouped according to similar topics.

If you miss a particular type of question more than once, start with these. I also recommend starting with problems that correspond to lower numbered test questions you've missed first and leaving the hard stuff (questions 50-60) for last. Some explanations may be a challenge for you, but don't fret. Some of my advice is primarily geared for those seeking perfection or close to it! Also remember, you don't have to do EVERY page in this book; prep smart and know yourself; do what you need and not what you don't.

PART ONE: ALGEBRA CORE

CHAPTER

1 BASIC ALGEBRA

SKILLS TO KNOW

- How to solve single variable equations
- How to solve two-variable equations with given values
- How to recognize the pattern "EQUATION–EXPRESSION" and plug in to solve
- How to simplify expressions & equations (PEMDAS, The distributive property, etc.)

NOTE: In this chapter, I assume you know Algebra basics already: the **distributive property,** how to **plug in for variables**, how to adhere to **the order of operations** and how to **solve single and two variable equations.** Below, I'll offer a few quick tips and walk you through examples of basic ACT style algebra problems to familiarize you with the way questions are worded. Need more help? Let us know and we'll update our book for future users: info@supertutortv.com.

SOLVING SINGLE VARIABLE EQUATIONS

Oftentimes on the ACT®, you'll need to know how to solve a basic algebraic equation. Sometimes you'll confront "Single Variable Equations" (i.e. $3x = 45 - x$ has a single variable, x). At other times, you'll see "Two Variable Equations" (i.e. $2x + 3y = 45$ has two variables, x and y), and be given the exact value for one of those variables. Either way, the path to the solution is similar: your job is to isolate the variable and solve for it by combining like terms and following the order of operations. Most of the time when students miss these questions, they're making careless errors or they're thrown off by awkward wording or an unfamiliar setup. When in doubt, **substitute** in any variables, **simplify** anything you have and reassess the situation! These two "S's" can get you through many seemingly challenging problems that are much easier than they look!

When $3x + 5y = 45$ and $y = 6$ what is the value of x?

To solve, simply plug in the value of y to get a single variable equation:

$$3x + 5(6) = 45$$

$$3x + 30 = 45$$

$$3x = 15$$

$$x = 5$$

SOLVING TWO VARIABLE EQUATIONS USING GIVEN VALUES

When $a = 6$ is the solution to the equation $3b - 1 = a + 10b$, what must b equal?

This problem isn't hard, it's just worded awkwardly. **WHEN IN DOUBT, plug in what you have:**

$$a=6$$
$$3b-1=a+10b$$ Substitute
$$3b-1=6+10b$$
$$-1=6+7b$$ Combine like terms
$$-7=7b$$
$$-1=b$$

MUST KNOW PATTERN: EQUATION-EXPRESSION

If $5+2p=21$, then $\frac{4}{3}p+\sqrt[3]{p}=?$

If you look at the $\frac{4}{3}p+\sqrt[3]{p}=?$ part of the problem, you may freak out. But don't! Find p first, then "plug in" the value of p you now have into the expression, $\frac{4}{3}p+\sqrt[3]{p}$.

I call this setup **EQUATION—EXPRESSION**. Here's how it works: You're given an equation. Then the test asks to evaluate some "expression" (some group of variables, monomials, or polynomials without an equals sign or with a question mark to the side of an equals sign) that is seemingly impossible to "solve" because it's simply not an equation. For these problems, always **simplify**—that first magical "S"—and then **substitute.** (We'll go deeper into Equation–Expression in Chapter 4 in this book, Manipulations.)

$$5+2p=21$$
$$2p=16$$
$$p=8$$

Great! Now we substitute 8 into the expression we're trying to solve:

$$\frac{4}{3}p+\sqrt[3]{p}$$ Substitute
$$\frac{4}{3}(8)+\sqrt[3]{8}$$
$$\frac{32}{3}+2$$

At this point, check the answers. Are they improper fractions? Mixed? Decimals? Let's say the answers are improper.

$$\frac{32}{3}+2\left(\frac{3}{3}\right)$$
$$\frac{32}{3}+\frac{6}{3}$$
$$\frac{38}{3}$$

If this problem still makes you panic, you can **use your calculator** if you can't remember cube roots or if you want to add fractions more easily. If you have a TI-83 or TI-84, you can hit the MATH key and then press "1" for "▷ FRAC" and then ENTER to turn any decimal solution into fraction form.

Remember the following tips to avoid common mistakes:

1. **Always distribute across items added or subtracted in parentheses**—don't distribute across multiplied or divided items:

$$\text{OKAY: } 3(x+7)=3x+21$$

$$\text{NOT OKAY: } 3\left(7\left(\frac{x}{y}\right)\right)\neq 21\left(\frac{3x}{y}\right)$$

Notice how the 2nd example "distributes" the 3 to the fraction being multiplied? Not cool.

2. **Keep track of your negative signs!** Always distribute them across added and subtracted elements when there is a negative in front of the parentheses!

$$\text{YES: } 5-(x+2)=5-x-2$$

Notice how the negative sign carries over the "$(x+2)$" to become "$-x-2$"?

Oftentimes students know this rule but somehow forget to apply it. If that's you, get in the habit of double checking yourself as you crunch numbers in algebra problems.

3. **A negative sign is not the same as a subtraction sign.** Sometimes I'll have students (incorrectly) write things like:

$$3x\cdot-7=5$$
$$3x=12$$
$$x=4$$

Oftentimes, they know that the -7 was being multiplied at some point, but because of their sloppy work—or because they use a tiny dot instead of parentheses—they end up off track and start thinking that they were subtracting the 7 instead of multiplying something by it. If you do this—be extra cautious! Put your elements in parentheses and write neatly.

4. **Two negatives make a positive. Whenever you have two negative signs in a row, ADD.** For example: $4--x=4+x$ and $5--6=11$.

5. **Write neatly and organized enough to keep track of your work—and above all WRITE!** So often students miss basic algebra questions because they're not writing down much work. You don't need to do all this in your head. When you do so you're far more likely to make careless errors. Thinking is often not much faster than writing. Use the pencil! It is your FRIEND!

6. **Remember the Order of Operations: PEMDAS (AKA GEMDAS).** Always perform mathematical calculations in the following order when simplifying, rather than simply working left to right: **"P" for Parentheses (or "G" for Grouping symbols), "E" for Exponents, "M" for Multiplication, "D" for Division, "A" for Addition, and "S" for Subtraction.** Remember multiplication and division can be done in either order (i.e. we can divide then multiply) and addition/subtraction also can be done in either order (i.e. we can subtract first then add). When the operations within parentheses are not easily simplified, we can distribute to remove parentheses, or we can leave what is in parentheses intact as we solve other portions.

1. Which of the following is a simplified form of the expression $-9(3-4y)+7+5y$?

A. $-y+10$
B. $41y-20$
C. $41y-34$
D. $-20y+41$
E. $-20y+10$

2. If $x=9$, $y=\frac{3}{4}$, and $w=-10$, what is the value of $\frac{2x^2y+w}{12yw-5x}$?

A. $-\frac{202}{270}$
B. $\frac{223}{270}$
C. $\frac{111.5}{270}$
D. 82
E. $-\frac{223}{270}$

3. If $x=20$, then which of the following is equal to 6065?

A. $15x+5$
B. $300x+15$
C. $15x^2+3x+5$
D. $15x^3+3x+5$
E. $15x^4+3x+5$

4. What is the value of $7\times4^{2y+x}$ when $x=-3$ and $y=3$?

A. .1093
B. 84
C. 448
D. 1372
E. 21,952

5. If $k-4=q$ and $k-7=p$, what is the value of $p-2q$?

A. 3
B. −3
C. $1-k$
D. $k-1$
E. $3k-11$

6. What value of x makes $\frac{2}{3}(x-3)+2x=21$ true?

A. $8\frac{3}{8}$
B. $8\frac{5}{8}$
C. $10\frac{2}{3}$
D. 18
E. 69

7. If $6+7x=27$, then $2x=$?

A. 6
B. 3
C. 12
D. 9
E. 15

8. What is the solution to the equation $4(3x-3)=4x+3$?

A. 0
B. −15
C. $\frac{15}{8}$
D. 7
E. 8

9. If $5x-7=3x+8$, then $x=$?

A. 4
B. $\frac{7}{2}$
C. 7
D. $-\frac{15}{2}$
E. $\frac{15}{2}$

10. When $4x+8y=24$ and $y=5$, what's the value of x?

A. −16
B. −4
C. −1
D. $\frac{1}{4}$
E. 8

11. If $7+3m=30$, then $\frac{9}{2}m=?$

A. 1.70
B. 7.67
C. 13.5
D. 34.5
E. 69

12. When $b=-2$ is a solution to the equation $2b-7=ab+10$, what must a equal?

A. $\frac{21}{2}$
B. 2
C. $\frac{1}{2}$
D. -2
E. -10

13. If $4x-5(x+2)=3$, what is the value of x^2-2x?

A. -221
B. -195
C. -143
D. 143
E. 195

14. What value of z will satisfy the equation $0.2(z-1230)=-z$?

A. 102
B. 205
C. 246
D. 308
E. 1025

15. What is the solution to the equation below?
$5(w+12)-7(2-3w)=9(w+4)-13$

A. $-\frac{9}{2}$
B. $-\frac{23}{17}$
C. $\frac{9}{12}$
D. $\frac{23}{17}$
E. $\frac{9}{2}$

16. If $\frac{x}{6}-\frac{x}{9}=\frac{2}{3}$, then $x=?$

A. $\frac{1}{15}$
B. $\frac{1}{2}$
C. 3
D. 6
E. 12

17. When $\frac{1}{2}x+\frac{1}{5}x=2$, what is the value of x?

A. $\frac{1}{7}$
B. $\frac{20}{7}$
C. $\frac{10}{7}$
D. 7
E. 20

18. How many ordered pairs (x,y) of real numbers satisfy the equation $3x+7y=63$?

A. 0
B. 1
C. 2
D. 3
E. Infinitely many

19. Which of the following is an equivalent form of $x-x-x+x(x+x+x)$?

A. $7x$
B. $2x^2$
C. $3x$
D. x^2+x
E. $3x^2-x$

ANSWER KEY

1. B **2. E** **3. C** **4. C** **5. C** **6. B** **7. A** **8. C** **9. E** **10. B** **11. D** **12. A** **13. E** **14. B**
15. B **16. E** **17. B** **18. E** **19. E**

ANSWER EXPLANATIONS

1. **B.** Distributing the -9, we get $-9(3)+(-9)(-4y)+7+5y=-27+36y+7+5y=-20+41y=41y-20$.

2. **E.** Plugging in all the values for the correct variables, we get:
$$\frac{2x^2y+w}{12yw-5x}=\frac{2(9)^2\left(\frac{3}{4}\right)+(-10)}{12\left(\frac{3}{4}\right)(-10)-5(9)}=\frac{2(81)\left(\frac{3}{4}\right)-10}{-12\left(\frac{3}{4}\right)(10)-45}=\frac{\left(\frac{243}{2}-\frac{20}{2}\right)}{-9(10)-45}=\frac{\frac{223}{2}}{-135}=\frac{223}{-270}.$$

3. **C.** We must plug in $x=20$ into each answer option until we get an expression that is equal to 6065. Plugging in $x=20$ for the equation in answer choice (A) gives us:
$$15(20)^2+3(20)+5=15(400)+3(20)+5=6000+60+5=6065.$$

4. **C.** Plugging in $x=-3$ and $y=3$, we get $7\cdot 4^{2(3)+(-3)}=7\cdot 4^{6-3}=7\cdot 4^3=7\cdot 64=448$.

5. **C.** Plugging in $k-4=q$ and $k-7=p$ into the expression $p-2q$, we get $(k-7)-2(k-4)$. Distributing the -2 (don't forget the negative!) gives us $k-7-2k+8$. Combining like terms, we get $-k+1$ which is $1-k$.

6. **B.** Distributing the $\frac{2}{3}$, we get $\frac{2}{3}x-\frac{2}{3}(3)+2x=21$. This simplifies into $\frac{2}{3}x+2x-2=21$. Combining like terms, we get $\frac{8}{3}x=23$. Now, to find the value of x, we multiply both sides by the reciprocal of $\frac{8}{3}$, which is $\frac{3}{8}$. So, $\left(\frac{3}{8}\right)\left(\frac{8}{3}\right)x=\left(\frac{3}{8}\right)23$. This simplifies into $x=\frac{69}{8}=8\frac{5}{8}$.

7. **A.** First solve for x, then worry about $2x$ later. Subtracting 6 on both sides of the equation, we get $7x=21$. Dividing both sides by 7, we find the value of $x=3$. Now, to find the value of $2x$, we multiply by 2 to get $2x=6$.

8. **C.** Distributing the 4 on the left side, we get $12x-12=4x+3$. Adding 12 on both sides, we get $12x=4x+15$. Now, subtracting $4x$ on both sides gives us $8x=15$. Finally, dividing both sides by 8, we get $x=\frac{15}{8}$.

9. **E.** Adding 7 to both sides of the equation, we get $5x=3x+15$. Subtracting $3x$ from both sides gives us $2x=15$. Finally, dividing both sides by 2, we get $x=\frac{15}{2}$.

10. **B.** Substituting in $y=5$ into $4x+8y=24$, we get $4x+8(5)=24$. Simplifying this gives us $4x+40=24$. Subtracting 40 on both sides gives us $4x=-16$, and dividing both sides by 4 gives us $x=-4$.

11. **D.** Subtracting 7 on both sides of the equation gives us $3m=23$. Dividing both sides by 3 gives us $m=\frac{23}{3}$. Now, to find the value of $\frac{9}{2}m$, we multiply the value of m by $\frac{9}{2}$, which is $\frac{9}{2}\left(\frac{23}{3}\right)=\frac{207}{6}=\frac{69}{2}=34.5$.

12. **A.** Plugging in $b=-2$ into the equation $2b-7=ab+10$, we get $2(-2)-7=a(-2)+10$. This simplifies to $-4-7=-2a+10$; $-11=-2a+10$; $-21=-2a$; $a=\frac{-21}{-2}=\frac{21}{2}$.

13. **E.** Distributing out the -5 on the left side of the equation gives us $4x-5x-10=3$. Combining like terms, we get $-x=13$, so $x=-13$. Plugging in $x=-13$ to the expression x^2-2x, we get $(-13)^2-2(-13)=169+26=195$.

14. **B**. Dividing both sides by 0.2 we get $z-1230=-\frac{z}{0.2}$. Recognizing that $-\frac{1}{0.2}$ is equivalent to $-\frac{5}{1}$, (using a calculator if necessary) we can rewrite the equation as $z-1230=-5z$. Adding $5z$ on both sides gives us $6z-1230=0$. Adding 1230 on both sides gives us $6z=1230$. Dividing both sides by 6, we get $z=\frac{1230}{6}=205$.

15. **B.** Distributing out the constants in the equation, we get $5w+60-14+21w=9w+36-13$. Combining like terms, we get $5w+21w-9w=-60+14+36-13$ which simplifies to $17w=-23$. Dividing both sides by 17 gives us $w=-\frac{23}{17}$.

16. **E.** Rewriting $\frac{x}{6}-\frac{x}{9}$ with the common denominator $=18$, we get $\frac{x}{6}\left(\frac{3}{3}\right)-\frac{x}{9}\left(\frac{2}{2}\right)=\frac{3x}{18}-\frac{2x}{18}-\frac{3x-2x}{18}=\frac{x}{18}$. So now we have $\frac{x}{18}=\frac{2}{3}$. Cross multiplying this equation gives us $3x=36$. Dividing both sides by 3 gives us $x=12$. For more on fractions, see Book 2's chapter 4.

17. **B.** We find the least common denominator of the fractions, which is the least common multiple of 2 and 5, to be 10. So we convert our fractions into $\frac{5}{10}x$ and $\frac{2}{10}x$. (You can also use your calculator to add the fractions if you like). Adding them together, we get $\frac{7}{10}x=2$. Multiply by the reciprocal of the fraction to get $x=2\left(\frac{10}{7}\right)=\frac{20}{7}$. For more on fractions, see Book 2's chapter 4.

18. **E.** Since the real numbers are infinite, there can be an infinite number of ordered pairs that satisfy the equation. For any value of x, we can always find a value of y in the real numbers that makes the equation true. As further proof we know that, graphically, this is a line, which infinitely extends in both directions with no discontinuities.

19. **E.** Use PEMDAS (Always perform mathematical calculations in the following **order of operations** when simplifying, rather than simply working left to right: "P" for Parentheses, "E" for Exponents, "M" for Multiplication, "D" for Division, "A" for Addition, and "S" for Subtraction.). Simplify what is in the parentheses to get $x-x-x+x(3x)$; then, as there are no exponents, multiply to get $x-x-x+3x^2$. Finally, sum all like terms to get $3x^2-x$.

CHAPTER

SYSTEMS OF EQUATIONS

SKILLS TO KNOW

- Solve a system of equations
- Set up word problems involving systems of equations
- Distinguish between systems with zero, one, or infinitely many solutions
- Solve for a given variable in a system of equations with single, infinite, or no solutions

SOLVE A SYSTEM OF EQUATIONS

Many students taking the ACT® know how to solve a system of equations, but often reviewing these problems can help students build speed and accuracy.

Remember there are two main ways to solve systems of equations: **Substitution** and **Elimination**.

SPEED TIP! Which method you choose to do the problem doesn't matter in terms of accuracy (you won't get a question wrong for choosing one method over another)—but sometimes one method will be faster than another—**knowing which is best in certain situations can help you speed up and finish the test in time.**

Use substitution: When one variable has "no" coefficient, or in other words, when the coefficient is technically equal to one; for example, the y in this expression has "no" coefficient: $2x+y$.

Use elimination: When two coefficients in two equations in front of the same variable match or are multiples of each other, such as the threes in the equations $3x+2y=8$ and $3x+5y=10$.

Substitution

What value of y solves the following system of equations?

$$x+6y=32$$
$$5x+y=24+2x$$

To substitute, we start by isolating a variable that has "no" coefficient whenever possible. Here, the first equation has "no" coefficient in front of the x and the second has "no" coefficient in front of the y. Thus, to substitute, you could really substitute either. BUT you can also be smart about which one you solve for so that you don't have to substitute back in at the end of the problem (Note: for scorers below a 22, the following "Isolate to Eliminate" explanation below may be overkill; if this is confusing, focus on the steps on the next page).

One issue many students have on these problems is that they may solve for and put x instead of y (what you actually need) as the answer. What if you could make sure you solve for the variable you actually need EVERY TIME? True, you can always plug in at the end, but with a little strategy, you can engineer which variable you solve for. Likewise, when you have three or more variables in a system of equations and need to reduce those down to fewer variables, you must know how to "keep" some variables while "eliminating" the other(s). For these problems, knowing how to target and eliminate a specific variable is essential.

The phrase to keep in mind when solving systems of equations with substitution is: **"ISOLATE to ELIMINATE."** We always eliminate the variable that is isolated, because that is what we substitute for. When we substitute, that isolated variable disappears.

Let's get back to this question. Because it is specifically asking for y, we need to keep the y and eliminate the other variable, x. To isolate the x we simply act as if we are variable bounty hunters. If you want to get rid of an enemy in a video game, you push that enemy into the corner—you isolate it—then eliminate it! That's what we'll do to x. The easiest variables to isolate are those without a coefficient, so we'll use the second equation, in which x has "no" coefficient:

STEP 1: ISOLATE! (Equation 1)

$$x+6y=32$$
$$x=-6y+32$$

Now we substitute x into the other equation and the x values will disappear. But before we do, we're going to combine like terms so we only have to substitute once.

STEP 2: SIMPLIFY (Equation 2)

$$5x+y=24+2x$$
$$y=24+2x-5x$$
$$y=24-3x$$

STEP 3: SUBSTITUTE

$x=-6y+32$	
$y=24-3(-6y+32)$	Substitute into equation 2 from equation 1
$y=24+18y-96$	Distribute the -3
$y=18y-72$	Simplify, combining like terms
$-17y=-72$	Subtract $18y$ from both sides
$y=\frac{72}{17}$	Divide both sides by -17

TIP: **Always substitute in with PARENTHESES!** If you don't, you may forget to distribute.

i.e. $y=24-3(-6y+32)$

Don't forget these!

What is the value of n in the solution to the following system of equations?

$$5n-m=26$$
$$3n+2m=39$$

A. 13 **B.** 9 **C.** 7 **D.** 5 **E.** -7

Here, m in the first equation has "no" coefficient, and we want to keep n so let's isolate that m! Remember, **ISOLATE TO ELIMINATE**!

STEP 1: ISOLATE! (Equation 1)

$$5n-m=26$$
$$5n=26+m$$
$$5n-26=m$$
$$m=5n-26$$

STEP 2: SUBSTITUTE (m into Equation 2)

$$3n+2m=39$$
$$3n+2(5n-26)=39$$
$$3n+10n-52=39$$
$$13n=91$$
$$n=7$$

Answer: **C**.

Elimination

Elimination involves stacking two equations and then adding or subtracting straight down after you've lined up your like terms.

If the following system has a solution, what is the x-coordinate of the solution?

$$2x+2y=58$$
$$3x-2y=27$$

A. 12 **B.** 17 **C.** 31 **D.** 34 **E.** 85

Here we want x, so we want to eliminate y. We're in luck because the coefficients of y match: y and $-y$ are opposites of each other, and if stacked, will "zero" out.

We set up a giant addition problem to eliminate the y terms:

$$\begin{array}{r} 2x+2y=58 \\ +\ 3x-2y=27 \\ \hline 5x+0y=85 \\ x=17 \end{array}$$

As you can see, the y's disappear, and we're left with the answer: **B**.
Problems aren't always so convenient—when coefficients don't match, we can manipulate the situation to make them match:

Solve for y:

$$2x+3y=16$$
$$4x-5y=21$$

Here, no coefficients match, and we have no variables with a coefficient of one, but the 2 and the 4 are multiples of each other. The idea is to get two coefficients that match in number but have opposite signs, so they cancel. We can multiply the first equation by -2 to make the resultant $-4x$ cancel with the $4x$ in the second equation.

$$-2(2x+3y=16)$$
$$\downarrow$$
$$(-2)2x+(-2)3y=(-2)(16)$$
$$\downarrow$$
$$\begin{array}{r} -4x-6y= -32 \\ +\quad 4x-5y=21 \\ \hline 0x-11y=-11 \\ y=1 \end{array}$$

TIP: The principle of **stacking equations** at work when we use elimination can also be used to solve problems creatively. Sometimes adding given equations together or subtracting them, without even trying to eliminate the variable(s), **can help you get to a correct answer faster.** To try this technique, see questions 10 and 11 in the problem set!

WORD PROBLEMS INVOLVING SYSTEMS OF EQUATIONS

Sometimes students get caught up on word problems that require systems of equations. Make sure you engage your mind and think through logically what each value represents. If you get stuck in setting up a word problem, make up numbers to help you understand what is what. In other words, imagine you know what each variable stands for.

At Montesquieu's Bistro, you can get 5 pastries and 2 espressos for \$21 (before tax). The price of a pastry is p dollars. The price of a pastry is equal to the price of 2 espressos. Which of the following systems of equations, when solved, gives the price, n dollars, of an espresso and the price, p dollars, of a pastry at Montesquieu's Bistro?

A. $\begin{cases} 5p+2n=21 \\ n=2p \end{cases}$ **B.** $\begin{cases} 5p+2n=21 \\ p=2n \end{cases}$ **C.** $\begin{cases} 2p+5n=21 \\ p=2n \end{cases}$ **D.** $\begin{cases} 5+2pn=21 \\ 2p=n \end{cases}$ **E.** $\begin{cases} 5n+2p=21 \\ n=2p \end{cases}$

Price of pastry $= p$
Price of espresso $= n$

To set this up, think about what's happening. Let's say pastries (p) are $\$1.50$ each. If we bought 5, that would be $5\times(\$1.50)$. Substitute p back in for the $\$1.50$ and you have $5p$. Now we need to do the same for espressos (n)—again if we knew the price we would multiply and get $2n$ as the cost of two espressos. We need the sum of these two to equal the total cost, 21:

$$5p+2n=21$$

For the second equation, we simply use translation—taking English and turning it into math. The price of a pastry (p) is equal to ($=$) the price of two espressos ($2n$):

$$p=2n$$

These two equations determine that the answer is **B.**

Word Problem Strategy Recap

Strategy 1: Make up numbers to make the problem more understandable so you can figure out what goes where.
Strategy 2: Turn English words into math equivalents.

ZERO, ONE, OR INFINITE SOLUTIONS

You'll need to know how to figure out whether **systems of linear equations** have **one, zero, or infinite solutions.** To figure out which condition is met, I'll walk you through two ways to solve. Advanced students may be able to solve some of these problems more quickly based on pattern recognition and quick slope calculations, but we'll keep it simple here.

METHOD 1: Solve the System

- **Solve by elimination or substitution and evaluate your final answer.**
- **When you finish your problem you'll have one of three conditions:**

One solution	Zero solutions	Infinite solutions
If you get a single x or y value, you have one solution	If you get a statement that is never true, you have no solutions.	If you get two values that always equal each other, you have infinite solutions.
Example: $x=6$ or $y=0$	**Example:** $5=6$ or $7=0$	**Example:** $5=5$ or $y=y$

If the following system of equations has a solution, what is the x – coordinate of the solution?

$$3y=12-x$$

$$3y=x-6$$

A. –9 **B.** –3 **C.** 1 **D.** 9 **E.** No solution exists

Here we can short cut and substitute in for $3y$; both equations are equal to $3y$, so we can make the right sides of both of the equations equal to each other.

Group Substitution Tip

Remember you can substitute whole expressions— you don't have to just isolate the single variable, you can isolate anything identical! Whenever you see elements in two equations that match, think substitution! This works especially well when you must eliminate more than a single variable. Here that means we can subtitute in for $3y$.

$$x-6=12-x$$
$$2x=18$$
$$x=9$$

Double check to ensure you solved for the variable the question asks for!
Answer: **D.**

If you get a single x or y value, you have one solution. Note: if you have nonlinear equations in your system, it's a great idea to double check answers for extraneous solutions and plug x back in to make sure y exists too. With linear equations that's not necessary, but it can be a good way to double check your work.

How many solutions exist in the system of equations below?

$$\frac{y}{2} = 8x + \frac{3}{2} \qquad \frac{16}{3}x = \frac{y-3}{3}$$

A. Zero solutions
B. One real solution
C. One imaginary solution
D. Two real solutions
E. Infinitely many solutions

Put the equations in two columns and simplify.

Multiply everything by 2	$\frac{y}{2} = 8x + \frac{3}{2}$	$\frac{16}{3}x = \frac{y-3}{3}$	Multiply everything by 3
	$y = 16x + 3$	$16x = y - 3$	

Now we substitute the left equation into the right one:

$$16x = y - 3$$
$$16x = (16x + 3) - 3$$
$$16x = 16x$$
$$x = x$$

Here we have something that equals itself: $x = x$. That statement is always true, so no matter what x or y equal, the equations overlap—that is, they are the same equation. If you get two equal quantities (i.e. $0 = 0$ or $5 = 5$), you have **infinite solutions.**

Answer: **E.**

What is the x – coordinate of the solution to the problem set below?

$$\frac{y}{4} = 2x + 3 \qquad \frac{y}{2} = 4x - 1$$

A. –4 **B.** –2 **C.** 2 **D.** 4 **E.** No solution exists

First, multiply the left equation by 4 and the right equation by 2 so the equations are in y-intercept form.

$$y = 8x + 12 \qquad y = 8x - 2$$

Then, set the equations equal to each other to solve for x:

$$8x + 12 = 8x - 2$$
$$12 = -2$$

Here we get two answers that are never equal. If you get something not true (i.e. $5 = 0$, $7 = 2$) you have no solutions (they are parallel lines). Parallel lines will never intersect, so we have **no real solutions.**

Answer: **E.**

METHOD 2: Put EVERYTHING in SLOPE-INTERCEPT FORM

- **Get both equations into slope-intercept form and compare!**

We can quickly look at the slopes and y-intercepts and determine the relationship according to the chart below:

One solution	Zero solutions	Infinite solutions
Different slopes (intercepts don't matter)	Same slopes (parallel lines), different y-intercepts	Same slope, same intercept
Example: $y=2x+1$ and $y=3x+7$	**Example:** $y=3x+2$ and $y=3x+4$	**Example:** $y=3x+2$ and $y=3x+2$

How many solutions exist in the system?

$$y=3x+\frac{4}{3} \qquad 3x=\frac{3y-4}{3}$$

Because the first equation is already in slope-intercept form ($y=3x+\frac{4}{3}$), we'll put the second one in that form too:

$$3x=\frac{3y-4}{3}$$
$$9x=3y-4$$
$$3y=9x+4$$
$$y=3x+\frac{4}{3}$$

Because these are the exact same equation, this set has infinite solutions! Every x value in one equation gives the same y in the other. Because a line is an infinite number of ordered pairs, the solution set is infinite: there are infinite points of intersection because the lines overlap and continue to infinity in both directions.

How many solutions exist for the system of equations?

$$\frac{y}{2}=4x-2 \qquad 2x=\frac{3y-9}{12}$$

Again, we make two columns and manipulate each equation to slope-intercept form.

$$\frac{y}{2}=4x-2$$
$$y=8x-4$$

$$2x=\frac{3y-9}{12}$$
$$24x=3y-9$$
$$8x=y-3$$
$$8x+3=y$$
$$y=8x+3$$

Same slope (8) and **different y-intercepts** (-4 and 3) indicate **parallel lines (no solution)**.

Answer: 0.

How many solutions exist for the following set of equations:

$$y = 3x + 1$$
$$y = 2x + 1$$

As you can see, these have different slopes. Thus, there will be one real solution.

Answer: 1.

For what value of a, if any, would the following system of equations have an infinite number of solutions?

$$a(y-6) = x \qquad \frac{y}{a} = x + 2$$

For both equations, isolate the y value to put them into slope-intercept form.

$$y - 6 = \frac{x}{a} \qquad \frac{y}{a} = x + 2$$
$$y = \frac{x}{a} + 6 \qquad y = ax + 2a$$
$$y = \frac{1}{a}x + 6$$

Now it's time for a technique called "matchy matchy." Line up matching portions of the equations and set them equal to each other. Remember for infinite solutions **every term must be identical in slope-intercept form!**

$$y = ax + 2a$$
$$y = \frac{1}{a}x + 6$$

We don't care much about the y's—line up the x terms and you'll see that the slopes need to be the same:

$$a = \frac{1}{a}$$
$$a^2 = 1$$
$$a = 1$$

Line up the last term and you'll see that $2a$ must be equal to 6.

$$2a = 6$$
$$a = 3$$

So we got that $a=3$ and $a=1$. That's not possible! So thus there are NO values of a that would make these equations have infinite solutions.

Answer: **No such value exists.**

1. For what value of z would the following system of equations be true for all real integers?

$$3x-2y=14$$
$$-12x+8y=8z$$

A. −56
B. −32
C. −14
D. −8
E. −7

2. What value of y solves the following system of equations?

$$3x+y+4=50$$
$$x+3y=50$$

A. 12
B. 13
C. 11
D. 49
E. 18.4

3. What is the x-coordinate of the solution of the following system, if the system has a solution?

$$5x-14y=47$$
$$2x+7y=53$$

A. 0
B. $2\frac{5}{7}$
C. $11\frac{1}{9}$
D. 17
E. The system has no solution.

4. Candice, Jill, and Kivo raised money for their school's golf team through a bake sale. They sold over-stuffed brownies for \$4.50 each, and they sold gourmet cupcakes for \$6 each. After selling 125 baked goods, they collected \$633 total. How much of the total did the trio collect from selling the cupcakes?

A. \$351
B. \$282
C. \$150
D. \$78
E. \$47

5. Tyler spends \$11.50 at Very Berry Frozen Custard on 2 large custards and 4 brownies. The price of each brownie is one-fifth the price of one large custard. Which of the following systems of equations, when solved, gives the price, b dollars, of a brownie and the price, c dollars, of one, large custard at Very Berry Frozen Custard?

A. $\begin{cases} 4c+2b=11.50 \\ b=\frac{1}{5}c \end{cases}$

B. $\begin{cases} 4c+b=11.50 \\ 4b=2c \end{cases}$

C. $\begin{cases} 4b+c=11.50 \\ 2b=4c \end{cases}$

D. $\begin{cases} 2c+4b=11.50 \\ b=\frac{1}{5}c \end{cases}$

E. $\begin{cases} 4c+b=11.50 \\ b=5c \end{cases}$

6. For what value of a would the following system of equations have an infinite number of solutions?

$$3x-7y=14$$
$$28y-12x=7a$$

A. −8
B. −2
C. −56
D. −7
E. −64

7. The solution to $ax=y$ is $x=-5$, and the solution to $ax+6=y$ is $x=3$. What is the value of a?

A. $-\frac{8}{6}$
B. $-\frac{6}{7}$
C. $-\frac{7}{9}$
D. $-\frac{3}{4}$
E. $\frac{7}{9}$

8. Given that $4x+3y=11$ and $3x+2y=13$, what is the value of $x-y$?

A. 5
B. −2
C. 36
D. 6
E. −36

9. If $x+y=5$, and $y-x=-7$, then $x^3+y^3=?$

A. 215
B. 217
C. −215
D. −217
E. 216

10. What is the value of c in the system of equations below?

$$3c-5d=a$$
$$2c+4d=-b$$

A. $a-b+d$
B. $a+b+d$
C. $-\left(\frac{a-b+d}{5}\right)$
D. $\frac{a+b+d}{3}$
E. $\frac{a-b+d}{5}$

11. Let $x+4y=12$ and $4x+2y=2.5$. What is the value of $5x+6y$?

A. 14.5
B. 9.5
C. 2
D. −9.5
E. −14.5

12. The solution of the system of equations below is the set of all (x,y) such that $3x+2y=12$. What is the value of w?

$$21x+14y=84$$
$$15x-wy=-6w$$

A. −10
B. −2
C. 3
D. 4
E. 5

13. Emily has printer paper and lined paper for her classroom. The reams of printer paper have 50 sheets per ream and cost \$10. The reams of lined paper have 75 sheets and cost \$12. Emily will order a total 45 reams of paper and her total cost is \$490. What system of equations gives the correct relationship between the p reams of printer paper and l reams of lined paper?

A. $\begin{cases} l+p=45 \\ 10l+12p=490 \end{cases}$

B. $\begin{cases} l-p=45 \\ 12l+10p=490 \end{cases}$

C. $\begin{cases} l+p=45 \\ 12l-10p=490 \end{cases}$

D. $\begin{cases} l-p=45 \\ 10l-12p=490 \end{cases}$

E. $\begin{cases} l+p=45 \\ 12l+10p=490 \end{cases}$

14. On opening, a high school play set a record by selling 630 tickets. They collected \$4350. If child tickets sold for \$5 and adult tickets sold for \$8, what is the difference between the number of adult and child tickets sold?

A. 230
B. 400
C. 200
D. 150
E. 170

ANSWER KEY

1. E **2. B** **3. D** **4. B** **5. D** **6. A** **7. D** **8. C** **9. A** **10. E** **11. A** **12. A** **13. E** **14. E**

ANSWER EXPLANATIONS

1. **E.** When you need to eliminate two variables out of three, try using group substitution or elimination. Multiplying the first equation $3x-2y=14$ by 4, we get $12x-8y=56$. Adding this equation to the second equation, we can cancel out the x and y variables.

$$\begin{array}{r} 12x-8y=56 \\ +(-12x+8y=8z) \\ \hline 0=56+8z \end{array}$$

Subtracting 56 on both sides and then dividing both sides by 8, we get $-56=8z$ and $-7=z$.

2. **B.** We can solve this problem using elimination. We want to find the value of y, so we wish to cancel out the x variable. Subtracting 4 from both sides of the first equation, we get $3x+y=46$. Multiplying the second equation by -3, we get $-3x-9y=-150$. Adding this to the first equation, we get:

$$\begin{array}{r} 3x+y=46 \\ +(-3x-9y=-150) \\ \hline -8y=-104 \end{array}$$

Dividing both sides of the equation by -8, we get $y=13$. You can also solve with substitution, isolating x.

3. **D.** We can solve this problem using elimination. We wish to find the value of x so we want to cancel out the y value. Multiplying the second equation by 2, we get $4x+14y=106$. Adding this to the first equation, we get:

$$\begin{array}{r} 5x-14y=47 \\ +4x+14y=106 \\ \hline 9x=153 \end{array}$$

Dividing each side of this equation by 9, we get $x=17$.

4. **B.** Let b be the number of brownies sold and c the number of cupcakes sold. The money made selling brownies can be expressed as $4.5b$ and the money made selling cupcakes can be expressed as $6c$. They sold a total of 125 baked goods, so $b+c=125$. They made a total of 633 dollars, so $4.5b+6c=633$. We now take the first equation and write b in terms of c. So, $b+c=125$ yields $b=125-c$. Now, we can plug in $b=125-c$ to the second equation $4.5b+6c=633$, giving us $4.5(125-c)+6c=633$. Distributing out 4.5, we get $562.5-4.5c+6c=633$. Combining like terms, we get $1.5c=70.5 \rightarrow c=47$. The total amount that they made from selling cupcakes alone is then $6c=6(47)=282$.

5. **D.** Let $c=$ the price of a custard and $b=$ the price of a brownie. The amount of money Tyler spends on custards can be calculated as c multiplied by the number of custards bought and likewise for brownies. We know that Tyler spent a total of \$11.50 on 2 custards and 4 brownies, so this information can be written as the equation $2c+4b=11.50$. We also know the price of each brownie is one-fifth the price of one custard. This means $b=\frac{1}{5}c$. Answer choice (D) matches our equations. If you have trouble, make up numbers to help you or review the lesson!

6. **A.** If a system of two equations has an infinite number of solutions, then the equations must be equal to each other. So, we wish to find the value of a such that the two equations are equal. We can use elimination to "solve the system." Multiplying the first equation by -4, we get $-12x+28y=-56$. At this point we don't actually need to use the elimination though as we see that the left side of this equation is already equal to the left side of the second equation. Now, we can play "matchy matchy" to set the right sides of the equations equal to each other. To find the value of a we write $-56=7a$ or $a=-8$. Alternatively, we could put both equations in slope intercept form and match up the corresponding pieces.

7. **D.** We can solve this problem using substitution and elimination. Plugging in $x=-5$ to $ax=y$, we get $-5a=y$. Plugging in $x=3$ to $ax+6=y$ is $3a+6=y$. Now, we have two new equations we want to use to solve for the value of a. Subtracting the second equation by the first, we get:

$$\begin{array}{r} -5a=y \\ -(3a+6=y) \\ \hline -8a-6=0 \end{array}$$

Adding 6 to both sides of the equation, we get $-8a=6$. Now, dividing both sides by -8, we get $a=-\frac{6}{8}=-\frac{3}{4}$.

8. **C**. We might first see if we can cleverly get $x-y$ by using elimination and stacking the equations:

$$\begin{array}{r} 4x+3y=11 \\ -(3x+2y=13) \\ \hline x+y=-2 \end{array}$$

Hmm, no such luck. We got $x+y=-2$ but that's not $x-y$. In any case, we can now more easily substitute to then solve for x and y. Remember the pattern from Chapter 1: Equation-Expression. First, we will solve for x and y using elimination or substitution, and then plug into the expression we need, $x-y$. Back to the result above, subtract y on both sides to get $x=-y-2$. Substituting this in for the value of x in $4x+3y=11$, we get $4(-y-2)+3y=11$. Distributing out the 4 and then simplifying this, we get $-4y-8+3y=11 \rightarrow -y=19$. So $y=-19$. Now we can solve for x by taking the equation $x+y=-2$ and plugging in -19 for y to get x-19=-2. After adding 19 to both sides we get x=17. Now we can calculate $x-y$ by plugging in our x and y values: 17-(-19)=36.

9. **A.** Remember the pattern from Chapter 1: Equation-Expression. First, we'll solve for x and y using elimination or substitution, and then plug into the expression we need. Let's use elimination. Adding two given equations we get:

$$\begin{array}{r} y+x=5 \\ +(y-x=-7) \\ \hline 2y=-2 \end{array}$$

Dividing both sides of the result by 2, we get $y=-1$. Now, we can plug $y=-1$ into the first equation to get $-1+x=5$. Adding 1 on both sides of this equation gives us $x=6$. Now we have the values of y and x, we can plug these in $x^3+y^3=(6)^3+(-1)^3=216-1=215$.

10. **E.** Look at the answer choices. All contain a, b and d. Thus we won't be able to simply isolate c in either single equation. We must combine them in some way. With trial and error we find that adding the equations together gives what we need:

$$\begin{array}{r} 3c-5d=a \\ +(2c+4d=-b) \\ \hline 5c-d=a-b \end{array}$$

To find the value of c, we add d to both sides of the equation and divide both sides by 5. This gives $c=\frac{a-b+d}{5}$.

11. **A.** We can solve quickly by seeing that we need 5x, while our coefficients are 1x and 4x, and we need 6y while our coefficients are 4y and 2y. Rather than solve for x and y, we can simply add our equations to get the expression we need.

$$\begin{array}{r} 4x+2y=2.5 \\ +(x+4y=12) \\ \hline 5x+6y=14.5 \end{array}$$

Thus our answer is A. 14.5. If that is too hard for you to see, you can use elimination, multiplying the second equation by 2: $2(4x+2y=2.5) \rightarrow 8x+4y=5$. Now, we subtract the first equation by this to get:

$$\begin{array}{r} 8x+4y=5 \\ -(x+4y=12) \\ \hline 7x=-7 \end{array}$$

Dividing $7x=-7$ by 7 gives us $x=-1$. Plugging $x=-1$ into the first equation: $(-1)+4y=12 \rightarrow 4y=13 \rightarrow y=\frac{13}{4}$.

To find $5x+6y$, we plug in $x=-1$ and $y=\frac{13}{4}$ to get $5(-1)+6\left(\frac{13}{4}\right)=-5+\frac{39}{2}=\frac{29}{2}=14.5$.

12. **A.** To quickly solve, we can make up any valid combination of (x,y) for $3x+2y=12$. Plugging in $y=0$ gives us $x=4$ and we can now plug these values into the second original equation $15x-wy=-6w$, giving us $60=-6w$. Dividing both sides by -10 leaves $w=-10$. Alternatively, recognize that to have a line as a solution set, the two equations must be identical; we could put all in slope intercept form and match up the pieces OR we could see the pattern of the coefficients as multiples of each other.

13. **E.** Let $p=$ reams of printer paper and $l=$ reams of lined paper Emily gets for her classroom. We are given that Emily gets a total of reams, so $p+l=45$. Then, the amount of money spent on printer paper is calculated as $10p$ since each ream of printer paper is $\$10$. Likewise, the amount of money spent on lined paper is calculated as $12l$ since each ream of lined paper is $\$12$. The total cost $\$490$ can then be represented as $10p+12l=490$. So, the two equations are $l+p=45$ and $10p+12l=490$.

14. **E.** Let $c=$ the number of child tickets sold and $a=$ the number of adult tickets sold. Then, the money made by selling child tickets is $5c$ and the money made from adult tickets is $8a$. The total amount of money, which is given, allows for this equation: $5c+8a=4350$. Our second equation, $c+a=630$, calculates the total number of tickets. Subtracting a on both sides of that equation, we get $c=630-a$. Substituting this value in for c in the first equation $5c+8a=4350$, we get $5(630-a)+8a=4350$. Distributing and simplifying, we get $3150-5a+8a=4350 \rightarrow 3a=1200 \rightarrow a=400$. Because $c=630-a$ and we now know $a=400$, $c=630-400$, which is 230. We can now find the difference between the number of adult and child tickets sold. It is $|a-c|=|230-400|=170$.

CHAPTER

3 FOIL AND FACTORING

SKILLS TO KNOW

- Multiplying polynomials (FOIL)/expanding expressions
- Factoring Monomials, Binomials, Trinomials (Quadratic Equations)
- The Zero Product Property/Solving Quadratics
- Special Products (Difference of Squares, Square of a Sum, Square of a Difference)
- Solve for "a," "h," or "k" in a quadratic equation by using factoring or FOIL
- "Rainbow" Distribution: expanding more complex products

FOIL

The expression $(5a+3)(a-4)$ is equivalent to:

A. $8a-4$ **B.** $5a^2-23a-12$ **C.** $5a^2+17a-12$
D. $5a^2-a-12$ **E.** $5a^2-17a-12$

In order to solve this problem, we need to use FOIL. **FOIL** is an acronym we use to describe **binomial expansion**: how we use the distributive property to solve for the product of a binomial expression (a two-term expression, such as $5a+3$) times another binomial (such as $a-4$). Each letter in FOIL represents the product of two terms. We find these four products and then add them together to find the total product of the binomials.

First—i.e. the FIRST TERMS: $(\mathbf{5a}+3)(\mathbf{a}-4)$
Outer—i.e. the OUTER TERMS: $(\mathbf{5a}+3)(a\mathbf{-4})$
Inner—i.e. the INNER TERMS: $(5a\mathbf{+3})(\mathbf{a}-4)$
Last—i.e. the LAST TERMS: $(5a\mathbf{+3})(a\mathbf{-4})$

This is the order in which we'll multiply the terms, creating four separate products. I draw a slightly sinister-looking smiley face to help me keep track of these four separate products:

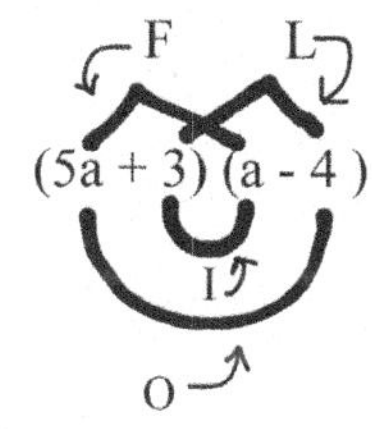

FIRST	OUTER	INNER	LAST
$(5a)(a)$	$(5a)(-4)$	$(3)(a)$	$(3)(-4)$
$5a^2$	$-20a$	$3a$	-12

Then we will add all four products together:

$$5a^2-20a+3a-12=5a^2-17a-12$$

The sum of each of the four products represented by F, O, I, and L is our answer: $5a^2-17a-12$.

Notice how the O and the I are "like terms" that can combine, in this case, $-20a$ and $3a$? This is true every time you FOIL a pair of binomials of the form $ax+b$ where x is any variable and a and b are constants. You will always combine the O term and the I term to get the center term in your final answer. In contrast, the first terms and last terms will not combine with any other terms.

Answer: $5a^2-17a-12$.

TIP: Remember to include the sign of terms like -4, or else you'll get the wrong answer! Negative signs always "hug" to the right: they travel with any number they are in front of when using the distributive property or FOIL.

FACTORING BASICS

Factoring Monomials

You'll need to be able to factor basic monomial elements out of any monomial, binomial, or polynomial.

For review:

A **monomial** is a single product such as $4x$, $7x^3$, or $8n^2$.
A **binomial** has two elements added together such as $4x+3$ or $5n^3+3n$.
A **polynomial** has multiple elements added together such as $5n^3+3n^2+7n+2$ or $5x^2+2x+4$.

These items can be factored by pulling out monomial factors.

All of the following monomials are factors over the integers of $24x^2y+16x^2y^3-8x^4y^2$ EXCEPT:

A. 8 **B.** $6x$ **C.** $2x^2$ **D.** $8x^2$ **E.** $4x^2y$

Here we need to understand what this means: we are looking for an answer choice that does not divide evenly into each "piece" of the original polynomial: $24x^2$, $16x^2y^3$, and $8x^4y^2$. Notice the word "EXCEPT!" We can use process of elimination by going through each answer, and checking whether the monomial answer choice divides evenly into each piece of our polynomial, i.e. if the monomial answer choice is a factor of each of $24x^2$, $16x^2y^3$, and $8x^4y^2$. I can keep track of each answer using Y for yes and N for no, so as not to be confused by the "EXCEPT." I'm looking for N (not a factor).

A. 8 goes into 24, 16 and -8 so that is a factor of each element. (Y)
B. 6 does go into 24—but not into 16 or -8—so this is NOT a factor of this polynomial. This is the answer. (N)
C. 2 goes into 24, 16 and -8 while x^2 goes into x^2, x^2, and x^4, so this is a factor of each element. (Y)
D. 8 goes into 24, 16 and -8 while x^2 goes into x^2, x^2, and x^4, so this is a factor of each element. (Y)
E. 4 goes into 24, 16 and -8 while x^2 goes into x^2, x^2, and x^4, and y goes into y, y^2, y^3. (Y)

Answer: **B**.

Factoring using FOIL (Quadratics with no leading coefficient):

You also need to know how to factor a typical quadratic expression into the product of two binomials. In other words, you need to know how to "undo" FOIL. These problems are simplest when the leading coefficient, i.e. the coefficient of x^2 (the number in front of x^2) is "invisible" or is 1. (Note: we don't actually write out the coefficient 1, i.e. x^2 has a coefficient of 1, or no leading coefficient, whereas $2x^2$ has a coefficient of 2.)

What is the factored form of the expression $x^2+5x-36$?

To solve this problem, we are essentially performing FOIL backwards.

In general, these types of problems will have solutions that look like this:

$(x+a)(x+b)$—what we start with when we FOIL.

Thinking back to FOIL, we can match up parts of the expression that relate to specific parts of the model solution, and FOIL the above model to see how the parts relate. Remember how the sum of the O and the I always form the middle term? That is an important fact when factoring:

Model: $(x+a)(x+b)$

FIRST	OUTER + INNER	LAST
x^2	$+5x$	-36
$(x)(x)$	$(ax)+(bx)$	$(a)(b)$

As you can probably see from the above, our simplest terms to deal with are always the first and the last—they have fewer variables and are less complicated, which means they are the best place to start.

Let's take that first term. We can see that these two terms equal each other. Because there is no leading coefficient (i.e. the coefficient is "one" in this case), this step is easy. We basically don't have to do anything.

$$x^2=x^2$$

Now let's look at our last terms:

$$-36=ab$$

If we set the model product equal to the "last" term in the original quadratic expression we get that the product of a and b is -36. Now we can start to understand what we'll need. We need two numbers that pair together to form -36.

Let's table this idea for a moment and look at our **middle term**.

From our "matchy matchy" setup we know:

$$5x=(ax)+(bx)$$

Factoring out the coefficients, I can simplify this to:

$$5x=(a+b)x$$

At this point, dividing by x (assuming it can be any number, not simply zero) I know 5 must equal $a+b$:

$$5=a+b$$

In short: we are looking for numbers that multiply to the last term ($\boldsymbol{ab}=-36$), and sum to the coefficient of the middle term ($\boldsymbol{a}+\boldsymbol{b}=5$). This sounds confusing, but the more you practice this process, the more you'll get the hang of it.

Thus to find the values of a and b:

STEP 1: Come up with factors of the last term, -36.
STEP 2: See if those factors sum to the middle term's coefficient, 5.

Step 1: Negative sign aside, we want a factor pair that forms 36. My factor rainbow below shows all the possible factor pairs of 36 (for more on factor rainbows, see Book 2's chapter on LCM & GCF):

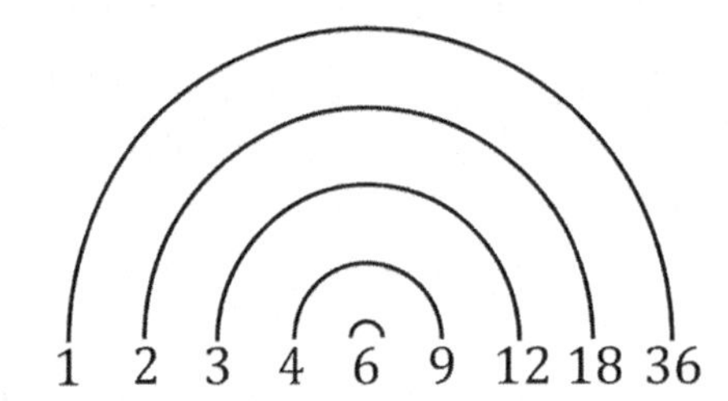

I also know that **to get a negative number as a product**, I'll need **one positive factor** and **one negative factor**. (If I wanted a positive product, both factors could be negative or both positive).

Now for Step 2: We are looking for a pair here such that when one of the pair is positive and one is negative, both sum to five. In other words, we want two factors in a pair that have a difference of 5.

Because 4 and 9 are five apart, let's try some combination of those:

$$4 \text{ and } -9$$

But $4+(-9)=-5$, not 5. We need positive 5.

If we flip the signs of the last pair to get -4 and 9, though we get $-4+9=5$, which works!

Thus, a and b are -4 and 9.

So, $ab=-4(9)=-36$ (The negative is very important!)

AND

$$-4+9=5.$$

Putting our numbers together, we get $(x-4)(x+9)$ as our answer.

TIPS:

1. To check your answer, FOIL it out and see if it matches the polynomial in the question.
2. You don't have to create an entire factor rainbow each time you factor. You can also list pairs of factors, or use your intuition given the idea of a sum and a product to deduce which factor pair might work best. Again, the more of these you do, the better you'll get. If you're ever stuck, though, a factor rainbow is a great way to ensure you haven't missed anything!

Answer: $(x-4)(x+9)$.

Factoring With Coefficients In Front Of The Largest Term

What is the factored form of the expression $3x^2+4x-4$?

The coefficient of 3 in front of x^2 makes this problem a little more complex than the last one. Still, we'll run a similar process, setting up a template to understand what we want:

$$(?x+a)(?x+b)$$

Let's skip straight to finding the coefficients of our first terms. We're going to assume as we factor that we are only looking for integer coefficients. Here we can be pretty certain that our first terms will be 3 and 1, since those are the only whole numbers whose product is 3.

$$(3x+a)(x+b)$$

Now, imagine using FOIL on this binomial. We can line up that FOIL product with our original expression:

F: $3x^2$	**O:** $3bx$	**PLUS**	**I:** ax	**L:** ab
$3x^2$		$+4x$		-4

The middle ones, O and I, sum to the middle term of our polynomial, $4x$, so $3bx+ax=4x$, or (to make things simpler) $3b+a=4$.

Just as we did before, we can look for pairs of numbers that create -4 and would create O and I terms that sum to $4x$. We just need to guess and check to find the answer, plugging into the equation $3b+a=4$. The difference here is that we aren't just adding a and b, the coefficient from the first term has an effect! Let's think of pairs of numbers whose product is -4.

-4 and 1: $3(-4)+1=-11$ **Incorrect**

-2 and 2: $3(-2)+2=-4$ **Closer**

2 and -2: $3(2)-2=4$ **Correct**

$b=2$ and $a=-2$ so if $(3x+a)(x+b)$, we can plug these in to form $(3x-2)(x+2)$.

If this method was confusing, just find the pairs that create -4 as a product, plug them in to the template above, and FOIL to see if it works; in other words, guess and check.

Answer: $(3x-2)(x+2)$.

TIP: Sometimes figuring out the first term is not so easy. If factoring a problem is very difficult, you can always use the **quadratic equation** to solve and then work backwards, subtracting each solution from x to find each factor. See the chapter 17 in this book on **Quadratics and Polynomials** for more on this equation.

ZERO PRODUCT PROPERTY

Now that we're up to speed on factoring, let's talk about how to solve an equation with factored parts.

Which of the following is NOT a solution of $(x-9)(x+2)(x-4)(x+7)=0$?

A. –9 **B.** 9 **C.** –2 **D.** 4 **E.** –7

To solve this, we need to understand the Zero Product Property.

ZERO PRODUCT PROPERTY

If $ab=0$, then $a=0$, $b=0$, or $a=b=0$.

TIP: Sometimes the ACT® calls answers to factored polynomials (such as this one) "roots" instead of solutions. Sometimes it will also call these answers "zeros."

If the product of any two or more items equals zero, then it follows that one of the items MUST be equal to zero. When we solve equations, we can use this idea to then simplify the entire equation down into further smaller equations, all of which are equal to zero:

$$(x-9)(x+2)(x-4)(x+7)=0 \text{ implies}$$
$$(x-9)=0 \text{ or } (x+2)=0 \text{ or } (x-4)=0 \text{ or } (x+7)=0$$

We can solve each equation and get all of the possible answers: $x=9$, $x=-2$, $x=4$, and $x=-7$. However, -9 is not a solution, so it is the answer.

Answer: **A.**

TIP: When a polynomial factors to a single binomial term to a power, for example, $(x-3)^2=0$, the polynomial has one real solution, because only one element can be set equal to zero. So **a perfect square factorization implies a single solution.** See **Quadratics and Polynomials** for more.

SPECIAL PRODUCTS

Certain patterns in math often emerge that we can memorize in order to speed up factoring. The idea with special products is to memorize the pattern and apply it, rather than using FOIL or traditional factoring methods.

The Difference of Squares

THE DIFFERENCE OF SQUARES

The product of the difference $a-b$ and the sum $a+b$ is equal to a squared minus b squared.

Pattern: $a^2-b^2=(a-b)(a+b)$

Why this works: Our middle term always cancels out:

Ex. 1: $(a-b)(a+b)$

$a^2+ab-ab-b^2$ FOIL

a^2-b^2 Simplify.

Ex. 2: Let $a=x$ and $b=3$:

$(x-3)(x+3)$

$x^2+3x-3x-3^2$

x^2-9

$9x^2-16=?$

To factor this, we apply this memorized pattern:

$$(a-b)(a+b)=a^2-b^2$$
$$\downarrow \quad \downarrow$$
$$9x^2-16$$
$$\downarrow \quad \downarrow$$
$$(3x)^2-(4)^2$$
$$a=3x,\ b=4$$

As you can see, this expression matches the second half of the above pattern—where $a=3x$ and $b=4$. Now that we know a and b, substitute in these values in the pattern's first half, and you're done:

$$(a-b)(a+b)$$
$$(3x-4)(3x+4)$$

Answer: $(3x-4)(3x+4)$.

The Square of a Sum

THE SQUARE OF A SUM

Pattern: $(a+b)^2=a^2+2ab+b^2$

Why this works:

$$(a+b)(a+b)$$
$$a^2+ab+ab+b^2$$
$$a^2+2ab+b^2$$

Example: Let $a=x$ and $b=3$:

$$(x+3)^2$$
$$(x+3)(x+3)$$
$$x^2+3x+3x+3^2$$
$$x^2+2(3)(x)+3^2$$
$$x^2+6x+9$$

$(4x+6)^2=?$

Apply the pattern: $(ax+b)^2=a^2x^2+2abx+b^2$

In our polynomial: $a=4x$; $b=6$

Plug in: $(4x)^2+2(4x)(6)+(6)^2=16x^2+48x+36$

Answer: $16x^2+48x+36$.

MISTAKE ALERT: Don't forget to distribute the exponent on the $4x$ term—you need to square both the 4 and the x!

The Square of a Difference

THE SQUARE OF A DIFFERENCE
Pattern: $(a-b)^2=a^2-2ab+b^2$

Why this works:

$$(a-b)(a-b)$$
$$a^2-ab-ab+b^2$$
$$a^2-2ab+b^2$$

Example: Let $a=x$ and $b=3$:

$$(x-3)^2=(x-3)(x-3)$$
$$x^2-3x-3x+3^2$$
$$x^2-6x+9$$

$(2x-4)^2=?$

In our polynomial:

$a=2x$, $b=4$

$a^2-2ab+b^2$

$$(2x)^2 - 2(2x)(4) + (-4)^2$$
$$= 4x^2 - 16x + 16$$

Answer: $4x^2 - 16x + 16$.

SOLVING FOR PARTS OF A QUADRATIC EQUATION

> Consider the equation $x^2 + 11x + b = 0$. One solution to this equation is -4. What is the value of b?

Here, we can use the idea of factors to set up a FOIL pattern and solve for b. We could also plug in the solution into x, set the equation equal to 0, and solve for b. Let's try the first way:

When we know a solution is equal to -4, we also know one of its factors will be $(x+4)$. We know the other factor will be x plus some other number, because the leading coefficient of x^2 is one:

$$x = -4 \rightarrow (x+4)(x+n)$$

Now we can FOIL this binomial product:

$$(x+4)(x+n) = x^2 + 4x + nx + 4n$$
$$= x^2 + (4+n)x + 4n$$

Now we play "matchy-matchy":

$$x^2 + (4+n)x + 4n$$
$$\downarrow \quad \downarrow$$
$$x^2 + 11x + b = 0$$

We can then solve for n and subsequently b:

$$4 + n = 11$$
$$n = 7$$
$$b = 4n$$
$$b = 4(7)$$
$$b = 28$$

Answer: 28.

Again, we can also solve this by plugging in -4 for x and setting the equation equal to zero:

$$x^2 + 11x + b = 0$$
$$4^2 + 11(-4) + b = 0$$
$$16 - 44 + b = 0$$
$$-28 + b = 0$$
$$b = 28$$

"RAINBOW" DISTRIBUTION: EXPANDING TERMS LARGER THAN BINOMIALS

$(x-2)^3 = ?$

When we have to expand more than a single product of two binomials, we must understand what is at play when we distribute terms. Some of you may have the cube of a binomial pattern (i.e. the way to expand the problem above) memorized. If so, awesome! You can solve this problem in a flash! But it's also important to know how to approach ANY type of more complex expansion problems. Plus, the cube of a binomial is rare on the ACT, so for most students it's not worth memorizing that pattern unless you already know it. Rainbow distribution, however, will get you not only through this problem, but any random complex expansion problem that might head your way.

Here, we can split this into pieces, first using FOIL to expand the product of the first two binomials:

$$(x-2)^2(x-2)=(x^2-4x+4)(x-2)$$

Now we use the distributive property. Thinking in terms of a "rainbow," we are going to take each term in $(x-2)$ and "rainbow" distribute it to each term in the quadratic (x^2-4x+4):

$$(x-2)(x^2-4x+4)$$

Now let's "rainbow" the -2 term to each element in the quadratic:

$$(x-2)(x^2-4x+4)$$

$$x(x^2-4x+4)+(-2)(x^2-4x+4)$$

$$(x^3-4x^2+4x)+(-2x^2+8x-8)$$

$$x^3-4x^2-2x^2+4x+8x-8$$

$$x^3-6x^2+12x-8$$

Answer: $x^3-6x^2+12x-8$.

What is the absolute value of the difference of the solutions for the equation $2x^2+5x=12$?

Here we can first solve for the solutions and then simply subtract those two numbers and find the absolute value of that result. Let's first factor:

$$2x^2+5x=12$$

$$2x^2+5x-12=0$$

$$(2x-3)(x+4)=0$$

Now we set each independent factor equal to zero using the Zero Product Property to find each solution:

$$2x-3=0 \qquad x+4=0$$
$$2x=3 \qquad x=-4$$
$$x=\frac{3}{2}$$

Now that we have two solutions, we subtract and find the absolute value of the difference:

$$\left|\frac{3}{2}-(-4)\right|=\left|\frac{3}{2}+4\right|=\left|5\frac{1}{2}\right|=5\frac{1}{2}$$

Answer: 5.5

For all values of $x>18$, which expression is equivalent to $\frac{3x^2+21x+18}{x^2+3x-18}$?

We also use factoring to simplify rational expressions. We have an entire chapter dedicated to rational expressions (Chapter 19 in this book), but below you can see how first factoring and then cancelling out can help to solve these problems. Note that whenever you cancel out common factors on the top and bottom, you must know that if these are potentially equal to zero you'll be eliminating a "hole" in the graph of the expression. I realize that may be confusing, but we'll talk more about that in chapter 22, Graph Behavior, in this book. Here, we don't need to worry about that because we know $x>18$, so no values in this domain would make the denominator zero. Between steps 2 & 3 we would guess and check for numbers whose product is 18 as we did in the previous problem until we found what works.

$$\frac{3x^2+21x+18}{x^2+3x-18}$$
$$\frac{(3x+?)(x+?)}{(x+6)(x-3)}$$
$$\frac{(3x+3)(x+6)}{(x+6)(x-3)}$$
$$\frac{(3x+3)}{(x-3)}$$

Answer: $\frac{3x+3}{x-3}$.

1. What is the magnitude of the difference of the solutions for the equation $(x+5)(x-1)=0$?

A. 3
B. 6
C. 5
D. 4
E. 2

2. Which value of x is a solution to the equation $x^2+7x-14=-26$?

A. 2
B. 5
C. −2
D. 3
E. −4

3. What is the complete factorization of $4x+8xy+16x^3y$?

A. $4x+8xy(1+2x^2)$
B. $4x(1+2y+4x^2y)$
C. $4x(2+y+2x^2y)$
D. $2x(2+4y+8x^2y)$
E. $x(4+8y+16x^2y)$

4. If $(x+9)$ is a factor of $3x^2+kx+36$, what is the value of k?

A. −31
B. 31
C. 9
D. −9
E. 15

5. The equation $x^2-16x+k=0$ has exactly one real solution for what value of k?

A. 16
B. −16
C. −64
D. 64
E. 8

6. Which of the following quadratic expressions has a solution at $n=4a$ and $n=-5b$?

A. $n^2+n(4a-5b)-20ab$
B. $n^2-n(4a-5b)-20ab$
C. $n^2+n(4a+5b)-20ab$
D. $n^2+n(4a-5b)+20ab$
E. $n^2-n(4a+5b)+20ab$

7. For the quadratic equation $3x^2+20x+K$, what value of K will yield $\frac{1}{3}$ and −7 as solutions for x?

A. 7
B. −7
C. 21
D. 12
E. −14

8. What expression is equivalent to $\left(\frac{2}{3}x-\frac{1}{2}y\right)^2$?

A. $\frac{4}{9}x^2+\frac{2}{3}xy+\frac{1}{4}y^2$
B. $\frac{4}{9}x^2-\frac{2}{3}xy-\frac{1}{4}y^2$
C. $\frac{4}{9}x^2+\frac{2}{3}xy-\frac{1}{4}y^2$
D. $\frac{4}{9}x^2-\frac{4}{3}xy+\frac{1}{4}y^2$
E. $\frac{4}{9}x^2-\frac{2}{3}xy+\frac{1}{4}y^2$

9. If $x^2-y^2=169$ and $x+y=13$, then $y=$?

A. 13
B. −13
C. 1
D. 0
E. 6

10. How many pairs of real numbers (x,y) satisfy $xy=5$ and $(x+y)^2=20$?

A. 0
B. 1
C. 2
D. 3
E. 4

11. What expression is equivalent to $(-4x-11)(x+2)$?

A. $(4x-11)(x+2)$
B. $(-4x+11)(-x-2)$
C. $(4x+11)(-x-2)$
D. $-(4x+11)(x-2)$
E. $(4x+11)(x+2)$

12. The trinomial x^2+8x-9 can be factored as the product of 2 linear factors in the form $(x+a)(x+b)$. What is the polynomial sum of these 2 factors?

A. $2x-8$
B. $2x+8$
C. $2x-9$
D. $2x+9$
E. $2x-1$

13. The expression $(4x+2)(x-3)$ is equivalent to:

A. $5x-1$
B. $5x+1$
C. $4x^2-6$
D. $4x^2-10x-6$
E. $4x^2+5x-6$

14. $(7a+2b)(3b-a)$ is equivalent to:

A. $11ab$
B. $7ab$
C. $21a^2-ab-2b^2$
D. $-7a^2+19ab+6b^2$
E. $7a^2+19ab+6b^2$

15. Which of the following is the factored form of the expression $3x^2-14x+8$?

A. $(x-2)(3x+4)$
B. $(x-4)(3x-2)$
C. $(x-4)(3x+2)$
D. $(x+4)(3x-2)$
E. $(x+2)(3x-4)$

16. Let a and b be real numbers. If $(a-b)^2=a^2-b^2$, which of the following must be true?

A. Either a or b is zero.
B. Both a and b are zero.
C. Both a and b are positive.
D. Both a and b are negative
E. $a=b$ or b is zero.

17. The trinomial x^2+x-20 can be factored as the product of 2 linear factors, in the form $(x+a)(x+b)$. What is the polynomial sum of these 2 factors?

A. $2x+1$
B. $2x-1$
C. $2x-9$
D. $2x+9$
E. $2x-20$

ANSWER KEY

1. B **2. E** **3. B** **4. B** **5. D** **6. B** **7. B** **8. E** **9. D** **10. C** **11. C** **12. B** **13. D** **14. D** **15. B** **16. E** **17. A**

ANSWER EXPLANATIONS

1. **B.** $(x+5)(x-1)=0$ implies (per the Zero Product Property) $(x+5)=0 \rightarrow x=-5$ or $(x-1)=0 \rightarrow x=1$. "Magnitude" is fancy for absolute value, so we find the difference of the possible x values $1-(-5)=1+5 \rightarrow 6$, or $-5-1=-6$. The absolute value of either is six, answer B.

2. **E.** To factor $x^2+7x-14=-26$, we first move everything to the same side (the left side) of the equal sign, so that the sum of the terms equals zero. By adding 26 on both sides, we get $x^2+7x+12=0$. Now, we factor the expression by finding two integers that add up to 7 and multiply to be 12. The numbers 3 and 4 work. We factor to $(x+3)(x+4)=0$. Thus, the values -3 and -4 are solutions to the equation. Only -4 is one of the choices given, so that is the answer.

3. **B.** Choice A is not a complete factorization because it only factors out elements of one added piece. We must have an overall product, so A is eliminated. Every term in the expression is a multiple of $4x$, so we can factor out $4x$ to get:

$$4x+8xy+16x^3y=4x\left(1+2y+4x^2y\right).$$

4. **B.** If $(x+9)$ is a factor of $3x^2+kx+36$, then that means $(x+9)(ax+b)=3x^2+kx+36$. Expanding $(x+9)(ax+b)$, we get $ax^2+9ax+bx+9b=3x^2+kx+36$. Comparing the two sides, we can conclude that $ax^2=3x^2 \rightarrow a=3$ and $9b=36 \rightarrow b=4$. Lastly, $9ax+bx=kx$, and plugging in $a=3$ and $b=4$, we get $9ax+bx \rightarrow 9(3)x+(4)x \rightarrow (27+4)x \rightarrow 31x=kx$. So, $k=31$. To move faster, we can also plug in $x=-9$, knowing that a factor creates a solution when it equals zero; set the expression equal to zero, plug in, and solve. Chapter 17 goes more in depth on this method, and we use it on the bottom of page 31 or in question 7 below.

5. **D.** $x^2-16x+k=0$ has exactly one real solution if it is a perfect square. So, we can write $x^2-16x+k=0$ in the form $(x+a)(x+a)=0$ where $a^2=k$ and $2ax=-16x$. Solving for a, we get $2ax=-16x \rightarrow 2a=-16 \rightarrow a=-8$. Now we can plug in $a=-8$ to solve for k. $a^2=k \rightarrow (-8)^2=k \rightarrow 64=k$.

6. **B.** If $n=4a$ and $n=-5b$ are two solutions to the equation, that means that when $n=4a$ or $n=-5b$, the equation equals zero. This means the equation can be written as $(n-4a)(n-(-5b)) \rightarrow (n-4a)(n+5b)$. FOIL this out, we get $n^2-4an+5bn-20ab=n^2-n(4a-5b)-20ab$. We go more in depth on this style of solving in Chapter 17.

7. **B.** This question is similar to number 4. Here let's solve by plugging in one of the given solutions to the equation $3x^2+20x+K=0$. We know the expression equals zero because that's what a quadratic solution is: a value that makes an expression equal zero. I'll plug in negative seven: $3(-7)^2+20(-7)+K=0$ Now solve for k. $147+(-140)+K=0 \rightarrow 7+K=0 \rightarrow K=-7$

8. **E.** Use FOIL or special products to get $\left(\frac{2}{3}x-\frac{1}{2}y\right)^2=\frac{4}{9}x^2-\frac{1}{3}xy-\frac{1}{3}xy+\frac{1}{4}y^2=\frac{4}{9}x^2-\frac{2}{3}xy+\frac{1}{4}y^2$.

9. **D.** x^2-y^2 is a difference of squares, so it can be written as $(x+y)(x-y)$. This gives us $(x+y)(x-y)=169$. Plugging in our given value that $(x+y)=13$, we get $13(x-y)=169 \rightarrow x-y=13$. Now we have two equations $x-y=13$ and $x+y=13$. We can solve for x and substitute: $x=13+y$. $(13+y)+y=13$. $13+2y=13$. $2y=0$. This can only be true if $y=0$.

10. C. We expand the binomial using special products (square of a sum) to $x^2+2xy+y^2=20$. Since $xy=5$, $y=\frac{5}{x}$. Plugging this in gives us: $x^2+2x\left(\frac{5}{x}\right)+\frac{25}{x^2}=20$. We can simplify this to: $x^2+\frac{25}{x^2}-10=0$. Multiplying both sides by x^2 gives us: $x^4-10x^2+25=0$. We can express this as $\left(x^2-5\right)^2=0$, and taking the square root of both sides gives us $x^2-5=0$. From this, it's simple to find that $x=\pm\sqrt{5}$, so there are 2 roots, each with their own respective y-values, $xy=5\rightarrow(y\sqrt{5})=5\rightarrow y=5/\sqrt{5}\rightarrow y=\sqrt{5}$ or $xy=5\rightarrow(y(-\sqrt{5}))=5\rightarrow y=5/-\sqrt{5}\rightarrow y=-\sqrt{5}$ so $y=\pm\sqrt{5}$, leading us to our answer of 2 pairs. Though this question relies on special products presented in this chapter, the problem is more on par with difficulty of questions in chapter 17, **Quadratics & Polynomials**. So if this was hard, don't worry, we haven't gotten to many skills it requires, yet! And be sure to complete chapter 17!

11. C. We can factor out the negative sign from $(-4x-11)(x+2)$ to get $-(4x+11)(x+2)$. Now, we can redistribute the negative sign to the second factor and get $(4x+11)(-x-2)$.

12. B. To factor x^2+8x-9, we must find two integers who have a difference of 8 and multiply to be -9. The numbers 9 and -1 work. We factor to get $(x+9)(x-1)$. If we sum the two factors, we get: $(x+9)+(x-1)\rightarrow x+9+x-1\rightarrow 2x+8$.

13. D. FOIL gives us $4x^2-12x+2x-6=4x^2-10x-6$.

14. D. FOIL gives us $21ab-7a^2+6b^2-2ab$. We rearrange this to $-7a^2+21ab-2ab+6b^2=-7a^2+19ab+6b^2$.

15. B. Our expression for $3x^2-14x+8$ will look like $(3x+n)(x+m)$. We need two numbers that multiply to 8 and sum to -14 when one of them is multiplied by 3. Whole number pairs that multiply to 8 are 8 and 1, -8 and -1, 4 and 2, and -4 and -2. Since they have to sum to a negative number, we can discard the positive sets, as the middle term coefficient, -14, is negative, leaving us with -8 and -1 and -4 and -2. If we sum the -8 and -1 pair while multiplying one of them by 3, we either get $-8+3(-1)=-8-3=-11$, or $-1+3(-8)=-1-24=-25$. Neither satisfies our equation. Using -4 and -2, we get either $-4+3(-2)=-4-6=-10$ or $-2+3(-4)=-2-12=-14$. The latter satisfies our equation. In order to have the -4 be multiplied by 3 when we FOIL, we have to put it in the opposite group from the $3x$. Thus, our solution will be $(x-4)(3x-2)$. You can also use the answers to help you brainstorm possibilities, and guess and check by expanding those options.

16. E. Applying FOIL to the left side, we get $a^2-2ab+b^2=a^2-b^2$. This simplifies to $-2ab=-2b^2\rightarrow ab=b^2\rightarrow a=b$ or $b=0$. Whenever you "divide" by a variable, it is also possible that that variable is equal to zero, and that would also be a solution. Choice E offers both these possibilities.

17. A. -20 is the product of -4 and 5 or -5 and 4. Only -4 and 5 sum to 1. Thus, x^2+x-20 can factor to $(x+5)(x-4)$. Adding these linear factors together gives us $x+5+x-4=2x+1$.

CHAPTER

MANIPULATIONS

SKILLS TO KNOW

- How to solve an equation "in terms of" any variable
- How to rework an equation or expression into another form or expression

Manipulations questions require you to rearrange equations or expressions using algebra. Often they won't ask you to solve down for a particular numeric answer, but rather will offer a variety of answer choices in equation or expression form. In other words, these questions typically have **VARIABLES in the ANSWER CHOICES.**

To save time, DO THESE PROBLEMS ALGEBRAICALLY!

Some teachers will tell you to "plug in numbers" for these problems. That way does work, and it's a must have tool in your problem solving toolbox, but it is sometimes time consuming. We'll show you how to do that for those times when the algebraic way is too confusing, but if you have trouble with timing, pay attention to the algebraic methods below.

The idea behind manipulation is that you need to make what you have look like what you need. Sometimes these problems can look scary—but don't be afraid. Instead remember your two S's and two F's:

- **S**implify
- **S**ubstitute
- **F**actor
- **F**OIL

As long as you apply these ideas, you'll find the answers.

Don't be overwhelmed! Most of the time there are variables in the answer choices. You're not being asked to solve for the impossible, but simply to rearrange things.

SIMPLIFY

Which of the following is the equation $9(x+y)=8$, solved for x?

A. $\frac{8-9x}{9}$ **B.** $\frac{8+9y}{9}$ **C.** $\frac{8-9y}{9}$ **D.** $\frac{8+y}{9}$ **E.** $\frac{8-y}{9}$

Here, we want x alone and the everything else (the numbers and the y) together—so don't distribute! Divide by 9 first and you'll save time. To speed up this process, think about which step will help you get numbers together and variables together the fastest. In other words, **combine like terms:**

$$9(x+y)=8$$
$$x+y=\frac{8}{9}$$
$$x=\frac{8}{9}-y$$

Now we have x alone, but if we look at the answers, nothing matches. What doesn't match? All the answers are single fractions. Thus we need to turn our answer into a single fraction:

$$x=\frac{8}{9}-\frac{9y}{9}=\frac{8-9y}{9}$$

Answer: **C.**

SUBSTITUTE

Often, you'll see a pattern in manipulations problems: an **EQUATION** (or two) then an **EXPRESSION** you need to solve for. We also discussed this idea in Chapter 1. The key here is that you will always **ISOLATE** some variable in the equation that you will then **SUBSTITUTE** into the expression. The equation part has an equals sign—the expression is what you are solving for. If you can break these problems into parts, seeing the equation half (usually first) and the expression you want (usually second) this strategy becomes easier to apply.

If $n-3=a$ and $n+8=b$, what is the value of $a-b$?

A. -11 **B.** 5 **C.** 11 **D.** $2n-5$ **E.** $2n+5$

Here our equations ($n-3=a$ and $n+8=b$) already have a and b isolated, so we can substitute straight into the expression $a-b$:

$$a-b=(n-3)-(n+8)$$
$$n-3-n-8=-3-8$$
$$=-11$$

Answer: **A.**

TIP: TRIAL AND ERROR IS OK! Don't give up! If you try substituting in for one variable and don't get anything similar to the answer choices, try substituting for the other.

FACTOR/FOIL

We cover these two strategies in Chapter 3 in this book. Just know you'll also sometimes need these strategies on questions with variables in the answer choices. See questions 9 and 16 in the previous chapter's problem set if you want to see examples of manipulations questions involving these steps.

MAKE UP A NUMBER

If you can't make progress algebraically, make up a number. Remember, this technique works when there are **variables** in the **answers**, but it is typically slower than working algebraically.

When $\frac{a}{b}+c=d$, and $b\neq 0$, $b=?$

A. $\frac{a}{d}-\frac{a}{c}$ **B.** $\frac{a+c}{d}$ **C.** $\frac{d}{a+c}$ **D.** $\frac{a}{d-c}$ **E.** $\frac{a}{d+c}$

TIP: Use distinct numbers and avoid ANY numbers already in the problem so you don't get confused!

First we make up three of the four variables, and then we solve for the variable we "need."

$$Let\ a=2;\ b=?;\ c=3;\ d=7$$

$$\frac{a}{b}+c=d$$

$$\frac{2}{b}+3=7$$

$$\frac{2}{b}=4$$

$$b=\frac{1}{2}$$

Now we plug in each value we made up above for a , c , and d and calculate the value of each answer choice until we find one that is equal to the value of b $(1/2)$.

A. $\frac{a}{d}-\frac{a}{c}=\frac{2}{7}-\frac{2}{3}=\frac{6-14}{21}=\frac{-8}{21}\neq\frac{1}{2}$

B. $\frac{a+c}{d}=\frac{2+3}{7}=\frac{5}{7}\neq\frac{1}{2}$

C. $\frac{d}{a+c}=\frac{7}{2+3}=\frac{7}{5}\neq\frac{1}{2}$

D. $\frac{a}{d-c}=\frac{2}{7-3}=\frac{2}{4}=\frac{1}{2}$ THIS ONE

E. $\frac{a}{d+c}=\frac{2}{7+3}=\frac{2}{10}=\frac{1}{5}\neq\frac{1}{2}$

Answer: **D.**

WARNING: When you make up numbers, you must plug in to EVERY lettered answer to be sure that you have the right answer. Occasionally two answers (or more) will appear to work with this method (i.e. above this would mean that two answer options would equal one half). If you get two answers that work, you must start over with new numbers. The only time not to check every answer would be if you have issues finishing on time. Still, to use this method and not check every answer presents a risk.

1. If $y \neq 0$, when $\frac{x^2}{y} = 5$, $25y^2 - x^4 = ?$

A. -25
B. -24
C. 0
D. 24
E. 25

2. If $x, y,$ and z are nonzero real numbers and $2xy - z = yz$, which of the following equations for x must always be true?

A. $x = zy + y + 2$

B. $x = 2yz - y$

C. $x = \sqrt{yz} - y$

D. $x = \frac{yz - z}{2y}$

E. $x = \frac{yz + z}{2y}$

3. For the equation $4x - 3a = -b$, which of the following expressions gives x in terms of a and b?

A. $\frac{3a - 4}{b}$

B. $\frac{3a - b}{4}$

C. $\frac{3a + 4}{4}$

D. $\frac{-b - 3a}{4}$

E. $3a - b - 4$

4. Which of the following is $3(m+n)^2 = 13$ solved for n?

A. $\pm\sqrt{\frac{13}{3}} - m$

B. $\pm\sqrt{\frac{13}{3}} + m$

C. $\pm\sqrt{\frac{3}{13}} + m$

D. $\pm\sqrt{\frac{13}{3} - m}$

E. $\pm\sqrt{\frac{3}{13}} - m$

5. If $x = 6z$ and $y = 15z$, which of the following is the relationship between x and y for each nonzero value of z?

A. $x = 2x - 1$

B. $y = 5y$

C. $x = y$

D. $x = \frac{2}{5}y$

E. $y = \frac{2}{5}x$

6. If $\sqrt{(14 - \sqrt{x})} = 3 - \sqrt{5}$, then $x = ?$

A. 0
B. 20
C. 25
D. 30
E. 180

7. For all nonzero $a, b,$ and c such that $2a = \frac{b}{c}$, which of the following *must* be equivalent to ab?

A. $\frac{c}{2a}$

B. $2ac^2$

C. $\frac{b^2}{2c}$

D. $\frac{a^2}{2c}$

E. $\frac{2c}{b}$

8. The relation between enthalpy and energy is $H = E + 8.31nT$, where H is the change in enthalpy, E is the change in energy, n is the change in moles, and T is the change in temperature. Which of the following expressions gives n in nonzero terms of H, E, and T?

A. $\frac{H - E}{8.31}$

B. $\frac{8.31T}{H - E}$

C. $\frac{H - E}{8.31T}$

D. $8.31E(H - T)$

E. $8.31T(H - E)$

9. Which of the following is *not* true for all (m,n) that satisfy the equation $\frac{n}{2}=\frac{m}{3}$ for $n, m \neq 0$?

A. $\frac{m}{2}=\frac{n}{2}$

B. $3n=2m$

C. $m=\frac{3}{2}n$

D. $n+m=\frac{5}{3}m$

E. $n \neq m$

10. If $x-y \neq 0$ and $\frac{3y+5x}{x-y}=\frac{5}{7}$, then $\frac{y}{x}=?$

A. $\frac{-15}{13}$

B. $\frac{15}{8}$

C. $\frac{5}{7}$

D. $\frac{12}{43}$

E. 3

11. If $\frac{2y-3x}{4y-5x}=\frac{3}{10}$, then $\frac{y}{x}=?$

A. $\frac{15}{8}$

B. $\frac{16}{30}$

C. $\frac{3}{10}$

D. $\frac{3}{16}$

E. 3

12. When $y=-x^3$, which of the following expressions is equal to $\frac{1}{y}$?

A. $\frac{1}{-x^3}$

B. x^{-3}

C. $\frac{1}{-x^{-3}}$

D. $\frac{1}{-x}$

E. $-x^3$

13. Given that A, B, C, and D are all positive real numbers satisfying $A^2=\frac{1}{2}B^2$, $C=D$, and $B=\sqrt{D}$, which of the following equations is NOT necessarily true?

A. $B^2=C$

B. $A=\sqrt{\frac{1}{2}D}$

C. $A=\frac{1}{2}D$

D. $A^2=\frac{1}{2}C$

E. $A=\frac{\sqrt{2}}{2}B$

14. For all real numbers x and y such that x is the quotient of y divided by 5, which of the following represents the difference of y and 5 in terms of x?

A. $x-5$

B. $\frac{x}{5}-5$

C. $5x-5$

D. $5(x-5)$

E. $\frac{x-5}{5}$

15. The area of a rectangle is A square units. The length is l units, and the width $5n+3$ units longer than l. What is n in terms of A and l?

A. $n=\dfrac{A-l^2-3l}{5l}$

B. $n=\dfrac{l^2+3l-A}{5l}$

C. $n=\dfrac{A-3l}{5l}$

D. $n=\dfrac{A-l-3}{5l}$

E. $n=\dfrac{A-3l}{5}$

ANSWER KEY

1. C **2. E** **3. B** **4. A** **5. D** **6. E** **7. C** **8. C** **9. A** **10. A** **11. A** **12. A** **13. C** **14. C** **15. A**

ANSWER EXPLANATIONS

1. **C.** We wish to write one variable in the terms of the other. For this problem, it is easier to write y in terms of x. We are given $\frac{x^2}{y}=5$, so multiplying by y on both sides and dividing by 5 on both sides gives us $\frac{x^2}{5}=y$. Now we can substitute in $y=\frac{x^2}{5}$ into the equation $25y^2-x^4$. This is equal to $25\left(\frac{x^2}{5}\right)^2-x^4=25\left(\frac{x^4}{25}\right)-x^4\rightarrow x^4-x^4=0$.

2. **E.** We wish to write x in terms of y and z, so our goal is to move the equation around so that x is on a side by itself. We do this by first adding z to both sides of the equation, giving us $2xy=yz+z$. Then, we divide both sides by $2y$, giving us $x=\frac{yz+z}{2y}$.

3. **B.** We wish to write x in terms of a and b, so our goal is to move the equation around so that x is on a side by itself. We do this by first adding $3a$ to both sides of the equation, giving us $4x=3a-b$. Then, we divide both sides by 4, giving us $x=\frac{3a-b}{4}$.

4. **A.** We wish to write n in terms of m, so our goal is to move the equation around so that n is on a side by itself. We do this by first dividing both sides by 3, giving us $(m+n)^2=\frac{13}{3}$. Then, taking the square root of both sides gives us $m+n=\pm\sqrt{\frac{13}{3}}$. Finally, subtracting m on both sides gives us $n=\pm\sqrt{\frac{13}{3}}-m$.

5. **D.** To solve this, we want to first write z in terms of y so we can plug in that expression into the variable z to evaluate x in terms of y. So, our first step is to write z in terms of y by dividing both sides of the second equation by 15. This gives us $\frac{y}{15}=z$. Now, we plug in $\frac{y}{15}$ for the z value in $x=6z$ to get $x=6\left(\frac{y}{15}\right)\rightarrow\frac{2y}{5}\rightarrow\frac{2}{5}y$.

6. **E.** We want to isolate x, so we first square both sides of the equation. This gives us $14-\sqrt{x}=\left(3-\sqrt{5}\right)^2$. Apply FOIL to the right side of the equation to get $14-\sqrt{x}=9-6\sqrt{5}+5$. Subtracting 14 on both sides, we get $-\sqrt{x}=-6\sqrt{5}$. Squaring both sides gives us $x=6^2\times5\rightarrow36\times5\rightarrow180$.

7. **C.** Dividing $2a=\frac{b}{c}$ by 2 on both sides, we get $a=\frac{b}{2c}$. To solve for the value of ab, we can plug in $a=\frac{b}{2c}$ into ab to get $ab=\frac{b}{2c}(b)$ or $\frac{b^2}{2c}$.

8. **C.** Subtracting E from both sides of the equation, we get $H-E=8.31nT$. Dividing both sides of the equation by $8.31T$, we get $\frac{H-E}{8.31T}=n$.

9. **A.** Multiplying the equation, we get $3n = 2m$, which eliminates answer choice (B). Dividing both sides of $3n = 2m$ by 2 gives us $\frac{3n}{2} = m$, which eliminates answer choice (C). Dividing both sides of $3n = 2m$ by 3 gives us $\frac{2m}{3} = n$, and adding m to both sides to this equation gives us $\frac{2m}{3} + m = n + m \rightarrow \frac{2m}{3} + \frac{3m}{3} = n + m \rightarrow \frac{5m}{3} = n + m$, which eliminates answer choice D. For all values of n and m not equal to zero, answer choice E is true. Lastly, Answer choice A is false for all values of n and m except when n and m both equal 0, but the problem states that $n, m \neq 0$. So, the only answer that is false is A.

10. **A.** Cross-multiplying the equation, we get $(3y + 5x)7 = (x - y)5$. Distributing the 7 and 5, we get $21y + 35x = 5x - 5y$. Now, we subtract $5x$ from both sides to get $21y + 30x = -5y$. Then, we subtract $21y$ from both sides to get $30x = -26y$. To find the value of we can divide both sides by x to get $30 = \frac{-26y}{x}$. Then divide both sides by -26, which is $-\frac{30}{26} = \frac{y}{x}$. $-\frac{30}{26}$ simplifies to $-\frac{15}{13}$.

11. **A.** Cross-multiplying the equation, we get $10(2y - 3x) = 3(4y - 5x)$. Distributing out the 10 and 3, we get $20y - 30x = 12y - 15x$. Combining like terms, we get $8y = 15x$. Dividing both sides by 8, we get $y = \frac{15x}{8}$. Then finally, to find the value of $\frac{y}{x}$, we divide both sides of the equation by x to get $\frac{y}{x} = \frac{15}{8}$.

12. **A.** Since $y = -x^3$, $\frac{1}{y} = \frac{1}{-x^3}$ by plugging in $y = -x^3$ in the denominator.

13. **C.** Since $B = \sqrt{D}$ and $C = D$, then by substitution, $B = \sqrt{C} \rightarrow B^2 = C$ so answer choice (A) is true. Since $A^2 = \frac{1}{2}B^2$ and $B = \sqrt{D}$, by substitution, $A^2 = \frac{1}{2}\sqrt{D}^2 \rightarrow A^2 = \frac{1}{2}D \rightarrow A = \sqrt{\frac{1}{2}D}$ so answer choice (B) is true. Since answer choice (B) is always true, answer choice (C) is not always true unless $A = 1$ so the correct answer is (C). Verifying that the rest of the answer choices are always true, we see that from answer choice (B) we know that $A = \sqrt{\frac{1}{2}D}$ and $D = C$, so $A = \sqrt{\frac{1}{2}C} \rightarrow A^2 = \frac{1}{2}C$ and answer choice (D) is true. Lastly, since $A^2 = \frac{1}{2}B^2$, taking the square root on both sides, answer choice (E) is true: $A = \frac{B}{\sqrt{(2)}} = \frac{\sqrt{2}}{2}B$.

14. **C.** x is the quotient of y divided by 5, which means $x = \frac{y}{5} \rightarrow 5x = y$. The difference between y and 5, using our new expression for y, is $5x - 5$.

15. **A.** The width of the rectangle is $l + 5n + 3$. The area of the rectangle, A, is $l(l + 5n + 3)$. Distributing gives us $l^2 + 5nl + 3l = A$. We isolate the term containing n: $5nl = A - l^2 - 3l$. Finally, isolate n: $n = \frac{A - l^2 - 3l}{5l}$.

CHAPTER

5 INTERCEPTS AND SLOPES

SKILLS TO KNOW

- How to find the slope of a line from two points or a graph.
- How to put any equation into slope-intercept form, and from that, derive the slope or y-intercept.
- How to find the equation of a line from two points or a point and a slope.
- How to find the x- or y-intercept(s) of linear (and non-linear) equations.
- How to use information about parallel or perpendicular lines to find the equation of a line.

When it comes to slopes, intercepts, and linear equations, you are expected to know just about everything you learned in Algebra 1—with less emphasis on actual methods of solving (you won't need point-slope form, for example) and vocabulary, but with full expectation of proficiency in the area of manipulation and problem solving.

SLOPE OF A LINE FROM TWO POINTS

SLOPE FORMULA

For points (x_1, y_1) and (x_2, y_2), $m = \frac{y_2 - y_1}{x_2 - x_1}$

Slope Shortcuts

1. If you're given a **GRAPH** instead of points, **pluck points off the graph** and you can still use this formula. OR simply count RISE over RUN but remember downhill lines are negative, and uphill lines are positive.

2. You could also **program this one in your calculator.** See SupertutorTV.com/BookOwners for programs and links to how to videos.

What is the slope of the line containing the points $(10,7)$ and $(14,19)$ in the standard (x, y) coordinate plane?

A. -3 **B.** $-\frac{1}{3}$ **C.** $\frac{1}{3}$ **D.** 3 **E.** $\frac{26}{24}$

To solve this one, simply plug in your numbers into the slope formula.

Let $(10,7) = (x_1, y_1)$ and $(14,19) = (x_2, y_2)$.

Then plug in and solve:

$$\frac{y_2 - y_1}{x_2 - x_1} = \frac{19-7}{14-10}$$
$$\frac{12}{4} = 3$$

Answer: **D**.

SLOPE-INTERCEPT FORM

You need to know what this means, and how to use it.

SLOPE-INTERCEPT FORM

$$y = mx + b$$

Where m is the **slope** of the line, and b is the **y-intercept** of the line at point $(0,b)$.

What is the slope-intercept form of $8x - 2y + 6 = 0$?

A. $y = -8x - 6$ **B.** $y = -8x + 6$ **C.** $y = 8x + 6$ **D.** $y = 4x + 3$ **E.** $y = 4x - 3$

If you're asked to put an equation in this form, you must isolate the "y" value.

$8x - 2y + 6 = 0$	Add $2y$ to both sides
$8x + 6 = 2y$	"Flip" the equation (put the y on the left)
$2y = 8x + 6$	Divide both sides by 2.
$y = 4x + 3$	

Answer: **D**.

Here's another example:

What is the slope of the line $\frac{y}{3} = \mathbf{x}$?

Here we can put the problem into slope-intercept form again by isolating the y value. Then once we get the equation in that form, we look for the number in the "m" position: that is our slope.

$$\frac{y}{3} = x$$
$$y = 3x$$
$$\downarrow$$
$$y = mx + b$$

Here we see the 3 is in the slope position, so it is our answer.

Answer: 3

EQUATION OF A LINE FROM TWO POINTS OR A POINT AND A SLOPE

Here you'll use the slope-intercept form to help again; Yes, you could memorize point-slope form, but for most students, it's easier to just memorize one equation (slope intercept) and plug in.

Which of the following is an equation for the line passing through $(2,2)$ and $(8,4)$ in the standard (x,y) coordinate plane?

A. $y=3x$ **B.** $y=3x-20$ **C.** $y=\frac{1}{3}x$ **D.** $y=\frac{1}{3}x+\frac{11}{3}$ **E.** $y=\frac{1}{3}x+\frac{4}{3}$

As you can see, the answers are all in slope-intercept form. Even if they weren't, we could still use this method and then manipulate the equation to the other form (i.e. standard form, etc.) later.

First, we find the slope:

$$m=\frac{y_2-y_1}{x_2-x_1}=\frac{4-2}{8-2}=\frac{2}{6}=\frac{1}{3}$$

Now we plug in that slope into $y=mx+b$:

$$y=\frac{1}{3}x+b$$

Finally, we plug in one of the pairs of numbers—it doesn't matter which pair you choose. I'll choose $(8,4)$. That means $x=8$ and $y=4$. I simply substitute these into the equation above since we know that they are a solution, and in the process, I solve for b:

$$4=\frac{1}{3}(8)+b$$

$$b=4-\frac{8}{3}=\frac{12}{3}-\frac{8}{3}=\frac{4}{3}$$

Upon finding $b=\frac{4}{3}$, I plug it back into our earlier equation:

$$y=\frac{1}{3}x+b$$

$$y=\frac{1}{3}x+\frac{4}{3}$$

Answer: **E**. At this point, if the answers are in a different form, you can manipulate the equation to match that form.

You can use simple calculator programs to help with problems such as this one.
See: SupertutorTV.com/BookOwners for more info.

X AND Y INTERCEPTS

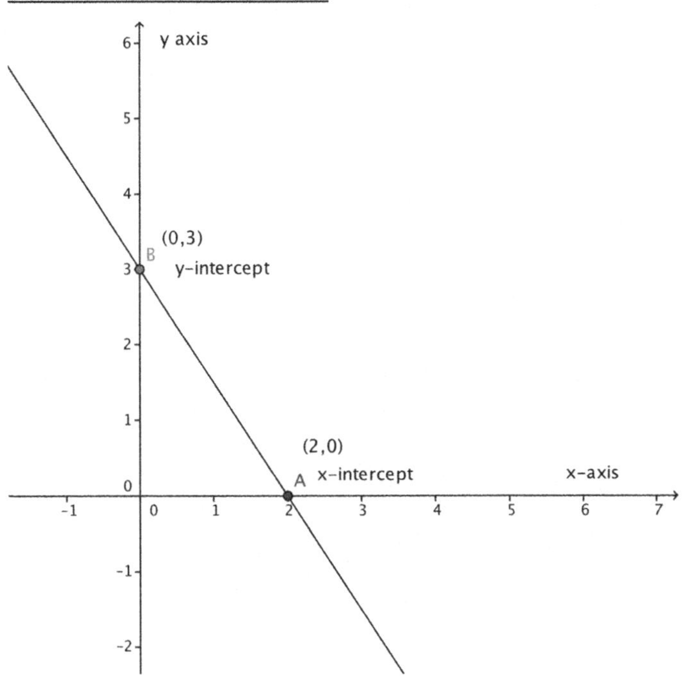

Intercepts

Intercepts are the points at which your line, shape, or other element crosses through the y-axis (y-intercept) or x-axis (x-intercept).

The most common error that students make on intercept questions is mixing up the y-axis and x-axis (i.e. when asked for an x-intercept, they put a y-intercept).

The memory trick that I use is to think about "y-i" and "x-i."

I let the "i" stand for IS: it is IS something, not nothing (0).

TIP: **If you are looking for a y-intercept, y-is the number, and $x=0$; i.e. (0,3) gives the y-intercept.**

If you are looking for an x-intercept, x-is the number, and $y=0$; i.e. (2,0) gives the x-intercept.

If you tend to confuse these, the other trick you can pull is to quickly sketch a graph—look at the y (rhymes with high—up and down) and x (a-"cross" (get it—x looks like a cross)—left to right) and figure out which you axis you want to intercept. The x-intercept is always on the x-axis and the y-intercept is always on the y-axis. In general, if you are getting coordinate geometry questions wrong, you probably need to draw more often. Little sketches are quick and often a great strategy!

One way to find **the y-intercept** from an equation is to put the equation into slope-intercept form , i.e. $y = mx + b$, and solve for b, which is by definiton the y-intercept.

What is the y-intercept of $y = 2x + 4$?

This equation is already in slope-intercept form, or $y = mx + b$. Thus $b = 4$, and b stands for the y-intercept, so the answer is 4.

Answer: 4.

But what if you need the **x-intercept**? Or if it's a pain to put the equation in slope intercept form? Or you have a quadratic or insane-looking polynomial!? In that case, plug in 0 for the appropriate variable (for y-intercept—plug in 0 for x; for x-intercept—plug in 0 for y) and solve for the remaining variable.

Find the y-intercept of $y = x^4 + x^3 + 2x - 7$.

Remember, y-intercept means y-IS the number—so $x = 0$.

Plug in 0 and you get:

$$y = 0^4 + 0^3 + 2(0) - 7$$
$$y = -7$$

You can also solve intercept problems using your **graphing calculator.** If you have an equation you can graph, plug it in, and then hit "trace." Look for the points where the function crosses the axis ($x = 0$ or $y = 0$).

In the (x, y) coordinate plane, the x-intercept of the line $y = -3x + 5$ is represented by:

A. -3 **B.** 3 **C.** 5 **D.** -5 **E.** $\frac{5}{3}$

If x "is" the number then y is zero. Plug in 0 for y and you get:

$$0 = -3x + 5$$
$$-5 = -3x$$
$$\frac{5}{3} = x$$

Answer: **E.**

A parabola with vertex $(-3,4)$ and an axis of symmetry at $y = 4$ crosses the y-axis at $(0, 4 + \sqrt{3})$. At what other point, if any, does the parabola cross the y-axis?

A. $(0, 4 - \sqrt{3})$ **B.** $(0, -4 - \sqrt{3})$ **C.** $(0, -4 + \sqrt{3})$

D. No other point **E.** Cannot be determined from the given information

We could simply sketch a graph knowing the parameters and use the idea of symmetry to find the point on the opposite side of about the line of symmetry, the point $(0, 4 + \sqrt{3})$.

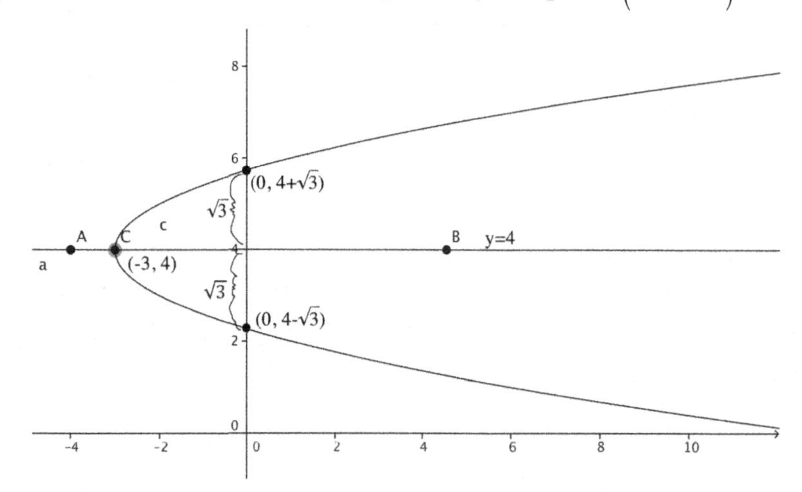

Since the line of symmetry is at 4 you're looking above and below that line by $\sqrt{3}$. So $\left(0,4-\sqrt{3}\right)$ is correct.

Answer: **A**.

This method is probably the fastest—but you may not see it.

If you don't see how to sketch out the graph, you could also memorize the standard vertex form of a parabola and use algebra, though this method is going to take much longer:

PARABOLA VERTEX FORM

$$y=a(x-h)^2+k \text{ OR } x=a(y-k)^2+h$$

You likely encounter with parabolas that are vertically oriented (the first equation) far more often, so this problem is a bit tough on that note alone.

We find out from the axis of symmetry at $y=4$ that it's a horizontally facing parabola. "y" is how "high" a line is—so this is a horizontal line across a consistent "height" ("high" and "y" rhyme if that helps!).

And thus, the parabola is symmetric about that horizontal line. So we need to use the 2nd equation.

We plug in $(-3,4)$ as our vertex at $\left(h,k\right)$ and $\left(x,y\right)$ our pair at $\left(0,4+\sqrt{3}\right)$ and can solve this algebraically:

	Plug in:
$x=a\left(y-k\right)^2+h$	$x=0$, $y=4+\sqrt{3}$, $h=-3$, and $k=4$
$0=a(4+\sqrt{3}-4)^2-3$	Be sure not to confuse equations or variables—h always follows the x, k always corresponds to the y.
$0=(y-4)^2-3$	The 4 and -4 cancel.
$0=3a-3$	Square the $\sqrt{3}$ to get 3.
$3=3a$	Add 3 to both sides.
$1=a$	Divide by 3.

Now that we know $a=1$, and, again $h=-3$ and $k=4$, we know the equation is:

$$x=1\left(y-4\right)^2-3$$

Now using this equation, plug in "0" for x and solve for the other point that crosses through the y-axis:

$0=(y-4)^2-3$

$3=(y-4)^2$ Add 3 to both sides.

$\pm\sqrt{3}=y-4$ Take the square root of both sides.

$4+\sqrt{3}=y$ or $4-\sqrt{3}=y$ Add 4 to both sides.

$(0,4-\sqrt{3})$ is the point we don't have yet, so that is the answer.

Another way to approach similar problems is to **graph on a graphing calculator** and trace to the intersection point between the parabola and the y-axis. However, in this problem, because of the horizontal orientation, you'd first have to work enough algebra to isolate the y value so it can be graphed, a time consuming feat. Sometimes, however, graphing can be a fast option depending on what information you're given.

Answer: **A.**

PARALLEL AND PERPENDICULAR LINES

Parallel lines share the same slope. Thus, if you know the slope or slope-intercept form of one line, and that another line is parallel to it, you can find its slope as well.

What is the slope of any line parallel to the x-axis in the (x,y) coordinate plane?

A. -1 **B.** 0 **C.** 1 **D.** Undefined

E. Cannot be determined from the given information

So this is a bit of a trick question, but let's think about what happens in a parallel line equation:

The "y" value (or height) of the equation is fixed. For example, $y=6$ is a line parallel to the x-axis. What is its slope? Well it is the same as $y=0x+6$, right? So the slope is 0 and the answer is **B**!

Answer: **B.**

I know it can be confusing to remember that horizontal lines have a slope of zero. Another way to think of this is with the slope equation—there is no rise, only run. That means the slope will be zero over some number.

Perpendicular lines have slopes that are opposite reciprocals of each other.

What is the slope of a certain line perpendicular to the line $4x+8y=20$ in the standard (x,y) coordinate plane?

A. -4 **B.** $-\frac{1}{2}$ **C.** $\frac{1}{2}$ **D.** 2 **E.** 4

$4x+8y=20$ First we must put the line into slope-intercept form.

$8y=20-4x$ Move x to the other side to isolate the variable y.

$y=\frac{20}{8}-\frac{4}{8}x$ Divide both sides by 8.

$y=-\frac{x}{2}+\frac{5}{2}$ Use commutative property to keep the x next to the equal sign.

$y=-\frac{1}{2}x+\frac{5}{2}$

Now, from slope-intercept form, we know the slope is $-\frac{1}{2}$.

Since perpendicular lines have slopes that are opposite reciprocals and the slope of this line is $-\frac{1}{2}$, the slope of the perpendicular line is 2.

Answer: **D**.

1. When $9x = 3y - 27$ is graphed in the standard (x,y) coordinate plane, what is the y-intercept?

A. 3
B. 6
C. 9
D. 12
E. 15

2. In the standard (x,y) coordinate plane, what is the x-intercept of the line expressed as $y = -\frac{x}{5} + 3$?

A. −15
B. −3
C. 3
D. 8
E. 15

3. What is the y-intercept of a line that contains the points $(5,-2)$ and $(3,2)$ in the standard (x,y) coordinate plane?

A. $\left(\frac{1}{2},0\right)$
B. $\left(0,-\frac{1}{2}\right)$
C. $(0,8)$
D. $(8,0)$
E. $(8,-8)$

4. What is the slope of the line containing the points $(15,8)$ and $(10,7)$ in the standard (x,y) coordinate plane?

A. $\frac{7}{3}$
B. $\frac{3}{5}$
C. $\frac{5}{3}$
D. 5
E. $\frac{1}{5}$

5. Points $A(-10,2)$ and $B(8,-5)$ lie in the standard (x,y) coordinate plane. What is the slope of $\overline{AB}$?

A. $-\frac{7}{18}$
B. $\frac{7}{18}$
C. $-\frac{3}{2}$
D. $\frac{3}{2}$
E. $\frac{2}{3}$

6. To check the slope of the roof of a house, an architect places an overlay of the standard (x,y) coordinate plane on the blueprint so the x-axis aligns with the horizontal on the blueprint. The line segment representing the side view of the roof goes through the points $(3,-2)$ and $(15,5)$. What is the slope of the roof?

A. $-\frac{1}{6}$
B. $-\frac{1}{4}$
C. 4
D. $\frac{7}{12}$
E. $\frac{12}{7}$

7. In the standard (x,y) coordinate plane, what is the slope of a line passing through the points $(-3,-4)$ and $(5,0)$?

A. −2
B. $-\frac{1}{2}$
C. $\frac{1}{2}$
D. $-\frac{7}{5}$
E. 2

8. What is the slope of the line through the points $(7,3)$ and $(12,9)$?

A. $\frac{10}{21}$

B. $\frac{12}{19}$

C. $\frac{5}{6}$

D. $\frac{6}{5}$

E. $\frac{21}{10}$

9. The graph of the line $6x = 4y - 20$ does NOT have any points in what quadrant(s) of the standard (x, y) coordinate plane below?

A. Quadrant I only
B. Quadrant II only
C. Quadrant III only
D. Quadrant IV only
E. Quadrants I and III only

10. For some real number k, the graph of the line of $y = (k-3)x + 13$ in the standard (x, y) coordinate plane passes through $(3,4)$. What is the slope of this line?

A. −9
B. −3
C. 0
D. 4
E. 9

11. Which of the following has the greatest slope value?

A. $y = 5x - 3$

B. $y = x + 12$

C. $y = 3x - \frac{5}{2}$

D. $2y + 18x = 15$

E. $3y = 2x - 10$

12. Lines a and b lie in the same standard (x, y) coordinate planes. The equation for line a is $y = 0.035x + 150$. The slope of line b is 0.01 less that that slope of line a. What is the slope of line b?

A. 0.0035
B. 0.025
C. 0.034
D. 0.135
E. 1.035

13. What is the slope of the line given by the equation $12x - 7y + 17 = 0$?

A. −7

B. $-\frac{12}{7}$

C. $-\frac{7}{12}$

D. $\frac{12}{7}$

E. 12

14. In the standard (x, y) coordinate plane, what is the slope of the line given by the equation $11x - 7y = 5$?

A. −7

B. $-\frac{11}{7}$

C. $\frac{7}{11}$

D. $\frac{11}{7}$

E. 11

15. What is the slope of the line represented by the equation $8y - 22x = 9$?

A. −22

B. $\frac{9}{8}$

C. $\frac{11}{4}$

D. 8

E. 22

16. When graphed in the standard (x,y) coordinate plane, the line $4x+5y-3=0$ has a slope of:

A. -4

B. $-\frac{4}{5}$

C. $\frac{4}{5}$

D. $\frac{5}{4}$

E. 4

17. The line with the equation $9x+5y=7$ is graphed in the standard (x,y) coordinate plane. What is the slope of this line?

A. $-\frac{9}{5}$

B. $-\frac{5}{9}$

C. $\frac{9}{5}$

D. $\frac{5}{9}$

E. $\frac{7}{5}$

18. For all $m \neq 0$, what is the slope of the line segment connecting $(-m,n)$ and $(m,-n)$ in the standard (x,y) coordinate plane?

A. 0

B. $-\frac{n}{m}$

C. $-\frac{m}{n}$

D. $2n$

E. Slope is undefined.

19. In the standard (x,y) coordinate plane, if the x-coordinate of each point on a line is 3 more than $\frac{1}{4}$ its y-coordinate, the slope of the line is:

A. -4

B. -3

C. $\frac{1}{4}$

D. 3

E. 4

20. What is the slope-intercept form of $3x-y+9=0$?

A. $y=-3x-9$

B. $y=-3x+9$

C. $y=9x+3$

D. $y=3x+9$

E. $y=3x-9$

21. The points $(-3,7)$ and $(0,9)$ lie on a straight line. What is the slope-intercept equation of the line?

A. $y=3x-9$

B. $y=\frac{2}{3}x+10$

C. $y=-\frac{2}{3}x+9$

D. $y=\frac{2}{3}x+9$

E. $y=-3x+7$

22. The slope of the line with the equation $y=mx+b$ is less than the slope of the line with the equation $y=nx+b$. Which of the following statements *must* be true about the relationship between m and n?

A. $m \geq n$

B. $m > n$

C. $m \leq n$

D. $m < n$

E. $m+.5 \leq n$

23. As part of a lesson on slopes and equations, Mr. Hurwitz rolled a barrel at a constant rate along a straight line. His students recorded the distance (d), in feet, from a reference point at the start of the experiment and at 4 additional times (t), in seconds.

t	0	1	2	3	4
d	12	14.5	17	19.5	22

Which of the following equations represents this data?

A. $d = t + 12$

B. $d = \frac{5}{2}t + 7$

C. $d = \frac{5}{2}t + 12$

D. $d = 12t + \frac{5}{2}$

E. $d = 14.5t$

24. What is the slope of any line parallel to the y-axis in the (x, y) coordinate plane?

A. -1

B. 0

C. 1

D. Undefined

E. Cannot be determined from the given information

25. If the graphs of $y = \frac{5}{3}x - 7$ and $y = ax + 12$ are parallel in the standard (x, y) coordinate plane, then $a = ?$

A. -12

B. $-\frac{3}{5}$

C. 0

D. $\frac{5}{3}$

E. 12

26. When graphed in the standard (x, y) coordinate plane, the graph of which of the following equations is parallel to the x-axis?

A. $x = -7$

B. $x = -7y$

C. $x = y$

D. $y = -7$

E. $y = -7x$

27. What is the slope of any line parallel to the line $6x + 7y = 5$ in the standard (x, y) coordinate plane?

A. -6

B. $-\frac{6}{7}$

C. $\frac{6}{5}$

D. $\frac{7}{6}$

E. 6

28. The table below contains coordinate pairs that satisfy a linear relationship. What does a equal?

x	y
-4	-11
-2	-8
0	-5
2	-2
7	a

A. $\frac{11}{2}$

B. 2

C. $-\frac{11}{2}$

D. 0

E. 3

29. Chris is planning a party for his friend. He receives the following prices from the restaurant:

Number of Guests	Price
30	$235
35	$250
40	$265
45	$280
50	$295

What equation, where x is the number of guests and y is the price in dollars, best fits the information in the table?

A. $y = 3x + 235$

B. $y = 3x + 145$

C. $y = 60x + 235$

D. $y = 30x + 235$

E. $y = 50x + 295$

30. What is the y-intercept of the line that contains the points $(-2,4)$ and $(3,1)$ in the standard (x,y) coordinate plane?

A. $\frac{14}{3}$

B. $-\frac{14}{3}$

C. $-\frac{14}{5}$

D. $-\frac{3}{5}$

E. $\frac{14}{5}$

31. What is the x-intercept of the line that passes through the point $(4,-7)$ and has a slope of $-\frac{1}{2}$?

A. 4

B. −10

C. 10

D. −5

E. 18

32. When $4x = 2y - 12$ is graphed in the standard (x,y) coordinate plane, what is the x-intercept?

A. −3

B. 3

C. 6

D. −6

E. −12

33. A parabola with vertex $(2,5)$ and an axis of symmetry at $x = 2$ crosses the y-axis at $(0,13)$. At what other point, if any, does the parabola cross the y-axis?

A. $(0,-13)$

B. $(4,-13)$

C. $(0,-9)$

D. No other point

E. Cannot be determine from the given information

34. The table below lists the number of volunteer organizations in a certain county in Nevada for the years 2001 through 2005. Which expression, using x as the number of years after 2001, best models the approximate number of volunteer organizations in that county?

Year	# Volunteer Orgs
2001	212
2002	217
2003	221
2004	226
2005	230

A. $\frac{9}{2}x + 212$

B. $\frac{9}{2}x + 2001$

C. $4x + 212$

D. $\frac{2}{9}x + 212$

E. $\frac{2}{9}x + 2001$

ANSWER KEY

1. C	**2. E**	**3. C**	**4. E**	**5. A**	**6. D**	**7. C**	**8. D**	**9. D**	**10. B**	**11. A**	**12. B**	**13. D**	**14. D**
15. C	**16. B**	**17. A**	**18. B**	**19. E**	**20. D**	**21. D**	**22. D**	**23. C**	**24. D**	**25. D**	**26. D**	**27. B**	**28. A**
29. B	**30. E**	**31. B**	**32. A**	**33. D**	**34. A**								

ANSWER EXPLANATIONS

1. C. We convert the equation into slope-intercept form: $9x=3y-27 \rightarrow 9x+27=3y \rightarrow 3x+9=y$ so $y=3x+9$. The y-intercept is the constant in the equation, 9.

2. **E.** To find the x-intercept, set y equal to 0 and solve for x : $0=-\frac{x}{5}+3$. This becomes $\frac{x}{5}=3$, so $x=15$.

3. C. We find the slope of the line as the change in y over the change in x : $\frac{2-(-2)}{3-5}=-2$. We put this into the point-slope form using one of the given points: $2=-2(3)+b$. From this, we can easily find that $b=8$. Since b represents the y-intercept, the y-intercept is $(0,8)$.

4. E. The slope of a line is the change in the y coordinate divided by the change in the x coordinate: $\frac{8-7}{15-10}=\frac{1}{5}$.

5. **A.** The slope of a line is the change in the y coordinate divided by the change in the x coordinate: $\frac{2-(-5)}{-10-(8)}=-\frac{7}{18}$.

6. **D.** The slope of the roof is equal to the change in the vertical direction divided by the change in the horizontal direction. Thus, the slope of the roof is equal to $\frac{5-(-2)}{15-3}=\frac{7}{12}$.

7. C. The slope is the change in y over the change in x : $\frac{-4-0}{-3-5}=\frac{-4}{-8}=\frac{1}{2}$.

8. D. The slope is the change in y over the change in x : $\frac{9-3}{12-7}=\frac{6}{5}$.

9. D. We change the formula into $4y=6x+20$, and divide by 4 on both sides to express it in slope-intercept form: $y=\frac{3}{2}x+5$. We can graph this to see that it does not pass through quadrant IV.

10. **B.** The line passes through the y-intercept, $(0,13)$, and $(3,4)$. The slope is equal to the change in the y direction divided by the change in the x direction: $\frac{13-4}{0-3} \rightarrow \frac{9}{-3} \rightarrow -3$.

11. A. By transforming each equation into slope-intercept form, we can easily compare the slopes by comparing the coefficients of x in each formula. A, B, and C are already in this form. (D) we can manipulate: $2y+18x=15 \rightarrow 2y=-18x+15 \rightarrow y=-9x+\frac{15}{2}$. We can also manipulate (E): $3y=2x-10 \rightarrow y=\frac{2}{3}x-\frac{10}{3}$. Now we see that (A) has the largest slope, of 5. The question doesn't ask for the steepest slope, so choice D, -9, is not the greatest slope though it may be steeper.

12. B. The slope of a is the coefficient of x, 0.035. The slope of b is $0.035-0.01=0.025$.

13. D. Isolate the term that contains y by moving it to the other side: $7y=12x+17$. Then, isolate y by dividing by $7: y=\frac{12}{7}x+\frac{17}{7}$. The slope is the coefficient of x in slope-intercept form: $\frac{12}{7}$.

14. **D.** Express the equation in slope-intercept form by isolating y. The equation becomes $7y = 11x - 5$, which simplified is $y = \frac{11}{7}x - \frac{5}{7}$. From this form we see that the slope is $\frac{11}{7}$.

15. **C.** Rewrite the equation and isolate y: $8y = 22x + 9$ becomes $y = \frac{22}{8}x + \frac{9}{8}$. Simplifying to $y = \frac{11}{4}x + \frac{9}{8}$, the slope is $\frac{11}{4}$.

16. **B.** Isolate y: $5y = -4x + 3$ becomes $y = -\frac{4}{5}x + \frac{3}{5}$. The slope is the coefficient of x : $-\frac{4}{5}$.

17. **A.** Isolate y: $5y = -9x + 7$ becomes $y = -\frac{9}{5}x + \frac{7}{5}$. The slope is the coefficient of x : $-\frac{9}{5}$.

18. **B.** The slope is the change in y over the change in x. Plugging into the slope formula yields $\frac{n-(-n)}{-m-m} = \frac{2n}{-2m} = -\frac{n}{m}$.

19. **E.** At every point on this line, $x = \frac{1}{4}y + 3$. This can be expressed as $y = 4x - 12$. The slope is 4.

20. **D.** Isolating y is simple in this problem: $y = 3x + 9$.

21. **D.** The y-intercept is given as $(0,9)$. We can partially fill out the slope-intercept form of the line as $y = mx + 9$. Using the slope formula, $m = \frac{9-7}{0-(-3)} = \frac{2}{3}$. Thus, $y = \frac{2}{3}x + 9$.

22. **D.** The slope of the first equation, m, is less than the slope of the second equation, n. Thus, $m < n$.

23. C. The d-intercept of the equation (note: the d-axis is vertical and the t-axis is horizontal) is 12, as given in the table. The slope is the change in d over the change in t. We can use any 2 sets of points from the table. The slope is $\frac{17-12}{2-0} = \frac{5}{2}$. In slope-intercept form, the equation of the line is $d = \frac{5}{2}t + 12$.

24. **D.** The slope of a vertical line is undefined. A line parallel to the y-axis in the (x,y) coordinate plain is vertical, so the slope of the line is undefined.

25. **D.** **Parallel lines share the same slope.** Because of the pattern inherent in slope-intercept form, we know the slope of the first line is the coefficient of its respective x, which is $\frac{5}{3}$. Because the second line is parallel to it, the slope of the second line must also be $\frac{5}{3}$. The coefficient of its x, a, is its slope, so $a = \frac{5}{3}$.

26. **D.** A horizontal line is parallel to the x-axis. The general equation for a horizontal line is $y = b$ where b is some constant. The only line that fits this general equation is $y = -7$.

27. **B.** The slopes of parallel lines are equal. We can find the slope by expressing the line in slope-intercept form by isolating y. The equation becomes $7y = -6x + 5$, which gives us $y = -\frac{6}{7}x + \frac{5}{7}$. The slope is $-\frac{6}{7}$.

28. **A.** The slope of a linear relationship is constant. From this, we can tell that $\frac{a-(-2)}{7-2} = \frac{-2-(-5)}{2-0}$. Simplifying $\frac{a+2}{5} = \frac{3}{2}$ gives us $a + 2 = \frac{15}{2}$. Isolating a: $a = \frac{15}{2} - 2 = \frac{11}{2}$.

29. **B.** If x is the number of guests and y is the price in dollars, then we want to look at the table and find a function that would describe y in terms of x. We notice that each x-value increases by 5 while each y-value increases by 15. This means the slope of the function is $\frac{rise}{run} = \frac{15}{5} = 3$. So, $y = 3x + b$. Plugging in the first given point $(30,235)$ for values of x and y, we get $235 = 30(3) + b \rightarrow 235 = 90 + b \rightarrow 145 = b$. So, the function that describes the given table is $y = 3x + 145$.

30. E. Find the slope given the two points: $m = \frac{y_2 - y_1}{x_2 - x_1} = \frac{1-4}{3-(-2)} = -\frac{3}{5}$. We can use the slope-intercept form and one of the points given to find b, which essentially is the y -intercept. If $y = mx + b$, then $4 = \left(-\frac{3}{5}\right)(-2) + b \rightarrow b = \frac{14}{5}$.

31. B. First, plug in all the values we know (the slope, m, and the x and y values) into slope intercept form, then solve for b to get the line's equation: $y = mx + b \rightarrow -7 = \left(-\frac{1}{2}\right)(4) + b \rightarrow b = -5$. Be careful, this is the y-intercept not the x-intercept! To find the x-intercept (x is the number, y is zero), now, plug in 0 for y: $0 = \left(-\frac{1}{2}\right)(x) - 5 \rightarrow 5 = -\frac{x}{2} \rightarrow x = -10$.

32. A. The x-intercept is the point where y is 0, so if we plug in 0 for y we get: $4x = 2(0) - 12 \rightarrow 4x = -12 \rightarrow x = -3$.

33. D. Because the parabola has a vertical axis of symmetry, it must either be facing up or down, and not left or right. Upward and downward facing parabolas can only have 1 y-intercept. You can sketch the parabola to confirm.

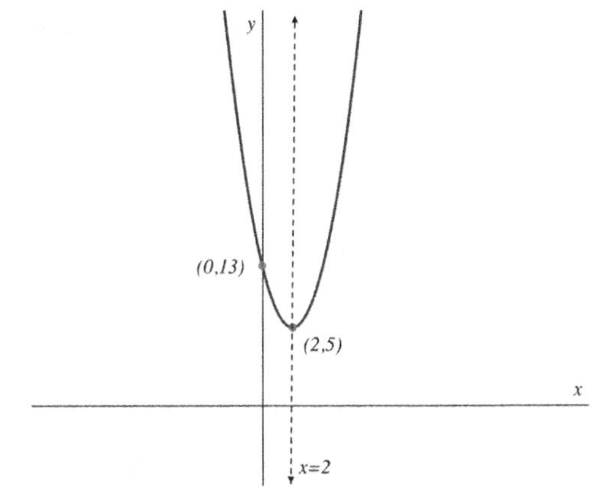

34. A. We can represent the data we have as coordinate points in the form (x,y) where $x =$ the number of years after 2001 and $y =$ the number of volunteer organizations. Then, we have the points $(0,212)$, $(1,217)$, $(2,221)$, $(3,226)$, and $(4,230)$. Now, we look for a pattern so we can represent these points with a function. From $(0,212)$ to $(1,217)$, the x -value increased by 1 and the y-value increased by 5. From $(1,217)$ to $(2,221)$, the x-value increased by 1 and the y-value increased by 4. From $(2,221)$ to $(3,226)$, the x-value increased by 1 and the y-value increased by 5. From $(3,226)$ to $(4,230)$, the x-value increased by 1 and the y-value increased by 4. So, we see a general pattern where the x-value increases by 1 and the y-value increases by 4 or 5. Taking the average of 4 and 5, we can say that each y-value increases by approximately 4.5. So, the slope of the equation would be $\frac{4.5}{1} = \frac{9}{2}$. The y-intercept would be the value of y where $x = 0$. From our points, we see this is the point $(0,212)$ so $y = 212$. The function with slope $\frac{9}{2}$ and y-intercept 212 is $y = \frac{9}{2}x + 212$. For more questions involving quadratic or polynomial graphs and intercepts, see **Coordinate Geometry** and **Quadratics and Polynomials** chapters in this book.

CHAPTER

6 DISTANCE AND MIDPOINT

SKILLS TO KNOW

- How to find the midpoint of a line (Midpoint Formula)
- How to find the distance between two coordinate points (Distance Formula)
- How to solve physical distance word problems

MIDPOINT

You may not remember midpoint from coordinate geometry, but don't worry, you can do these problems without a complicated formula by using a little common sense to memorize the idea.

The midpoint is essentially the **average** of the x values and the **average** of the y values.

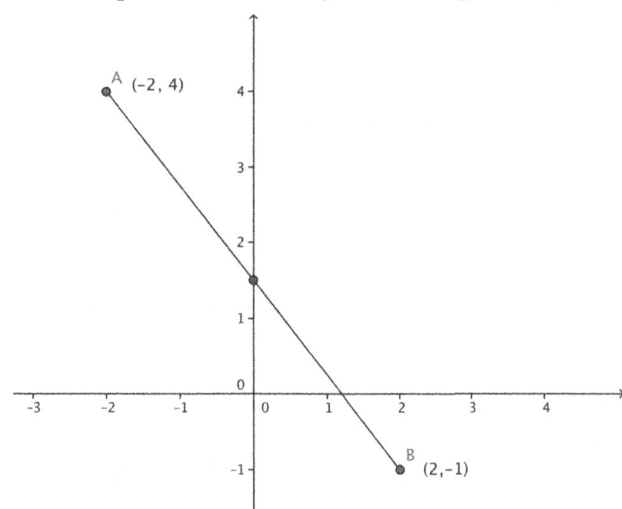

Average of the x-values:

$$\frac{(-2+2)}{2}=x$$
$$\frac{0}{2}=x$$
$$0=x$$

Average of y-values:

$$\frac{(4+-1)}{2}=y$$
$$\frac{3}{2}=y$$
$$1.5=y$$

Answer: $(0,1.5)$

That's right. It's that easy. If you have two points you add the x's together and divide by two, and then add the y's together and divide by two. Above you take the 2 and -2 and find halfway between—that's the average of 2 and -2, which is 0. Same with the y—find the halfway point or average between 4 and -1, which is 1.5.

THE MIDPOINT FORMULA

The midpoint of two coordinate points (x_1,y_1) and (x_2,y_2) is:

$$\left(\frac{(x_1+x_2)}{2},\frac{(y_1+y_2)}{2}\right)$$

TIP: Some questions on the ACT® that deal with midpoint will NOT give you the end points, but rather may give you one end point, the midpoint, and ask for the other endpoint. Always read the questions carefully! Use algebra to solve after setting up an equation.

In the standard (x,y) coordinate plane, the midpoint of a line segment is $(5,7)$ and an endpoint of that segment is located at $(9,-3)$. If (x,y) are the coordinates of B, what is the value of $x+y$?

Knowing the midpoint and an endpoint, we write a formula to solve for the missing endpoint (x,y):

$$\left(\frac{(x_1+x_2)}{2},\frac{(y_1+y_2)}{2}\right)= \text{(midpoint } x \text{ value, midpoint } y \text{ value)}$$

$$\left(\frac{(x+9)}{2},\frac{(y-3)}{2}\right)=(5,7)$$

$$\frac{(x+9)}{2}=5 \qquad \frac{(y-3)}{2}=7$$

$$x+9=10 \qquad y-3=14$$

$$x=1 \qquad y=17$$

$$x+y=1+17$$

$$x+y=18$$

Answer: 18.

THE DISTANCE FORMULA

To find the distance between two coordinate points, think about the Pythagorean Theorem. If you wanted to find distance between A & B in the picture below, you could draw a right triangle, find the physical distances for both horizontal and vertical legs, and then use the Pythagorean Theorem:

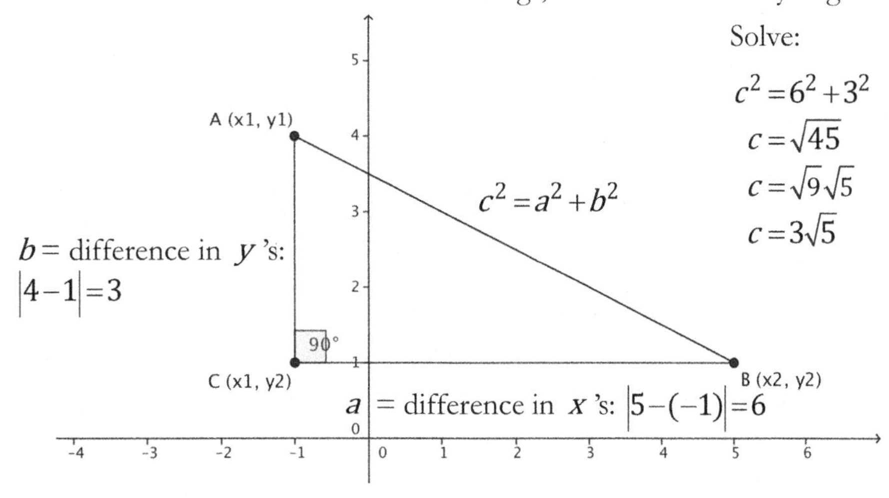

TIP: SIMPLIFYING RADICALS

For some problems in this chapter, you'll need to simplify radicals. However, **first look at the answer choices as sometimes ACT won't require you to reduce your radical to lowest term.**

To simplify radicals, remember, $\sqrt{ab}=\sqrt{a}\sqrt{b}$. Factor out any perfect squares under the radical and break them apart into separate radicals to simplify. For example:

$$\sqrt{36}=\sqrt{(9)(4)}=\sqrt{9}\sqrt{4}=(3)(2)=6$$

$$\sqrt{50}=\sqrt{(25)(2)}=\sqrt{25}\sqrt{2}=5\sqrt{2}$$

As long as you remember that **the distance formula is essentially the Pythagorean theorem,** you can rely on that idea to solve. If you prefer, however, there's a formula you can memorize. You can even program that formula into your calculator (see SupertutorTV.com/BookOwners for more info).

THE DISTANCE FORMULA

Given two points, (x_1,y_1) and (x_2,y_2), the distance between them is:

$$d=\sqrt{(x_2-x_1)^2+(y_2-y_1)^2}$$

What is the distance in coordinate units between the points $(5,4)$ and $(10,7)$ in the standard (x,y) coordinate plane?

The easiest way to approach this, if you get confused by formulas or tend to make careless errors, is to sketch it out. Then count out the difference from 4 to 7 and from 5 to 10—voila! You get 3 and 5.

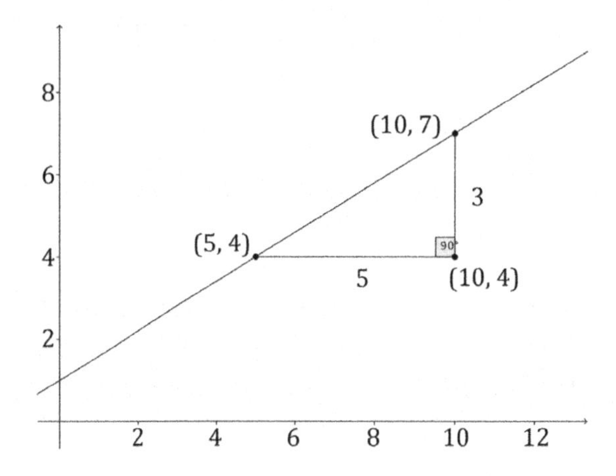

Enter the Pythagorean Theorem: $a^2+b^2=c^2$

Square each, and you get: $9+25=34=c^2$

Solve for c and you get: $\sqrt{34}$

Now sometimes on the test things won't be that easy.

(x,y) and $(-4,2)$ are 10 units away from each other in the standard coordinate (x,y) plane. If y is eight less than x, than which of the following could be the value of x?

A. 4 **B.** –10 **C.** –2 **D.** 6 **E.** 8

You could do this problem by sketching it out and using the answers, but the most reliable method is algebraic. First, translate the "y is eight less than x" into $y=x-8$. Then substitute $x-8$ for y into (x,y). Now you have $(x,x-8)$ as the point you are looking for.

Plug in to the ordered pair: $(x,x-8)$ and your original pair, $(-4,2)$ into the distance formula:

$$d=\sqrt{(x_2-x_1)^2+(y_2-y_1)^2}$$

$$10=\sqrt{(-4-x)^2+(2-(x-8))^2}$$

$100=(-1(4+x))^2+(2-x+8)^2$	Square both sides, factor out -1, distribute negative sign to $(x-8)$
$100=x^2+8x+16+100-20x+x^2$	Expand: $(a-b)^2=a^2-2ab+b^2$ (special product); Commutative property
$0=2x^2-12x+16$	Simplify
$0=x^2-6x+8$	Divide both sides by 2
$0=(x-4)(x-2)$	Factor
$0=(x-4)$ *or* $0=(x-2)$	Apply Zero Product Property
$x=4$ *or* 2	Solve for x

Because only 4 is an available answer (A), it is the correct answer.

Answer: **A.**

PHYSICAL DISTANCE WORD PROBLEMS / SPATIAL ARRANGEMENTS

The ACT sometimes includes word problems that make you chart out physical distances. These could be scenarios involving north, south, east and west distances or line segments involving points such as PQRST in a row. Regardless, draw a picture, mark all distances you know, if helpful assign coordinates to your points (i.e. east is positive x movement right and south is negative y movement), and then set up algebraic equations if necessary to solve for what you need.

The points P, Q, R, S, and T lie in that order on a straight line. The midpoint of $\overline{PR}$ is Q, and the midpoint of $\overline{PT}$ is S. The length of $\overline{PQ}$ is n inches, and the length of $\overline{ST}$ is $3n+4$, what is the length of $\overline{RS}$?

A. n **B.** $n+4$ **C.** $2n+4$ **D.** $3n+4$ **E.** 4

Draw out a picture to represent all this information, and you should get something like this:

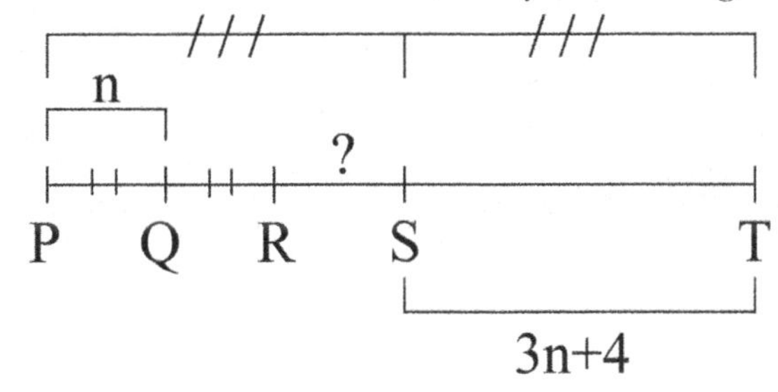

We can see that $PQ=QR$ and these are each equal to n. We also can see that $PQRS$ is equal to $3n+4$ as PS equals ST. Now turn this into an algebraic equation: $2n+RS=3n+4$. Simplifying, we get that $RS=n+4$, our answer, B.

1. In the standard (x, y) coordinate plane, approximately how many coordinate units is $(4,5)$ from $(9,-2)$?

A. 9
B. 10
C. 7
D. 4
E. 6

2. What is the distance, in coordinate units, between points $P\ (-4,-6)$ and $Q\ (1,3)$ in the standard (x, y) coordinate plane?

A. $3\sqrt{2}$
B. $3\sqrt{10}$
C. $\sqrt{106}$
D. 10
E. 12

3. What is the distance, in coordinate units, between points $P\ (6,2)$ and $Q\ (3,-7)$ in the standard (x, y) coordinate plane?

A. $9\sqrt{10}$
B. $3\sqrt{10}$
C. 9
D. 10
E. 3

4. On a map of Truetown in a standard (x, y) coordinate plane, where 1 coordinate unit represents 1 block, the firehouse is at $(-7,2)$ and the police station is at $(3,-4)$ What is the straight-line distance, in blocks between the firehouse and the police station?

A. 8
B. 4
C. $2\sqrt{34}$
D. 136
E. $2\sqrt{5}$

5. A middle school is located 2 miles south and 3 miles east of the city library. A high school is located 4 miles north and 2 miles west of the library. Which of the following is the shortest distance, in miles, between the middle school and the high school?

A. $\sqrt{65}$
B. 12
C. $\sqrt{11}$
D. $\sqrt{61}$
E. $\sqrt{51}$

6. Tomas is standing 50 feet due west of a picnic table at his local park, and 120 feet due south of a water fountain outside a city recreation center. How far apart, in feet, are the picnic table and the water fountain?

A. 130
B. 170
C. 70
D. $10\sqrt{17}$
E. $\sqrt{119}$

7. In the pentagon $ABCDE$ shown in the standard (x, y) coordinate plane below, what is the distance, in coordinate units, from the midpoint of $\overline{AC}$ to the midpoint of $\overline{ED}$?

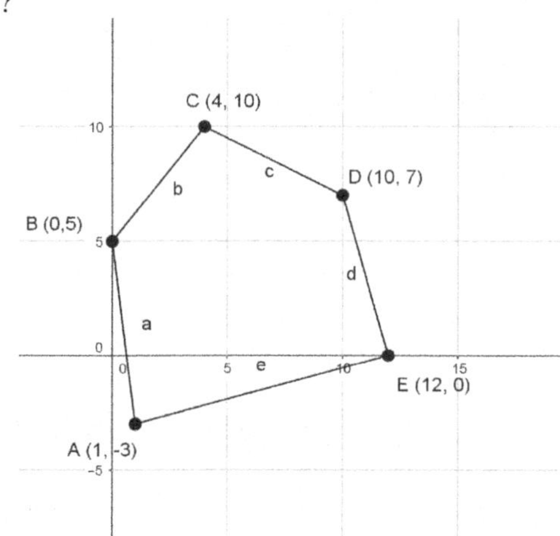

A. 7
B. 8.5
C. 10.8
D. 11.4
E. 12

8. Margaret, Alice, and Trevor start walking from the same point in the center of the schoolyard. If Margaret walks 5 feet due south, and 6 feet due east, Alice walks 7 feet due north and 8 feet due west, and Trevor walks to the point half way between the Margaret and Alice, how far from the original starting point is Trevor, in feet?

A. 2
B. 1
C. $\sqrt{2}$
D. $\sqrt{85}$
E. 13

9. The Rosa Parks Community Center is planning a community jobs fair. Several booths lettered **A, B, C, D,** and **E** are to be placed at the center according to the representation on the standard (x,y) coordinate plane below, in which each coordinate unit represents 1 yard. If the general information booth is to be placed directly halfway between booths **C** and **E**, what is the approximate distance from the general information booth to booth **B**?

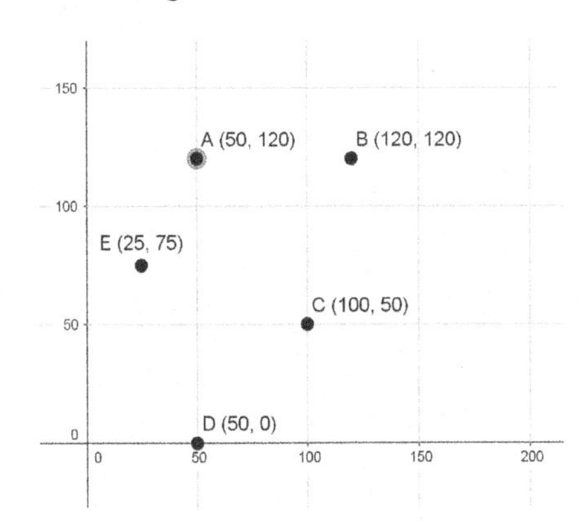

A. 63.6
B. 81.3
C. 88.3
D. 115
E. 130

10. A line segment in the standard (x,y) coordinate plane has endpoints $(-5,8)$ and $(1,2)$. What is the x-coordinate of the midpoint of this line segment?

A. -3
B. -2
C. $\frac{3}{2}$
D. 0
E. 5

11. What is the midpoint of the line segment with endpoints $(-3,10)$ and $(10,-28)$ in the standard (x,y) coordinate plane?

A. $\left(-\frac{7}{2},9\right)$
B. $\left(\frac{7}{2},-9\right)$
C. $(4,18)$
D. $\left(\frac{13}{2},-9\right)$
E. $(13,-18)$

12. The point $(6,-5)$ is the midpoint of the line segment in the standard (x,y) coordinate plane joining the point $(9,1)$ and the point (a,b). Which of the following is (a,b)?

A. $(-3,11)$
B. $(3,7)$
C. $(3,8)$
D. $(3,-11)$
E. $(7.5,-2)$

13. In the standard (x,y) coordinate plane, the midpoint of a line segment is $(19,2)$ and an endpoint of that segment is located at $(1,-3)$. If (x,y) are the coordinates of B, what is the value of $x+y$?

A. 44
B. 37
C. 30
D. 7
E. -30

14. In the standard (x,y) coordinate plane, $(8,15)$ is halfway between $(5z, a-3)$ and $(3z, z+1)$. What is the value of z?

A. 0
B. 2
C. 3
D. 4
E. 6

15. Points $P, Q, R,$ and S lie on a line in the order given. The midpoint of $\overline{PR}$ is Q. The length of $\overline{PQ}$ is b and the length of $\overline{PR}$ is 16 inches, and the length of $\overline{RS}$ is $3b-12$ inches. What is the approximate length, in inches, of $\overline{QS}$?

A. 8
B. 12
C. 20
D. 26.67
E. 30

16. Points $A, B, C,$ and D lie on the real number line as shown below. The length of $\overline{AD}$ is 60 units; $\overline{AC}$ is 39 units long; and $\overline{BD}$ is 32 units long. How many units long, if it can be determined, is $\overline{BC}$?

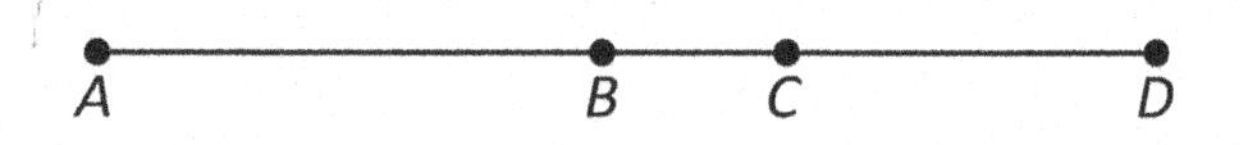

A. 10
B. 11
C. 12
D. 13
E. Cannot be determined from the given information.

17. Points P, Q, R, S and T lie on a line in the order given. The coordinate of R is 0, Q is the midpoint of $\overline{PR}$ and S is the mid-point of $\overline{PT}$. $\overline{RS}$ is 4 units longer than $\overline{PQ}$ and $\overline{ST}$ is 8 units longer than double the length of $\overline{QR}$. What is the coordinate of S?

A. 4
B. 8
C. 12
D. 20
E. Cannot be determined from the given information.

For Questions 19-21, refer to the figure below.

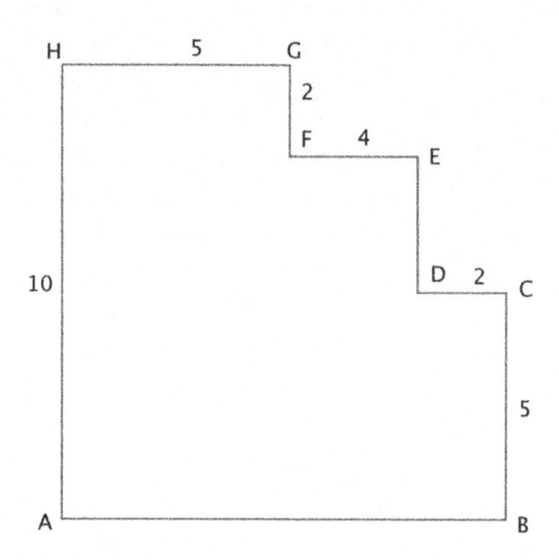

The shape above has only right angles. In a coordinate plane (not shown), $A=(0,0)$ and all other points are the given distances in coordinate units. Point J (not shown) is half-way between A and E.

18. What is the perimeter of the figure?

A. 28
B. 31
C. 39
D. 42
E. 45

19. What are the coordinates of J?

A. $(4,4.5)$
B. $(4.5,4)$
C. $(5,5)$
D. $(8,9)$
E. $(9,8)$

20. Approximately how far, in coordinate units, is J from B?

A. 6
B. 6.7
C. 7
D. 7.6
E. 8.2

21. In pentagon $ABCDE$ shown in the standard (x,y) coordinate plane below, what is the distance, in coordinate units, from the midpoint of $\overline{AC}$ to the midpoint of $\overline{BD}$?

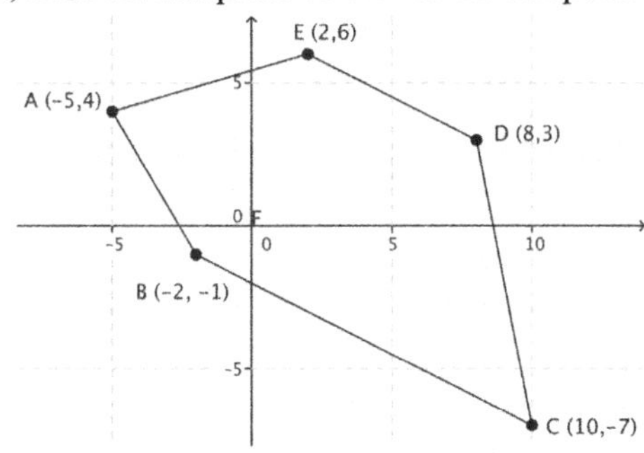

A. $\frac{\sqrt{2}}{2}$
B. $\frac{\sqrt{26}}{2}$
C. $\frac{\sqrt{122}}{4}$
D. $\frac{\sqrt{146}}{2}$
E. 6.5

22. Sarah is planning on throwing a birthday party at her local park. In the figure below, the possible locations for placing a bounce house are represented as points $A, B, C, D,$ and E on the standard (x,y) coordinate plane, where each coordinate unit represents 1 meter. The water fountain is exactly between B and E.

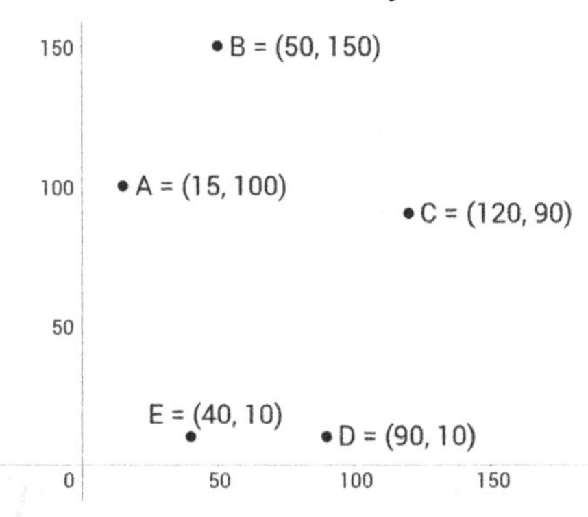

How many meters will the water fountain be from A?

A. 30
B. $\sqrt{931.25}$
C. $\sqrt{1000}$
D. $\sqrt{1125}$
E. $\sqrt{1300}$

23. In the standard (x,y) coordinate plane shown below, what is the distance in the y direction, in units, from point X to point Y?

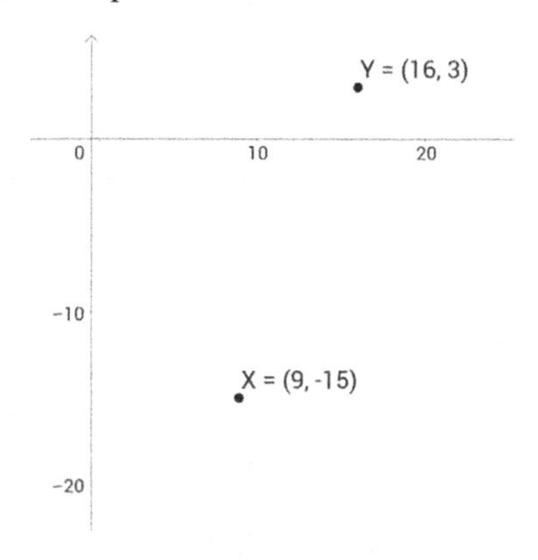

A. 7
B. 18
C. $\sqrt{373}$
D. -12
E. -18

24. What is the distance, in coordinate units, from $(3,4)$ to $(9,-3)$ in the standard (x,y) coordinate plane?

A. $\sqrt{17}$
B. $\sqrt{37}$
C. $\sqrt{39}$
D. $\sqrt{41}$
E. $\sqrt{85}$

25. In the standard (x,y) coordinate plane, what is the distance between the points $(x, x-3)$ and $(3x, x+5)$?

A. $2x+8$
B. $\sqrt{68}$
C. $2\sqrt{x^2+16}$
D. $\sqrt{4x^2+4}$
E. $2x+2$

ANSWER KEY

1. A **2. C** **3. B** **4. C** **5. D** **6. A** **7. B** **8. C** **9. B** **10. B** **11. B** **12. D** **13. A** **14. B**
15. C **16. B** **17. B** **18. D** **19. B** **20. D** **21. B** **22. E** **23. B** **24. E** **25. C**

ANSWER EXPLANATIONS

1. **A.** We can solve this problem using the Pythagorean theorem. The horizontal and vertical distances between the two points make up the two legs of our triangle. We find the horizontal distance as the difference between the x coordinates: $9-4=5$, and the vertical distance as the difference between the y coordinates: $-2-5=-7$. We plug 5 and -7 into our formula to get $5^2+(-7)^2=c^2$ where c is the distance we are looking for. Simplify into $25+49=c^2$ which becomes $c^2=74$. Thus, $c\approx 8.6$ which rounds to 9.

2. **C.** We solve this problem the same way we solved question 1, only this time the answers are exact. Plug in the difference between x values: $1-(-4)=5$ and the difference between y values: $3-(-6)=9$ into the Pythagorean theorem: $5^2+9^2=c^2$. $c^2=25+81=106$. Thus, $c=\sqrt{106}$.

3. **B.** We solve this problem exactly how we solved #2. Plug in the difference between the x values: $6-3=3$ and the difference between the y values: $2-(-7)=9$ into the Pythagorean theorem: $3^2+9^2=c^2$. $c^2=9+81=90$. $c=\sqrt{90}=\sqrt{9}\sqrt{10}=3\sqrt{10}$.

4. **C.** This problem is just like the ones in the easy section; we are simply looking for the distance between two points. The difference between the x coordinates is $3-(-7)=10$ and the difference between the y coordinates is $-4-2=-6$. Plug into the Pythagorean theorem: $10^2+(-6)^2=c^2\rightarrow c^2=100+36=136\rightarrow c=\sqrt{136}=\sqrt{4}\sqrt{34}=2\sqrt{34}$.

5. **D.** Sketch the word problem to gain a clearer understanding. The total North-South distance between the middle school and high school is $4+2=6$ miles. The total East-West distance is $2+3=5$ miles. Plug these values into the Pythagorean theorem to get $5^2+6^2=c^2\rightarrow c^2=25+36=61\rightarrow c=\sqrt{61}$.

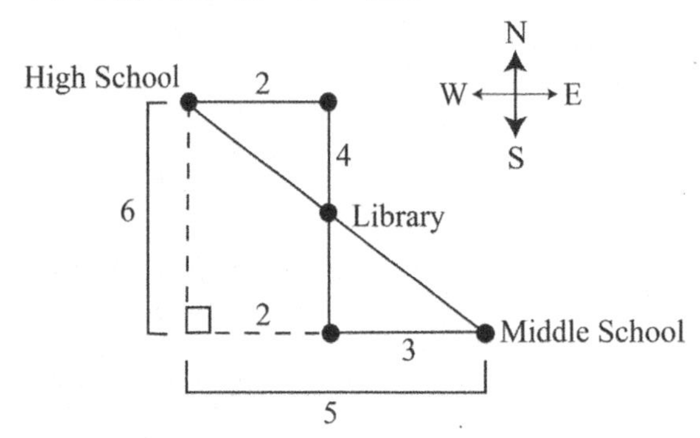

6. **A.** We can see from a sketch of the problem that the water fountain, the picnic table, and Tomas form a right triangle with legs of lengths 120 and 50. This triangle is similar to the right triangle whose sides are the Pythagorean triple 5-12-13. Thus, we can deduce that the hypotenuse of the larger triangle is 130. (See Triangles in Book 2 for more Pyth. triples)

7. **B.** The midpoint between A and C is $\left(\frac{1+4}{2},\frac{-3+10}{2}\right)=(2.5,3.5)$. The midpoint between D and E is $\left(\frac{10+12}{2},\frac{7+0}{2}\right)=(11,3.5)$. The distance between $(2.5,3.5)$ and $(11,3.5)$ is only the difference between the x values, since the y values are equal. The distance is $11-2.5=8.5$.

8. **C.** If we define the starting point as the origin $(0,0)$, we can express the position of Margaret as $(6,-5)$ and the position of Alice as $(-8,7)$. We find the position of Trevor as the midpoint: $\left(\frac{6-8}{2},\frac{-5+7}{2}\right)=\left(\frac{-2}{2},\frac{2}{2}\right)=(-1,1)$. We find the distance between Trevor and the origin using the Pythagorean theorem: $(-1)^2+1^2=c^2.c^2=1+1=2$. $c=\sqrt{2}$.

9. **B.** The coordinates of the general information booth are $\left(\frac{25+100}{2},\frac{50+75}{2}\right)=(62.5,62.5)$. The distance between this and B, $(120,120)$ is given by $\sqrt{(120-62.5)^2+(120-62.5)^2}=\sqrt{57.5^2+57.5^2}=\sqrt{3306.25+3306.25}=\sqrt{6612.5}\approx 81.3$.

10. **B.** Plug the coordinates into the midpoint formula: $\left(\frac{-5+1}{2},\frac{8+2}{2}\right)=\left(\frac{-4}{2},\frac{10}{2}\right)=(-2,5)$. The x-coordinate of this point is -2.Or, average the x-values given (you don't need to solve for the y values).

11. **B.** Plug the coordinates into the midpoint formula: $\left(\frac{-3+10}{2},\frac{10+(-28)}{2}\right)=\left(\frac{7}{2},\frac{-18}{2}\right)=\left(\frac{7}{2},-9\right)$.

12. **D.** Plug the coordinates into the midpoint formula: $\left(\frac{a+9}{2},\frac{b+1}{2}\right)=(6,-5)$. By multiplying the x and y coordinates on either side by 2, $(a+9,b+1)=(12,-10)$. We simplify this to $(a,b)=(3,-11)$. Note that when solving, we do not add or subtract the same value from the x and y coordinates, only one or the other.

13. **A.** We solve this the same way we solved the question before. Plug in the values given into the midpoint formula: $\left(\frac{x+1}{2},\frac{y+(-3)}{2}\right)=(19,2)$. This becomes $(x+1,y-3)=(38,4)$. Thus, $(x,y)=(37,7)$. $x+y=37+7=44$.

14. **B.** Plug in the values given into the midpoint formula: $\left(\frac{5z+3z}{2},\frac{a-3+z+1}{2}\right)=(8,15)$. This becomes $(5z+3z,a-3+z+1)=(16,30)$. $(8z,a+z-2)=(16,30)$. We can match up the x values to form the equation $8z=16$, which yields $z=2$.

15. **C.** We know that $\overline{PR}=2\overline{PQ}$, since Q is the midpoint of $\overline{PR}$. $\overline{PR}=16=2b$, so $b=8$. $\overline{RS}=3b-12=3(8)-12=12$. $\overline{QR}+\overline{RS}=\overline{QS}$. $\overline{QS}=8+12=20$.

16. **B.** $\overline{AC}=39$. $\overline{BD}=32$. $\overline{AC}+\overline{BD}-\overline{BC}=60$, which means $30+32-\overline{BC}=60$. Solve to find that $\overline{BC}=11$. You can also make up variables for each of the smaller segments, and then create a system of equations to solve if you can't see that subtracting the overlap from the total sum will give you your answer. I can set $AB=x$, $BC=y$ and $CD=z$. I know $x+y=39$ and $y+z=32$. I also know $x+y+z=60$. I can add the first two equations to get $x+y+y+z=39+32$. I simplify to $x+2y+z=71$. I know that is 11 more than $x+y+z$ (which equals 60) and the only difference is that it includes one more y. Thus y must be equal to 11. This is similar to Venn Diagram problems covered in Word Problems, Chapter 11 in this book. See that section for more on this alternate method.

17. **B.** Drawing this out, we get: 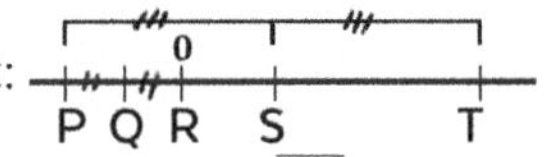

. Instead of getting jumbled up with our points, let's label the $\overline{PQ}$ (which is the same as $\overline{QR}$ because Q is the midpoint) n. We are told that $\overline{RS}=4+n$, so we can say that $\overline{PS}=n+n+(4+n)$. We also know that $\overline{ST}=2\overline{QR}+8$, which can also be written as $2n+8$. Because we know that $\overline{PS}=\overline{ST}$, we can set these equations equal to each other: $n+n+(4+n)=2n+8$.

Simplifying this, we have $3n+4=2n+8 \rightarrow n=4$. Because we're looking for the coordinate of S, we are really looking for the distance of $\overline{RS}$, since we know R is at 0. $\overline{RS}=4+n \rightarrow 4+4=8$. The coordinate of S is 8.

18. D. The sum of horizontal and vertical lengths on opposite sides of this shape are equal. (See Polygons chapter Book 2 for more on this rule). As an example, notice how in this figure, $\overline{GE}, \overline{ED}, and\ \overline{CB}$ correspond exactly to sections of $\overline{AH}$ to the left. Thus, the perimeter is equal to 2 times the vertical length of the shape, 10, plus 2 times the horizontal length of the shape, $5+4+2$. The perimeter is equal to $2(10)+2(5+4+2)=2(10)=2(11)=20+22=42$. (For more questions like this, see Polygons, Book 2).

19. B. The vertical lengths $\overline{GE}$, $\overline{ED}$, and CB are equal to 10. $GF=2$ and $CB=5$, so by substitution, $2+ED+5=10$. Simplifying, $ED=3$. The horizontal distance between A and E is equal to $HG+FE=5+4=9$. The vertical distance between A and E is equal to $BC+ED=5+3=8$. Since A is at $(0,0)$, the coordinates of E are $(9,8)$. Point J is the midpoint of the two, so it is at $\left(\frac{0+9}{2},\frac{0+8}{2}\right)=(4.5,4)$.

20. D. B is 11 units to the right of $(0,0)$, giving it the coordinates $(11,0)$. The distance between J and B is given by the Pythagorean theorem. J is at $(4.5,4)$. We can plug that in as:

$$\sqrt{(11-4.5)^2+(0-4)^2}=\sqrt{6.5^2+(-4)^2}=\sqrt{42.25+16}=\sqrt{58.25}\approx 7.6.$$

21. B. The midpoint of $\overline{AC}$, given by the midpoint formula, is $\left(\frac{-5+10}{2},\frac{4+(-7)}{2}\right)=\left(\frac{5}{2},\frac{-3}{2}\right)$. The midpoint of $\overline{BD}$ is $\left(\frac{-2+8}{2},\frac{-1+3}{2}\right)=\left(\frac{6}{2},\frac{2}{2}\right)=(3,1)$. The distance between the midpoints is:

$$\sqrt{\left(3-\frac{5}{2}\right)^2+\left(1-\left(-\frac{3}{2}\right)\right)^2}=\sqrt{\left(\frac{1}{2}\right)^2+\left(\frac{5}{2}\right)^2}=\sqrt{\frac{1}{4}+\frac{25}{4}}=\sqrt{\frac{26}{4}}=\frac{\sqrt{26}}{2}.$$

22. E. The water fountain is at the midpoint between B and E: $\left(\frac{40+50}{2},\frac{10+150}{2}\right)=\left(\frac{90}{2},\frac{160}{2}\right)=(45,80)$. The distance between this point and A is $\sqrt{(45-15)^2+(80-100)^2}=\sqrt{900+400}=\sqrt{1300}$.

23. B. Keep in mind that distance is always a positive quantity. The distance in the y direction is $3-(-15)=18$. Be sure to read carefully, this question only asks for y-value distance, NOT total distance. You only need to measure vertical movement.

24. E. Plug the coordinates into the distance formula: $\sqrt{(9-3)^2+(-3-4)^2}=\sqrt{6^2+(-7)^2}=\sqrt{36+49}=\sqrt{85}$.

25. C. We find the distance between two points using the Pythagorean theorem, as expressed in the formula $\sqrt{(x_2-x_1)^2+(y_2-y_1)^2}$. Plugging in the points from the equation, we get $\sqrt{(3x-x)^2+((x+5)-(x-3))^2}=\sqrt{(2x)^2+(8)^2}=\sqrt{4x^2+64}=\sqrt{4}\sqrt{x^2+16}=2\sqrt{x^2+16}$.

CHAPTER

7 INEQUALITIES: CORE

SKILLS TO KNOW

- Flip the sign when multiplying/dividing by a negative
- How to graph inequalities on a number line
- How to "pluck points"

THE BASICS

Inequalities are **just like equations**—with one big exception. If you're multiplying or dividing both sides of the inequality by a negative number, you must "flip" the sign to the other direction:

$-3x > 9$ Divide by negative three

$\frac{-3x}{-3} < \frac{9}{-3}$ Flip the sign!

$x < -3$ Simplify

INEQUALITIES ON A NUMBER LINE

You'll need to know how to graph a basic inequality on a number line and properly shade. Remember **open circles** refer to **less than/greater than** and **closed circles** refer to **less than or equal to/greater than or equal to.** For example, in the line below, $-5 < x \leq 2$ because the point at -5 is an open circle, meaning the exact point -5 is not included in the solution, while the point at 2 is a closed, black shaded circle, meaning that 2 is included in the solution set.

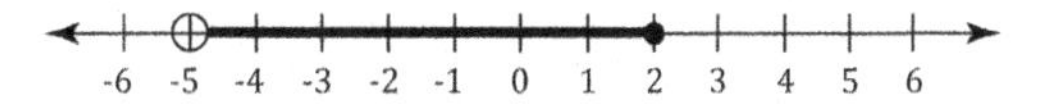

Which of the following number line graphs shows the solution set of $-6(x+2)+1 \geq 4-x$?

A. -6 -5 -4 -3 -2 -1 0 1 2 3 4 5 6

B. -6 -5 -4 -3 -2 -1 0 1 2 3 4 5 6

C. -6 -5 -4 -3 -2 -1 0 1 2 3 4 5 6

D. -.2 -6 -5 -4 -3 -2 -1 0 1 2 3 4 5 6

E. -6 -5 -4 -3 -2 -1 0 1 2 3 4 5 6

$$-6(x+2)+1 \geq 4-x \quad \text{To simplify, distribute the multiplier, } -6$$

$$-6x-12+1 \geq 4-x \quad \text{Group like terms together}$$

$$-6x+x \geq 4+12-1 \quad \text{Simplify}$$

$$-5x \geq 15 \quad \text{Divide by negative five}$$

$$\frac{-5x}{-5} \leq \frac{15}{-5} \quad \text{Flip the sign!}$$

$$x \leq -3 \quad \text{Simplify}$$

Answer: **B.**

WHEN IN DOUBT, PLUCK POINTS

If you're ever unsure of how to solve an inequalities problem, plucking points is a great way to solve if you have graphed answers on number lines. Take advantage of the fact that the ACT® is multiple choice!

Remember, every part of the shaded line must make the inequality statement true, so if more than one seems possible to you, find a point that is only on one of the two number lines in question and test that point. Then, use process of elimination.

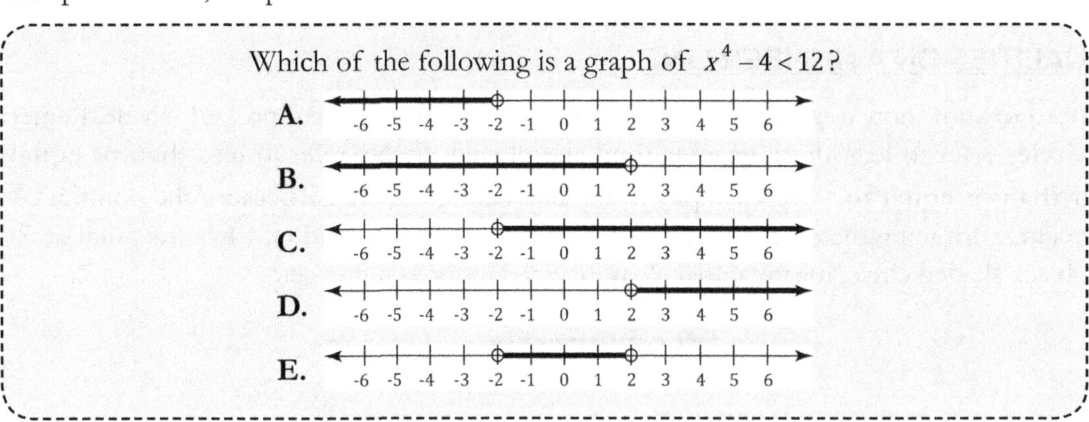

As you can see, all the answer choices are some combination of an inequality that involves 2 or negative 2. To do the actual math on this problem is probably a waste of time: you're going to be comparing x to some version of 2 or -2. Instead, let's **pluck points** to find the answer.

Zero (0) is an answer to A, C, and E. Let's test that first, plugging in 0 for x.

$$0^4-4<12$$
$$-4<12 \text{ : TRUE}$$

Because this statement is true, we can eliminate D. Now let's check 6, part of the solution for C.

$$6^4-4<12$$

At this point, I know the term on the left is going to be pretty big ($6\times6\times6\times6$ is 36 squared!). This won't be true.

Now let's check -6, part of the solution for choices A and B.

$$(-6)^4 - 4 < 12$$

This is going to give the same result as choice C. This will not be true, as the sign will go away when we take this number to the 4th power.

Only choice E remains. It is correct.

Answer: **E**.

1. Which of the following is a solution for the inequality $\frac{9}{5}a+3>\frac{3}{4}a-7$?

   A. $a>-\frac{5}{3}$

   B. $a>-\frac{210}{20}$

   C. $a>-\frac{200}{21}$

   D. $a<-\frac{200}{21}$

   E. $a<-\frac{210}{20}$

2. Which of the following is equal to the inequality $5n-21<13+2n$?

   A. $n<\frac{34}{3}$

   B. $n<-\frac{8}{3}$

   C. $n>-\frac{8}{3}$

   D. $n>\frac{34}{3}$

   E. $n<\frac{8}{3}$

3. The inequality $8(y-4)>7(y+2)$ is equivalent to which of the following?

   A. $y<46$

   B. $y>46$

   C. $y>6$

   D. $y>18$

   E. $y<34$

4. Which of the following is equal to $\frac{4}{2-x}-8>0$?

   A. $x>2$

   B. $\frac{3}{2}<x<2$

   C. $x<\frac{3}{2}$

   D. $x>\frac{3}{2}$

   E. $x>\frac{3}{2}$ or $x>2$

5. Which of the following is equivalent to $(|a|-7)^3\geq 8$?

   A. $a\geq 7$ or $a\leq -7$

   B. $a\geq 8$ or $a\leq -8$

   C. $a\geq 15$ or $a\leq -15$

   D. $a\geq 3$ or $a\leq -3$

   E. $a\geq 9$ or $a\leq -9$

6. For what values of n is $\frac{1}{4}n-9>\frac{5}{2}n$?

   A. $n>-4$

   B. $n<-4$

   C. $n<4$

   D. $n<36$

   E. $n>46$

7. What is the smallest integer value x that satisfies the inequality $\frac{x}{20}>\frac{13}{23}$?

   A. 10

   B. 11

   C. 12

   D. 13

   E. 14

8. Which of the following graphs shows the solution set for the inequality $7x+2>9$?

A.

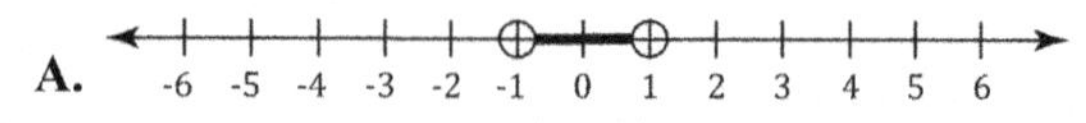

B. -6 -5 -4 -3 -2 -1 0 1 2 3 4 5 6

C. -6 -5 -4 -3 -2 -1 0 1 2 3 4 5 6

D.

E.

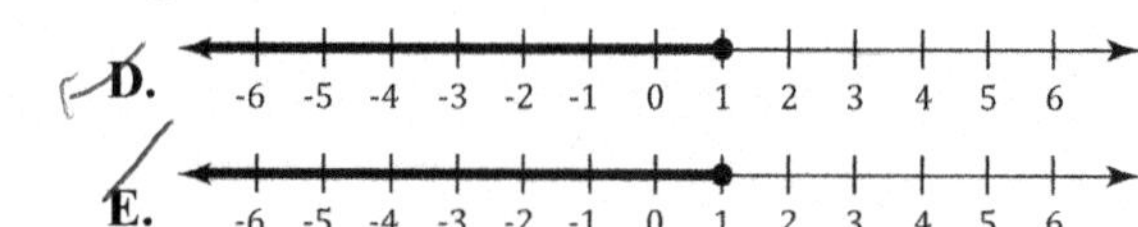

9. Which of the following is the graph of the solution set for the inequality $11-\frac{x}{2}\leq 9$?

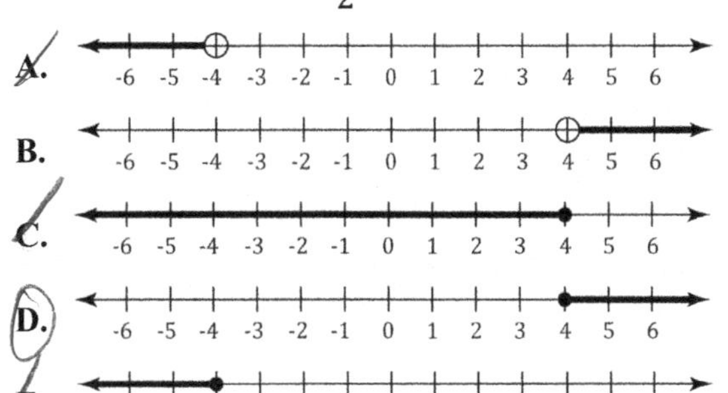

10. When 3 times x is increased by 13, the result is less than 4. Which of the following is a graph of the real numbers x that satisfy this relationship?

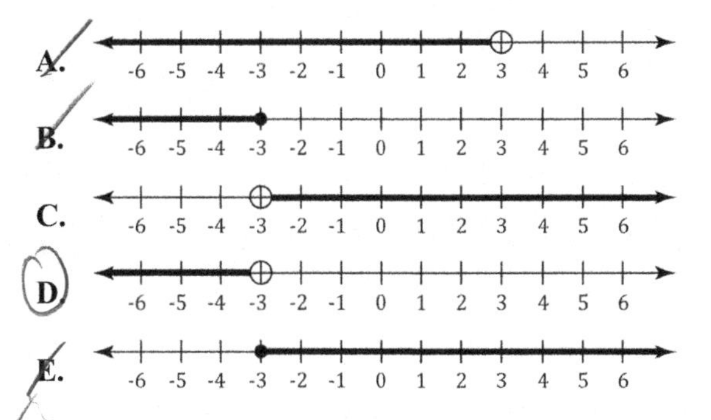

11. Which of the following inequalities represents the graph shown below on the real number line?

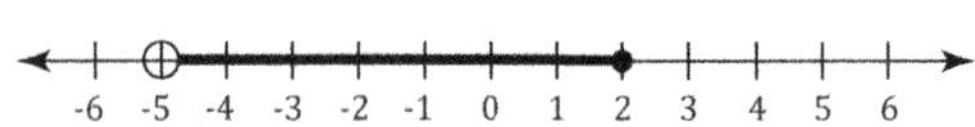

A. $-5\leq x\leq 2$
B. $-5<x<2$
C. $-5<x\leq 2$
D. $x<-5$ and $x\geq 2$
E. $x\leq -5$ and $x>2$

12. The number lined graphed below is the graph of which of the following inequalities?

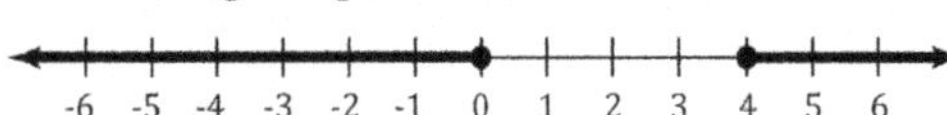

A. $x\geq 0$ and $x\leq 4$
B. $x>4$ and $x<0$
C. $x>0$ or $x<4$
D. $x<0$ or $x>4$
E. $x\leq 0$ or $x\geq 4$

13. Which of the following graphs illustrates the solution set for the system of inequalities $3x-25\geq -15$ and $-4x+10<-10$?

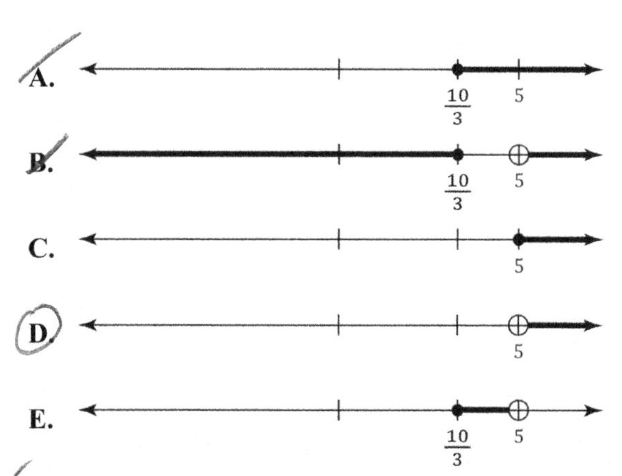

14. Which of the following number line graphs shows the solution set for x of $x^2<16$?

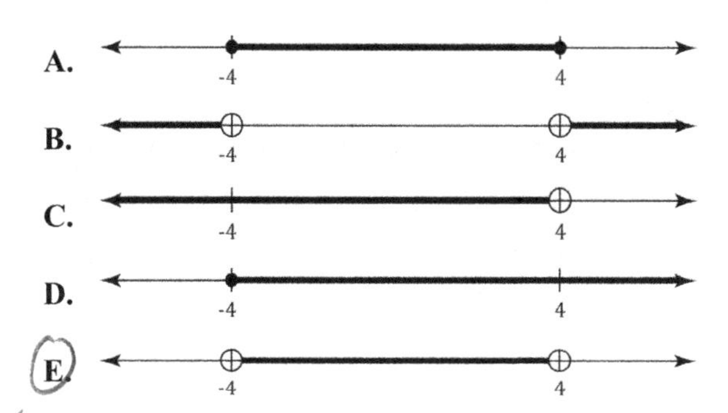

15. Given real numbers $a, b, c, d,$ and e such that $b<a$, $e<d$, $a<c$, and $d<b$, which of these numbers is the least?

A. a
B. b
C. c
D. d
E. e

ANSWER KEY

1. C **2. A** **3. B** **4. B** **5. E** **6. B** **7. C** **8. C** **9. D** **10. D** **11. C** **12. E** **13. D** **14. E** **15. E**

ANSWER EXPLANATIONS

1. **C.** Adding 7 on both sides, we get $\frac{9}{5}a+10>\frac{3}{4}a$. Subtracting $\frac{9}{5}a$ on both sides, we get $10>\frac{3}{4}a-\frac{9}{5}a \rightarrow 10>\frac{3}{4}\left(\frac{5}{5}\right)a-\frac{9}{5}\left(\frac{4}{4}\right)a \rightarrow 10>\frac{15}{20}a-\frac{36}{20}a \rightarrow 10>-\frac{21}{20}a$. Now, dividing both sides by $-\frac{21}{20}$, and flipping the sign, we get $10\left(-\frac{20}{21}\right)<a \rightarrow -\frac{200}{21}<a$ or $a>-\frac{200}{21}$.

2. **A.** Adding 21 on both sides, we get $5n<34+2n$. Subtracting $2n$ on both sides, we get $3n<34$. Dividing by 3 on both sides, we get our answer, $n<\frac{34}{3}$.

3. **B.** Distributing the constants on both sides of the inequality, we get $8y-8(4)>7y+7(2) \rightarrow 8y-32>7y+14$. Adding 32 on both sides, we get $8y>7y+46$. Subtracting both sides by $7y$, we get $y>46$.

4. **B.** Adding 8 on both sides, we get $\frac{4}{2-x}>8$. The denominator is an expression that could be either negative or positive, depending on the value of $2-x$. Thus, when we multiply by $2-x$ on either sides, we get $4<8(2-x)$ if $x<2$ (which would make the expression positive) and $4>8(2-x)$ if $x>2$ (which would make the expression negative, and this requires the sign to change directions). Distributing the 8 on the right side, we get $4<16-8x$ or $4>16-8x$. Subtracting 16 on both sides, we get $-12<-8x$ or $-12>-8x$. Lastly, dividing each side by -8 gives us $\frac{3}{2}<x$ when $x<2$, or $\frac{3}{2}>x$ when $x>2$. The second statement gives us no solution because no number can be simultaneously less than $\frac{3}{2}$ and greater than 2. Thus, our answer is $\frac{3}{2}<x<2$.

5. **E.** Taking the cube root of both sides of the equation, we get $|a|-7\geq 2$. Adding 7 on both sides gives us $|a|\geq 9$. This means $a\geq 9$ or $a\leq -9$.

6. **B.** Subtracting $\frac{1}{4}n$ on both sides, we get $-9>\frac{5}{2}n-\frac{1}{4}n \rightarrow -9>\frac{9}{4}n$. Dividing both sides by $\frac{9}{4}$, we get $-9\left(\frac{4}{9}\right)>n \rightarrow -4>n$ or $n<-4$.

7. **C.** Rewriting the fractions with a common denominator, we get $\frac{x}{20}\left(\frac{23}{23}\right)>\frac{13}{23}\left(\frac{20}{20}\right) \rightarrow \frac{23x}{460}>\frac{260}{460}$, which simplifies to $23x>260$. Dividing both sides by 23, we get $x>11.3$ and the smallest integer value that is greater than 11.3 is 12.

8. **C.** Subtract 2 from both sides and divide by 7 to get $x>1$. x is greater than, not greater than or equal to, so the circle in open or unfilled, since $x\neq 1$.

9. **D.** Subtract 11 from both sides and multiply both sides by -2. Remember to switch the direction of the sign because we are multiplying both sides by a negative number. $11-\frac{x}{2}\leq 9 \rightarrow -\frac{x}{2}\leq -2 \rightarrow x\geq 4$. Because $x\geq 4$, x can equal 4, so the circle on 4 is filled.

10. D. Translating our inequality into numbers, $3x+13<4$. Now simplify: $3x+13<4 \rightarrow 3x<-9 \rightarrow x<-3$. x is less than but *not* equal to 3, so our bubble is empty.

11. C. Looking at the graph, we see that the thick line starts at -5 and ends at 2, so x can be any value in between the two, and that the bubble is unfilled at -5 but filled at 2, meaning that x *cannot* be -5 but *can* be 2. As an inequality, this looks like $-5<x\leq 2$.

12. E. Looking at the graph, we see that the thick line that is our x spans values that are less than 0 and greater than 4 and that the circles at 0 and 4 are filled, which means that x could potentially be 0 or 4. Than means that our x values are less than or equal to 0 *or* greater than or equal 4: $x\leq 0$ or $x\geq 4$. It is logically impossible for a number to be less than or equal to 0 *and* greater than or equal 4, which is why we say *or*.

13. D. First simplify the inequalities given: $3x-25\geq -15 \rightarrow x\geq \frac{10}{3}$ and $-4x+10<-10 \rightarrow -4x<-20 \rightarrow x>5$. The solution of the system is the intersection (AKA overlap) of the two inequalities: $x\geq \frac{10}{3} \cap x>5$, which simplifies to $x>5$, because anything that is greater than 5 will always be greater than $\frac{10}{3}$, and both conditions must be true.

14. E. To solve this inequality, we can approach in few ways: 1. Pluck points OR turn the inequality into an equation, solve for the significant or "hinge" points, and test each region between the significant points. Solving this way, we take the square root of both sides $\sqrt{x^2}=\pm\sqrt{16} \rightarrow x=\pm 4$. (Often the answer choices will tell you these significant points so you can skip this step and just start by testing regions or plucking points, too). Now we can take the three regions and test a point in each: $x<-4, 4<x<-4, x>4$ Plugging in one number in each of these ranges respectively (I simply make up some numbers), I see:

$(-6)^2=36<16$ FALSE $\quad 2^2<16$ TRUE $\quad 5^2<16$ FALSE

Thus I need to shade between the two points. Because my symbol is less than/greater than NOT less than or equal to/ greater than or equal to, I know I need open circles so choice E is correct.

2. We can solve by thinking about the problem and solving algebraically. When we take the square root, it is the same as when we multiple or divide by a negative when the variable is negative, so for any negatives we'll flip the sign on our inequality: $\sqrt{x^2}<\pm\sqrt{16} \rightarrow x<\sqrt{16}$ and/or $x>-\sqrt{16}$. Solving both inequalities we see that $x<4$ and $x>-4$, which can also be written as $-4<x<4$, which is represented by the bottommost graph because that graph shows the set of numbers between but not including -4 and 4.

15. E. We cannot relate the first two inequalities given, but the first and the third combined is $b<a<c$. Because $e<d$ and $d<b$, we get $e<d<b<a<c$. Thus, e is the least element.

CHAPTER

INEQUALITIES: ADVANCED

SKILLS TO KNOW

- How to graph/shade inequalities in a coordinate plane
- How to think abstractly about math involving inequalities
- How to solve "must be" and "could be" true problems with inequalities
- How to solve absolute value inequalities (see: **Chapter 14, Absolute Value**)

INEQUALITIES ON A COORDINATE PLANE

These can seem complicated, but they don't have to be. A few tips to remember:

1. You can always "pluck points."
Most of the time on a multiple choice question you'll be given a system of inequalities and every picture will have the same lines—only the shading will be different. As such, you can always choose test points, plug them in, and use the process of elimination to then eliminate choices. $(0,0)$ is almost always the easiest point to test.

2. Remember—get y isolated. If the y is **greater** than the line, **shade above the line.** If y is **less than** the line, **shade below.** Even if you're in function notation, $f(x)$, the same rule applies. This idea can also be applied to parabolas. Shade up or down from the line according to the inequality sign.

The shaded portion of one of the following graphs is the set of all x and y such that $-1 < x + y < 1$. Which one?

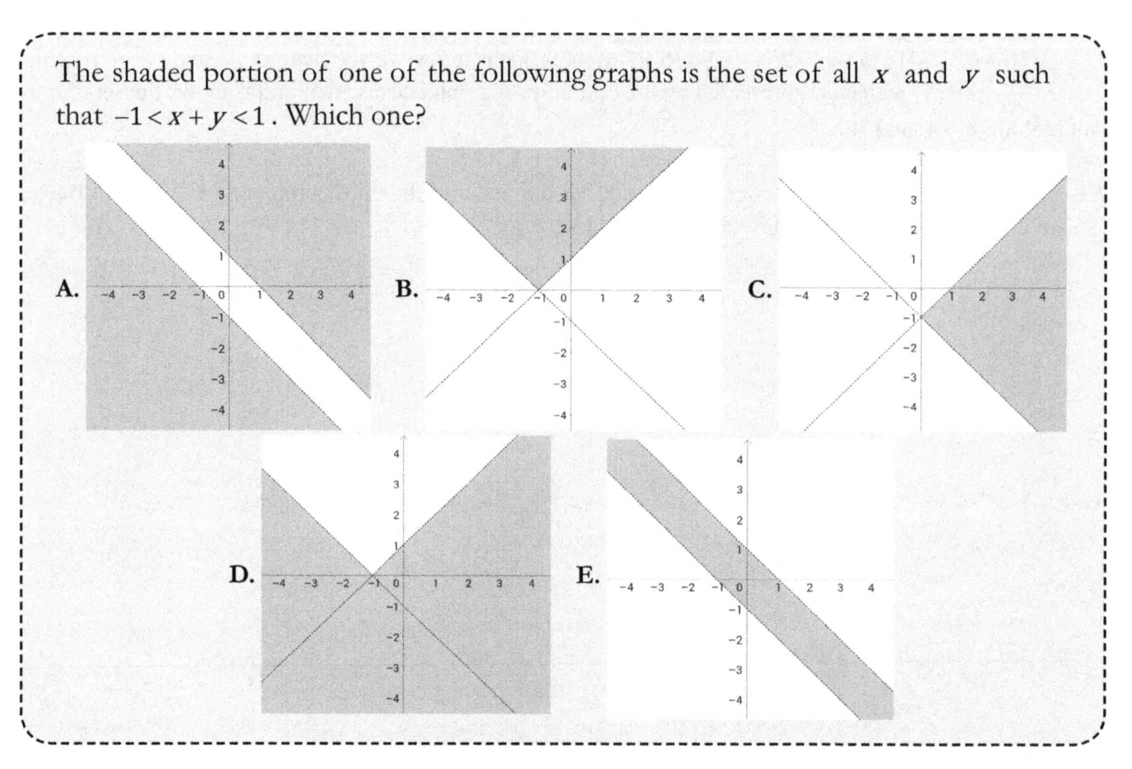

First we need to split this into two inequalities and solve for each.

$-1<x+y<1$ can become two:

$-1<x+y$ and $x+y<1$.

Isolate the y value for each:

$$-1<x+y \qquad x+y<1$$
$$-x-1<y \qquad y<-x+1$$
$$y>-x-1$$

Now let's think about what these mean. The first of these ($y>-x-1$) says to shade "up" (above the y) from the line $-x-1$: that line is going to have a negative slope and a y-intercept of -1. Choices (A) through (E) above all have that line, but only (A), (C), & (E) shade exclusively above this line.

The second of these inequalities, $y<-x+1$, indicates that we should shade "under" the line $y=-x+1$. That line has a y-intercept of 1 and a slope of -1. Of the remaining choices, only (A) and (E) include this line, and only (E) shades below it, so that is our answer.

We could also make other observations that would help us reach this same conclusion: both lines have the same slope, we're looking for parallel lines—clearly (A) & (E). Then pluck point (0,0) to see that it should be shaded. Again, (E) is correct.

Answer: **E.**

What is the graph of the system:

$$f(x)>3(x-2)^2+1$$

$$g(x)\le-\frac{x}{2}+4$$

Given the inequality signs, we shade above the parabola and below the line. We can also look to ensure we have the right parabola and line. From vertex form, we know the parabola vertex is (2,1) and that it faces upwards because the coefficient of the x term is positive (3). We also know that the line has a y-intercept of 4 and a slope of -1/2 given the slope intercept form of the inequality. See Graph Behavior chapter for more on determining graph features from equations (or inequalities) alone.

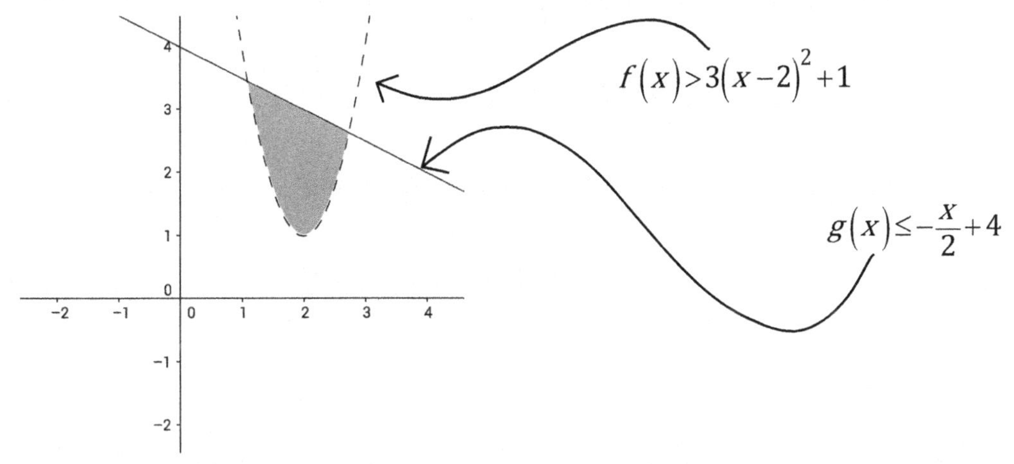

3. Use a DOTTED line with graphs of inequalities using the signs > and < (as in $f(x)$ above). Use a SOLID line with graphs of inequalities using the signs ≤ and ≥ (as in $g(x)$ above).

4. You always have your calculator.
The TI-84 series and TI-Nspire (note: only TI-Nspire non-CAS is permitted for the ACT®) can graph equations or inequalities to help you.

COULD BE TRUE/MUST BE TRUE

Oftentimes, questions that have inequalities in them are testing your understanding of basic math concepts or number behavior, such as how positive or negative numbers react in different situations. (I also cover similar questions to this in Book 2: Chapter 2, Properties of Numbers). When you have these problems:

1. Try to simplify algebraically first.
2. Plug in numbers if you get stuck.

Let p and q be numbers such that $|p| \leq 0 < q$. Which of the following expressions *must be* true for all such p and q?

A. $q \div p < 0$ **B.** $p + q < 0$ **C.** $pq = 0$ **D.** $p - q > 0$ **E.** $p + q < q$

Let's start with the first part of this inequality:

$$|p| \leq 0$$

Now let's think about what this means: the distance between p and 0 is at most 0. If there is no distance between these values, then p must be equal to 0. What's more, absolute values can only be positive or 0, meaning p must be at least zero but also less than or equal to zero, so p will have to be 0:

$$p = 0$$

From that conclusion, we can look at the answer choices and see which requires p to equal 0:
(A) Mathematically impossible—you can't divide anything by 0.
(B) Doesn't work—q must be positive and p is 0—that makes a positive number.
(C) From the zero product property: $pq = 0$ implies that either p or q must equal 0. If p equals 0, then this expression must be true. P is 0, so this expression is true.
(D) If p is 0 and we subtract a positive number q, the result will be negative, not greater than zero.
(E) Substituting in 0 for p we get q is less than q, an expression that makes no sense.

Answer: **C.**

Another way to attempt that problem, if we can't synthesize the ideas above, is to make up some numbers. Let's try $p = 5$:

$$|5| \leq 0 < q$$

That doesn't work—5 isn't less than or equal to zero! What about -5?

$$|-5| = 5$$

That's still not less than or equal to zero! You can keep trying numbers—but a good rule of thumb is to always try from the following list:

LIST OF NUMBERS TO TRY
(for must be true/could be true problems)

1. A negative number (-5)
2. A positive number (4)
3. Zero
4. One
5. Negative one
6. A positive fraction $(1/4)$
7. A negative fraction $(-1/9)$
8. A huge number $(1{,}000{,}000{,}000$ *etc.*$)$
9. A very small number $(-1{,}000{,}000$ *etc.*$)$

Here, by trying #3, we'll get to zero, see that it works and would continue on as we did above.

1. How many different integer values of d satisfy this inequality $\frac{2}{15}<\frac{4}{d}<\frac{6}{13}$?

A. 20
B. 22
C. 29
D. 18
E. 21

2. For which values of n is $n>n^2$?

A. $n<-1$
B. $n>1$
C. $0<n<1$
D. $-1<n<0$
E. No real numbers

3. Which of the following inequalities is representative of the statement below? 35 is greater than $\frac{1}{3}$ the difference between 10 and the reciprocal of a number p. (Two numbers are reciprocals if their product equals 1).

A. $35>\frac{1}{3}\left(10-\frac{1}{p}\right)$
B. $35+\frac{1}{3}>10-\frac{1}{p}$
C. $35>\frac{10}{3}-\frac{1}{p}$
D. $35>\frac{1}{3}-10\left(\frac{1}{p}\right)$
E. $35>\frac{1}{3}\left(\frac{1}{10-p}\right)$

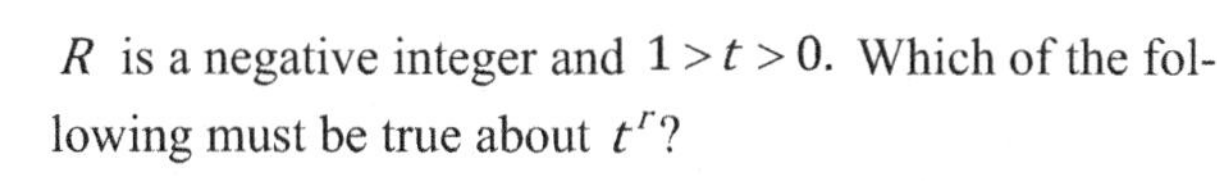

4. R is a negative integer and $1>t>0$. Which of the following must be true about t^r?

A. $t^r \leq 1$
B. $t^r < 1$
C. $t^r > 1$
D. $t^r \geq 1$
E. $t^r = 1$

5. A and b are two numbers such that $0<b<a$. All of the following statements are true except:

A. $a+1>b+1$
B. $a^2>b^2$
C. $\frac{1}{b}>\frac{1}{a}$
D. $\frac{1}{a}>\frac{1}{b}$
E. $-b>-a$

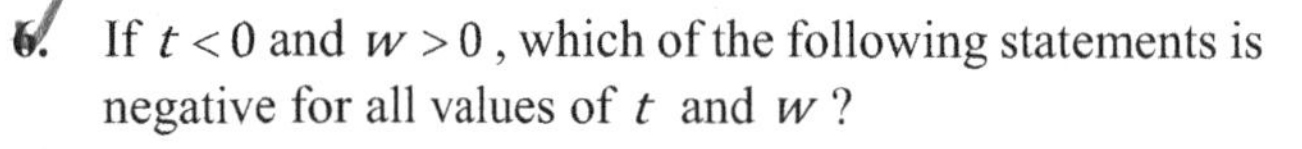

6. If $t<0$ and $w>0$, which of the following statements is negative for all values of t and w?

A. t^2+w
B. tw^2
C. w^2t^2
C. $t+w$
E. $w-t$

7. Which of the following gives the solution set for the system of inequalities below?

$$n \geq 10$$
$$9-3n \leq 0$$

A. $n \geq 3$
B. $n \geq 10$
C. $10 \geq n \geq 3$
D. $n \leq 3$ or $n \geq 10$
E. $n \leq 3$

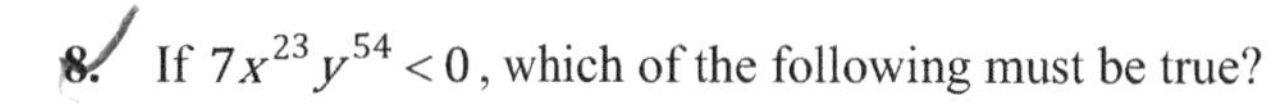

8. If $7x^{23}y^{54}<0$, which of the following must be true?

A. $x<0$
B. $y>0$
C. $x>y$
D. $y<0$
E. $x<y$

9. The variables x, y, z are defined by the following statements.

1. y is 8 more than x
2. z is at least 4 less than y

Which of the following inequalities expresses the relationship between x and z?

A. $x \geq z-4$
B. $x \geq -z-4$
C. $x \leq z-4$
D. $x \leq -z-4$
E. $x \geq z+12$

10. If g is a variable between 0 and 1, not inclusive, which of the following is the largest?

A. g^2
B. g^3
C. g^4
D. g^5
E. $\sqrt[3]{g}$

11. For positive real numbers x, y, z such that $\frac{1}{2}x = y\sqrt{5} = \frac{z\sqrt{5}}{7}$, what is the correct ordering of x, y, z?

A. $z > x > y$
B. $y > x > z$
C. $y > z > x$
D. $x > z > y$
E. $z > y > x$

12. If p, q, r are real numbers such that $p^6q^9r^{10} > 0$, which of the following must be greater than 0?

A. pr
B. pq
C. rq
D. p^2q
E. pq^2

13. When $-8 \leq a \leq -7$ and $2 \leq b \leq 3$, what is the largest value of $a-b$?

A. -10
B. -9
C. -11
D. -6
E. -5

14. The Congress of a certain country has 267 members. If more than $\frac{2}{3}$ of the members of the Congress must vote yes on a certain bill to pass it, which of the following expressions shows the necessary number of yes votes, x, for a bill to pass?

A. $x = 178$
B. $x > 178$
C. $x < 179$
D. $x < 89$
E. $x < 177$

15. A line contains the points $P, Q, R,$ and S. Point R is between P and S. Point Q is between Point R and Point S. Which of the following inequalities must be true?

A. $PQ > PR$
B. $PR > QR$
C. $PR < RS$
D. $RQ > RS$
E. $QS > RS$

16. Whenever $(1-x)(x-5) > 0$, which of the following expressions *always* has a negative value?

A. $-x+5$
B. $x-1$
C. $-2x+5$
D. x^2-6x+2
E. $(x-3)(x-1)$

17. If $a \neq b$, what are the real values of a that make the following inequality true?

$$\frac{ab-b}{2b-4b} < 0$$

A. $a = 1$
B. $a < 1$
C. $a > 1$
D. all positive real numbers
E. all negative real numbers

18. If j and k can be any integers such that $4j-2k=12$ and $k<8$, which of the following is the solution set for j?

A. $j < 7$
B. $j > 0$
C. $j > 7$
D. $j < 0$
E. $j > 1$

19. For real numbers a and b such that $0 < ab < a < b$, which of the must be true?

A. $a > 1$ and $b > 1$
B. $a < 1$ and $b < 1$
C. $a < 1$ and $b > 1$
D. $ab > a+b$
E. $ab > a-b$

20. Whenever $\frac{a}{b} < \frac{a}{c} < \frac{a}{d} < 1$ is true for some positive integers $a, b, c,$ and d, which of the following, if it can be determined, must be the order of $\frac{b}{a}, \frac{c}{a},$ and $\frac{d}{a}$?

A. $1 < \frac{d}{a} < \frac{c}{a} < \frac{b}{a}$

B. $\frac{d}{a} < \frac{c}{a} < \frac{b}{a} < 1$

C. $1 < \frac{b}{a} < \frac{c}{a} < \frac{d}{a}$

D. $\frac{b}{a} < \frac{c}{a} < \frac{d}{a} < 1$

E. $1 < \frac{c}{a} < \frac{d}{a} < \frac{b}{a}$

21. Given that $a > b + 1$ and $(a + b) < (a^2 - b^2)$, then all possible values for the sum of a and b could be most specifically described as:

A. Zero
B. Positive
C. Negative
D. Between 0 and 1
E. Between 0 and −1

22. Which of the following is the solution statement for the inequality $14x - 3 \leq 9x + 4$?

A. $x \leq \frac{7}{5}$

B. $x \leq \frac{5}{7}$

C. $x \leq \frac{1}{5}$

D. $x \leq \frac{1}{3}$

E. $x \leq \frac{7}{3}$

23. Tickets for the Annual Poetry Slam at Van Buren High School are \$8 for adults and \$4 for students. In order to make a profit, at least \$200 must be collected from ticket sales for the show. One of the following graphs in the standard (x,y) coordinate plane, where x is the number of adult tickets sold and y is the number of student tickets sold represents all possible combinations of ticket sales necessary to make a profit. Which graph is it?

A.

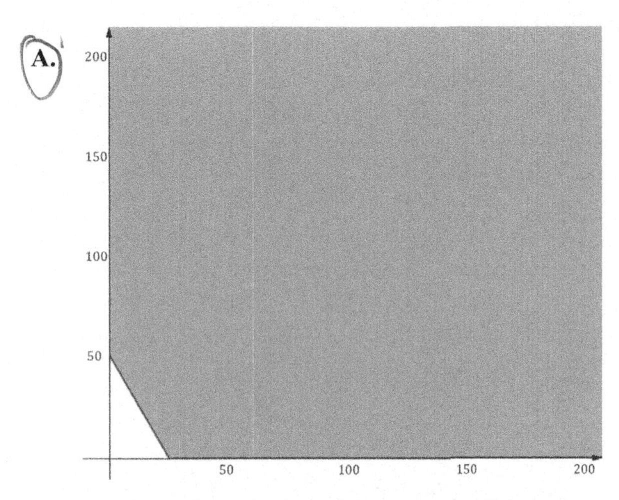

B.

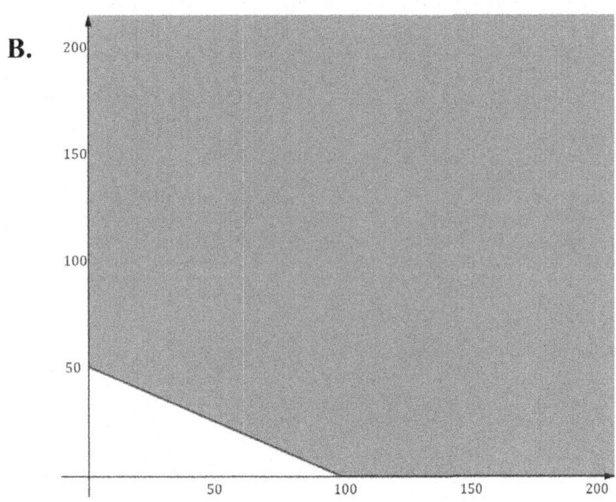

C.

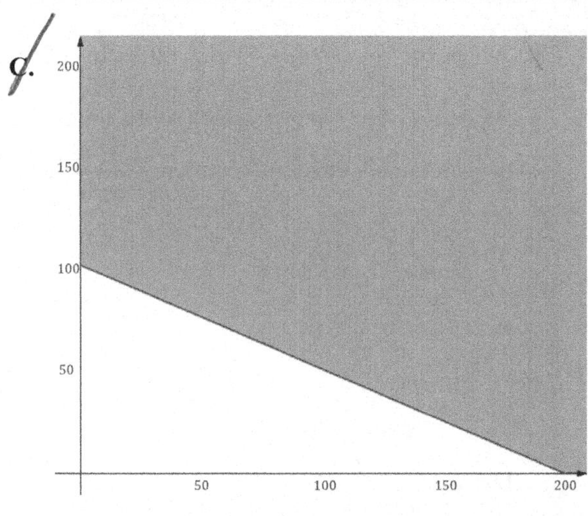

D.

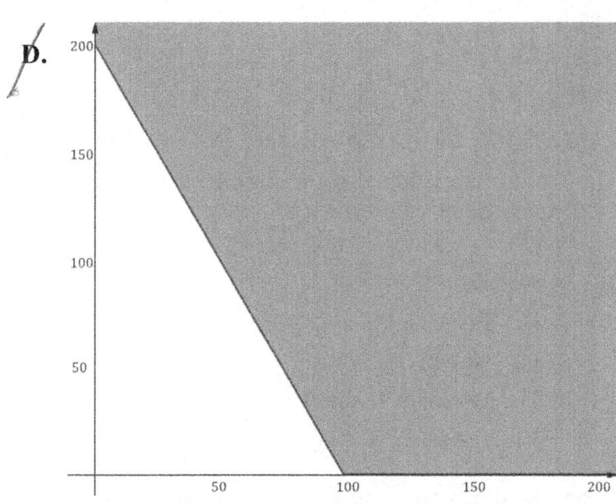

E.

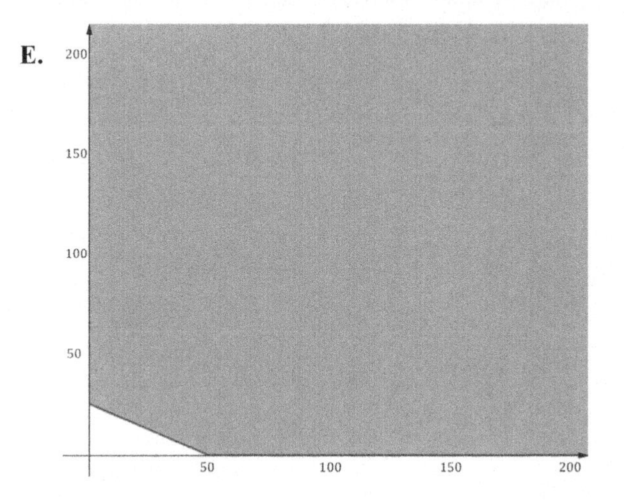

24. One of the following graphs in the standard (x, y) coordinate plane is the graph of $y > ax + b$ for some positive a and positive b. Which graph?

A.

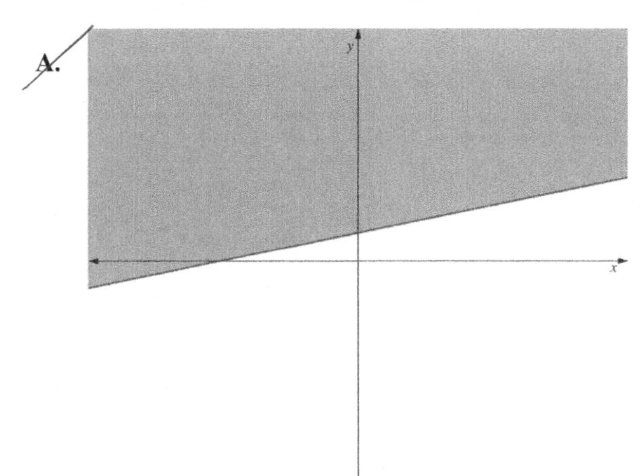

B.

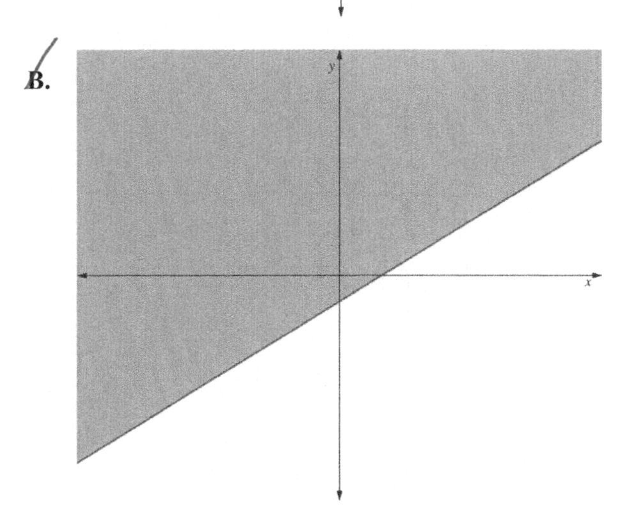

C.

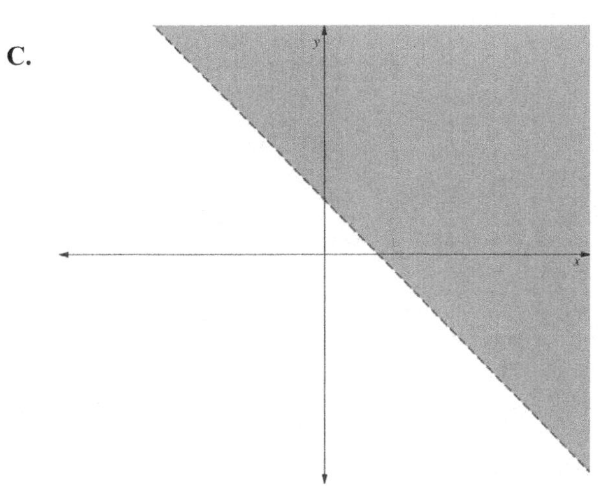

D.

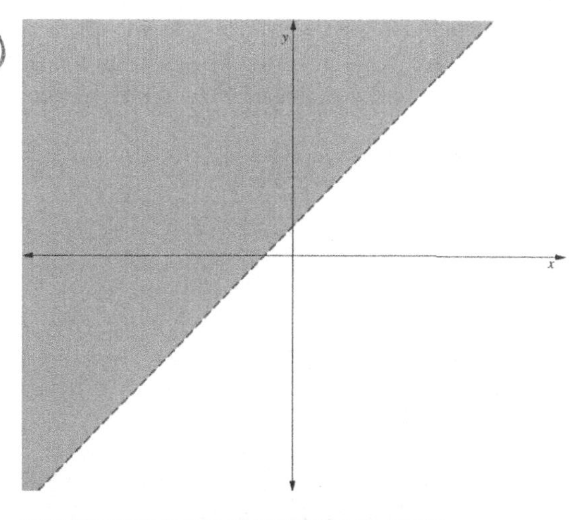

E.

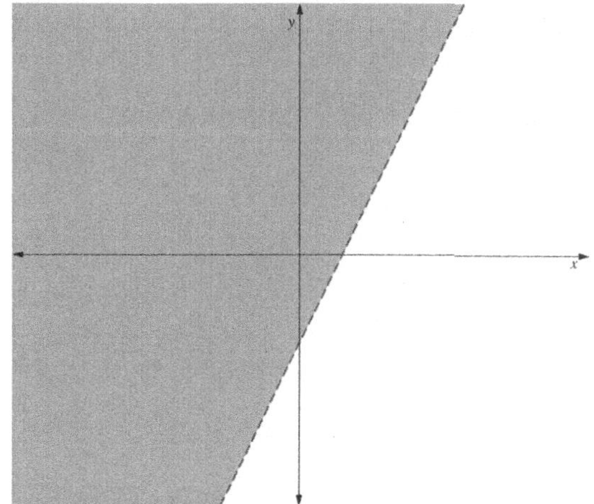

25. One of the following inequalities, where both constants a and b are positive real numbers, is graphed in the standard (x, y) coordinate plane below. Which inequality is it?

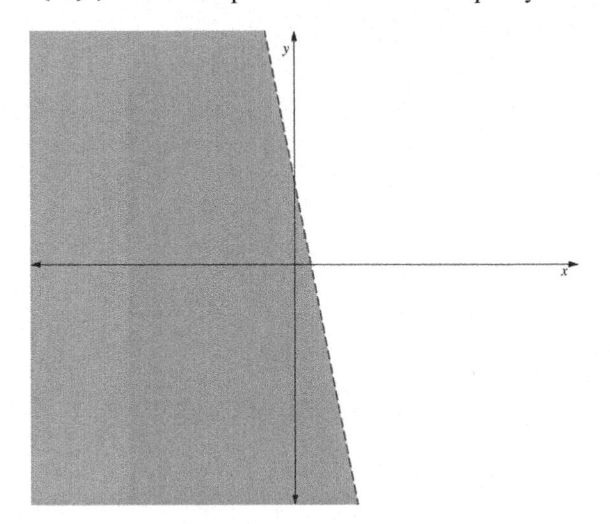

A. $y \geq ax + b$
B. $y > -ax + b$
C. $y < ax - b$
D. $y < -ax + b$
E. $y \leq ax + b$

26. The shaded region in the standard (x, y) coordinate plane below is bounded by a parabola and a line. The shaded region and its boundary is the solution set of which of the following systems of inequalities?

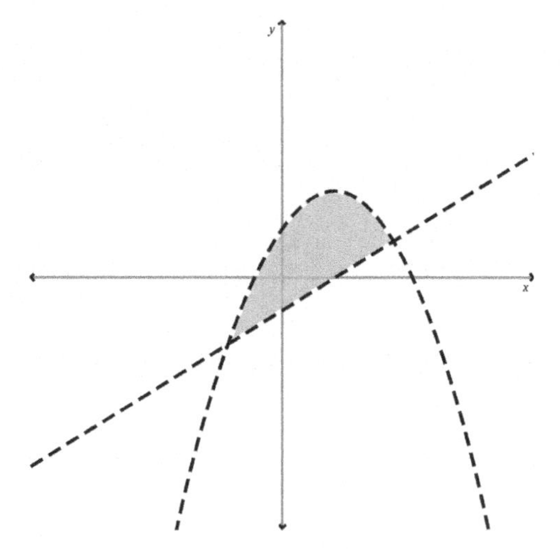

A. $\begin{cases} y \leq \left(\dfrac{x}{2} - 2\right)^2 + 6 \\ y \geq \dfrac{x}{2} - 2 \end{cases}$

B. $\begin{cases} y > \left(\dfrac{x}{2} - 2\right)^2 + 6 \\ y < \dfrac{x}{2} - 2 \end{cases}$

C. $\begin{cases} y \leq \left(\dfrac{x}{2} - 2\right)^2 + 6 \\ y < \dfrac{x}{2} - 2 \end{cases}$

D. $\begin{cases} y < \left(\dfrac{x}{2} - 2\right)^2 + 6 \\ y > \dfrac{x}{2} - 2 \end{cases}$

E. $\begin{cases} y > \left(\dfrac{x}{2} - 2\right)^2 + 6 \\ y > \dfrac{x}{2} - 2 \end{cases}$

27. The graph of the equations $y_1 = x^5 - \frac{x}{2} + 1$ and $y_2 = \frac{x}{2} + 1$ are shown in the standard (x, y) coordinate plane below. What real values of x, if any, satisfy the inequality $x^5 - \frac{x}{2} + 1 > \frac{x}{2} + 1$?

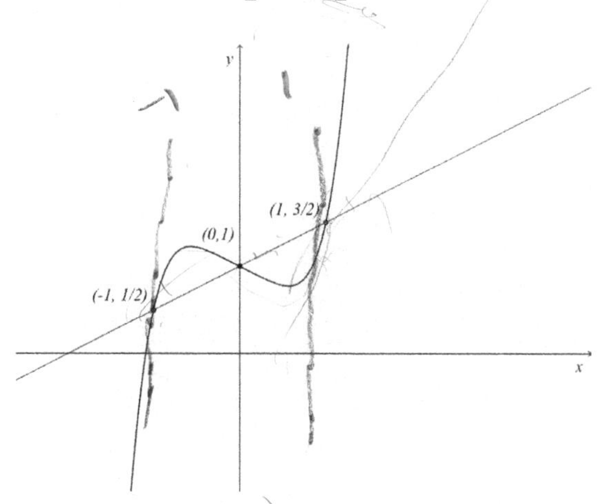

A. $-1 < x < 1$
B. $-1 < x < 0$ and $x > 1$
C. $-1 < x < 0$
D. $x < -1$ and $0 < x < 1$
E. $0 < x < 1$

28. Which of the graphs below best represents the system of inequalities below?

1. $2 \le x \le 20$

2. $1 \le y \le 10$

3. $y \le x - 2$

4. $y \ge -7x + 50$

A.

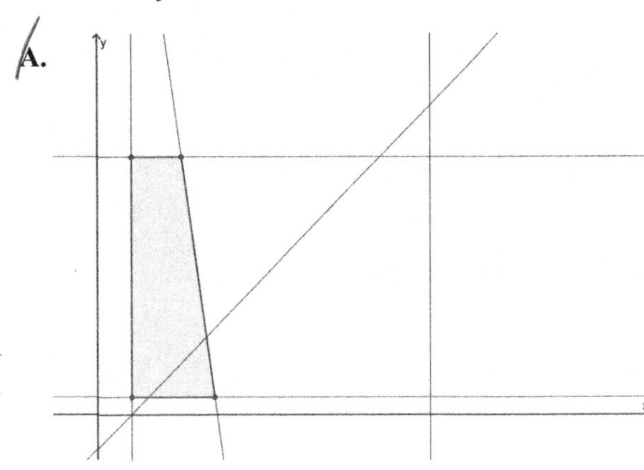

B.

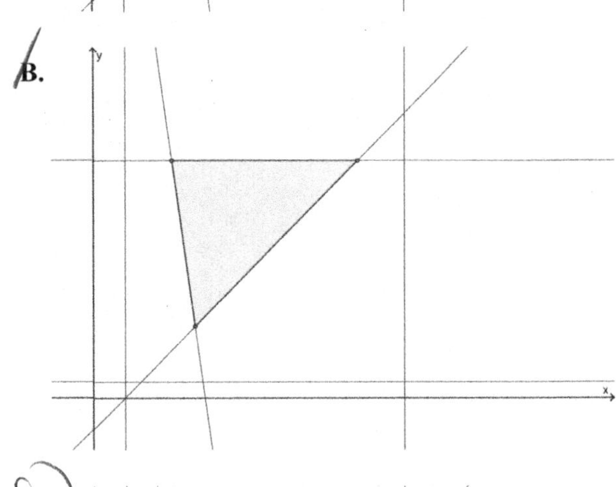

C.

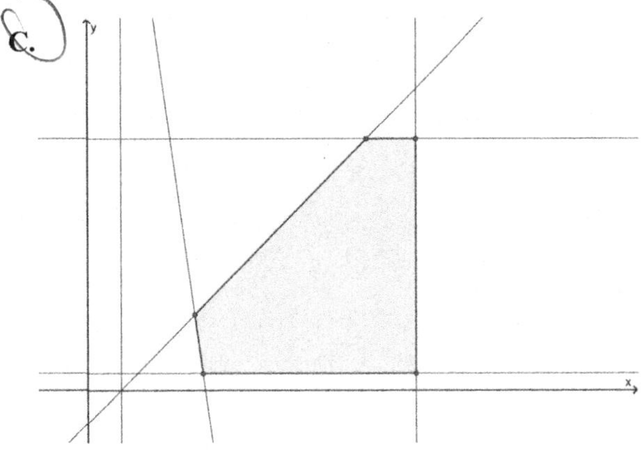

D.

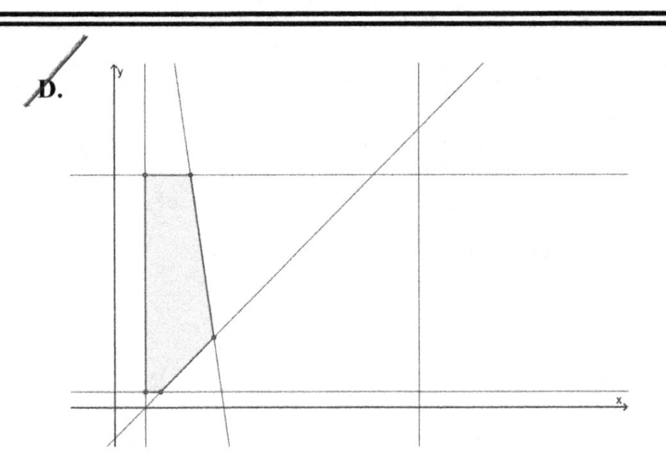

E.

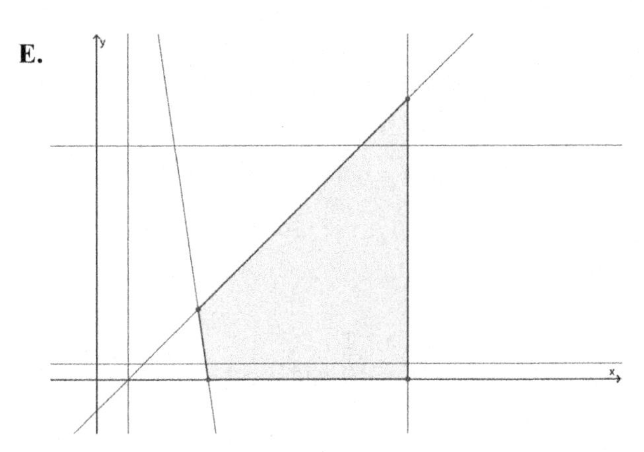

29. Which of the following graphs in the standard (x, y) coordinate plane represents the solution set of the inequality $|x - y| < 1$?

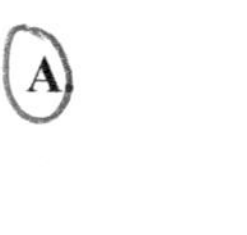

A.

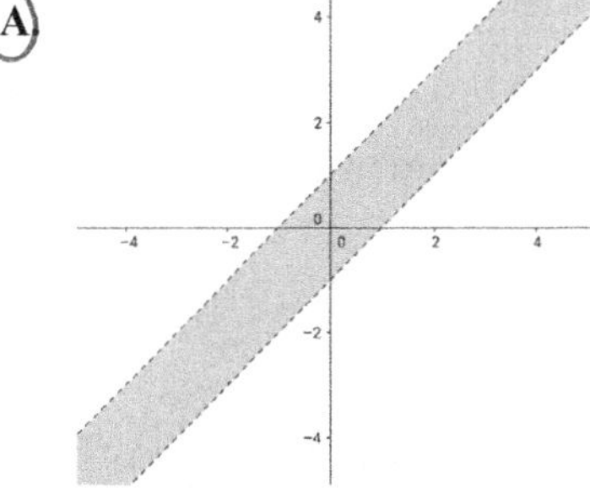

B.

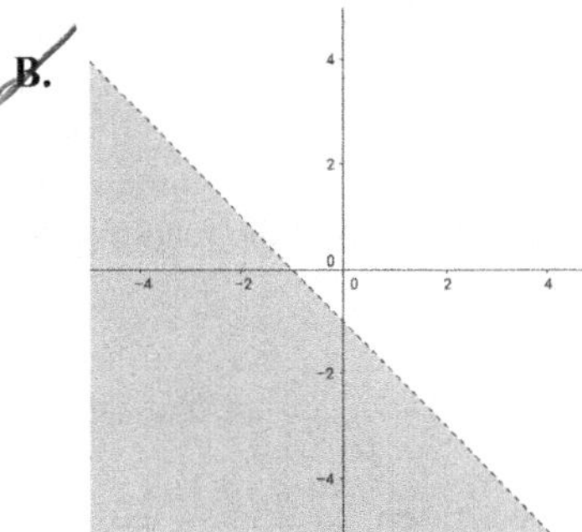

C.

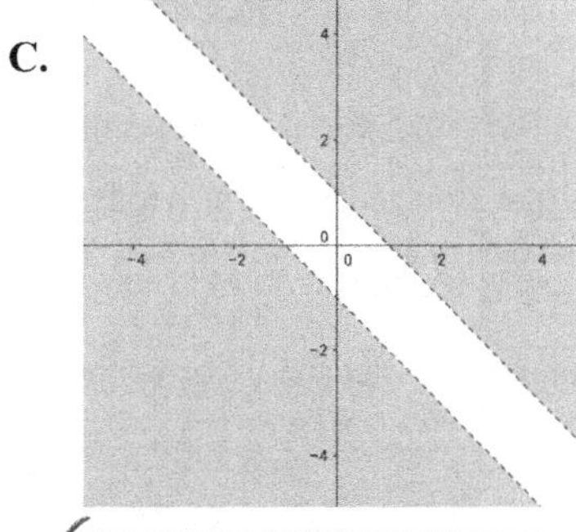

D.

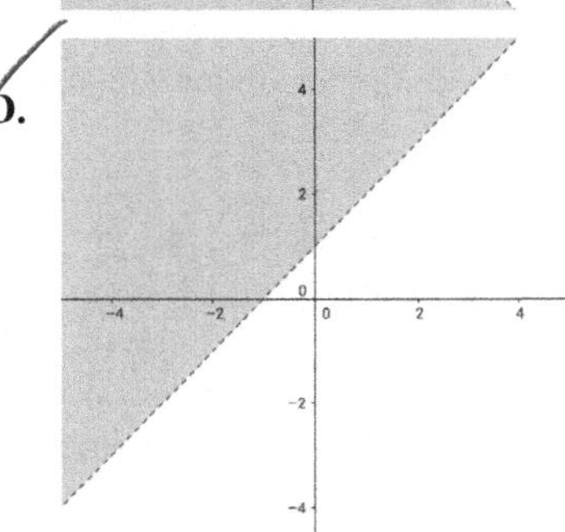

E.

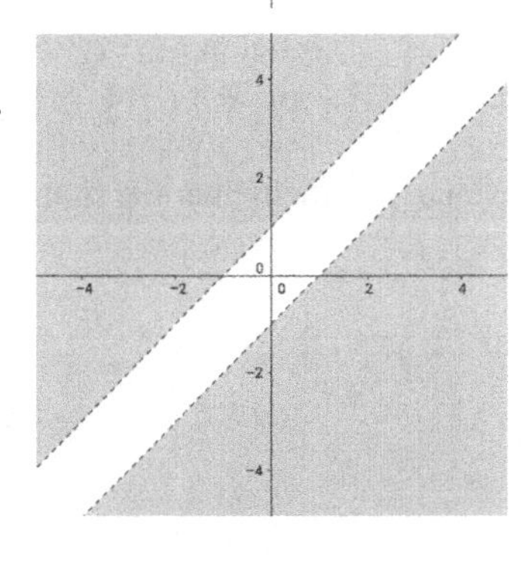

30. Which of the following systems of inequalities is represented by the shaded region of the graph below?

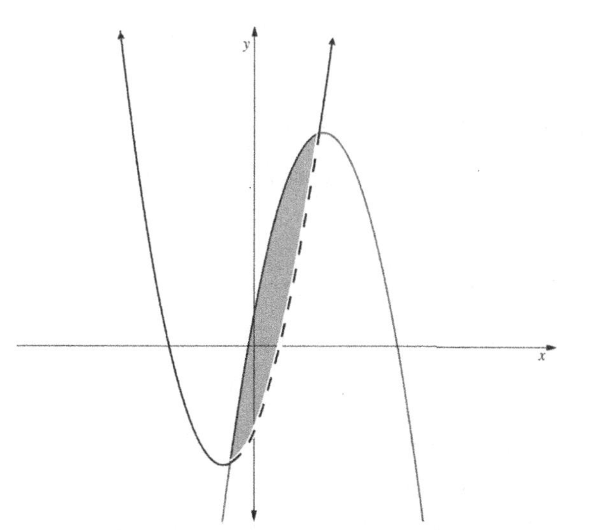

A. $y > x^2 + 2x - 3$ and $y < -x^2 + 5x + 1$

B. $y > x^2 + 2x - 3$ and $y \leq -x^2 + 5x + 1$

C. $y \leq x^2 + 2x - 3$ and $y < -x^2 + 5x + 1$

D. $y > x^2 + 2x - 3$ and $y > -x^2 + 5x + 1$

E. $y \geq x^2 + 2x - 3$ and $y \leq -x^2 + 5x + 1$

ANSWER KEY

1. E	**2. C**	**3. A**	**4. C**	**5. D**	**6. B**	**7. B**	**8. A**	**9. C**	**10. E**	**11. A**	**12. D**	**13. B**	**14. B**
15. A	**16. D**	**17. C**	**18. A**	**19. B**	**20. A**	**21. B**	**22. A**	**23. A**	**24. D**	**25. D**	**26. D**	**27. B**	**28. C**
29. A	**30. B**												

ANSWER EXPLANATIONS

1. **E.** Multiplying the inequality by d, we get $\frac{2}{15}d<4<\frac{6}{13}d$. Now splitting the inequality into two separate ones, we have $\frac{2}{15}d<4$ and $4<\frac{6}{13}d$. Dividing $\frac{2}{15}d<4$ by $\frac{2}{15}$ we get $d<4\left(\frac{15}{2}\right)\rightarrow d$. Dividing $4<\frac{6}{13}d$ by $\frac{6}{13}$, we get $\frac{26}{3}<d$. Now, combining the two inequalities, we have $\frac{26}{3}<d<30$. $\frac{26}{3}\approx 8.67$. So, the integer values that would satisfy the inequality are $9\le d\le 29$. This gives us $29-9+1=21$ possible integers. Notice that we must add 1 after subtracting the boundaries because we include both boundaries in our count.

2. **C.** The square of a fraction between zero and 1 is smaller than the fraction itself because when the denominator is larger than the numerator, the denominator increases more than the numerator when squared. This makes the resulting fraction smaller.

3. **A.** The reciprocal of a number p is written $\frac{1}{p}$. The difference between 10 and $\frac{1}{p}$ is $10-\frac{1}{p}$. $\frac{1}{3}$ of that difference is $\frac{1}{3}\left(10-\frac{1}{p}\right)$. Finally, the inequality states that 35 is greater than the expression we've written, so $35>\frac{1}{3}\left(10-\frac{1}{p}\right)$.

4. **C.** The easiest solution is to make up numbers that adhere to the given parameters. Set t=1/2 and r=-2 and plug into t^r : $(\frac{1}{2})^{-2}=((\frac{1}{2})^2)^{-1}=(\frac{1}{4})^{-1}=4$ Since 4 is greater than 1, the answer is C. Note: this question requires knowledge of exponent rules **power of a product** $(ab)^c=a^cb^c$ and the **negative exponent rule:** $a^{-b}=\frac{1}{a^b}$ (covered in Ch 15).

5. **D.** Since a and b are positive numbers, $\frac{1}{a}<\frac{1}{b}$ because the larger the denominator, the smaller the value of the fraction. So, since $a>b$, $\frac{1}{a}<\frac{1}{b}$. This makes answer choice (D) the only false statement.

6. **B.** w^2 is positive and t is negative. tw^2 is a negative number multiplied by a positive number, so it will always result in a negative value.

7. **B.** Adding $3n$ to both sides of the second equation, we get $9\le 3n$ or $3n\ge 9$. Dividing both sides by 3, we get $n\ge 3$. So, our two inequalities are $n\ge 3$ and $n\ge 10$. The set of solutions that satisfies these inequalities is $n\ge 10$ because if n is greater than 10, it will automatically be greater than 3.

8. **A.** Any number to an even power is positive, so y^{54} is positive. If $7x^{23}y^{54}$ is given to be negative, we know that x must be negative because it is the only possible term that could make the product negative. So, $x<0$.

9. **C.** Statement #1 can be written as $y=x+8$. Statement #2 can be written as $z\ge y-4$. Plugging in $y=x+8$ into $z\ge y-4$ we get $z\ge x+8-4$ or $z\ge x+4$ Subtracting 4 on both sides, we get $z-4\ge x$ or $x\le z-4$.

10. **E.** g is a positive fraction, so the higher the power it is raised to, the smaller its value, and the higher the root taken, the larger its value. So, $\sqrt[3]{g}$ would yield the largest value.

11. **A.** Because the expressions set equal, the variable with the smallest coefficient is the largest. Rewriting $\frac{z\sqrt{5}}{7}$ as $\frac{z\sqrt{5}}{\sqrt{49}}$, It

is easy to see that $\frac{\sqrt{5}}{\sqrt{49}} < \frac{1}{2}$. Putting the coefficients in the correct order, we have $\frac{\sqrt{5}}{\sqrt{49}} < \frac{1}{2} < \sqrt{5}$. This means the order of the variables is $z > y > x$.

12. D. Since $p^6q^9r^{10} > 0$ we know that $q > 0$ since it is the only variable risen to an odd power. Thus, the expression p^2q is guaranteed to be greater than zero because we know $q > 0$ and a square number such as p^2 is always positive.

13. B. The largest value of $a - b$ would be the difference between the largest possible value of a and the smallest possible value of $b: -7 - 2 = -9$.

14. B. The number of yes votes, x, must be greater than $\frac{2}{3}$ of 267. This can be mathematically written as $x > \frac{2}{3}(267)$ or $x > 178$.

15. A. Drawing the points on a number line according to the description given, we have P R Q S. We can see that $PQ > PR$ because $PQ = PR + RQ$.

16. D. First we need to find the set of numbers for which the inequality is true. We can do this by setting up a number line. We know that the expression equals zero at $x = 1$ and $x = 5$, so we can split up the number line into three main regions to test. If we plug in numbers less than 1 or numbers greater than 5 into the expression for x, the expression is negative, so our inequality is not satisfied. If $1 < x < 5$, the expression is positive, so the inequality is satisfied, and we now know which values of x we are looking at. Next, we need to find the expression that is negative for all these values of x. The only expression given that is negative while $1 < x < 5$ is $x^2 - 6x + 2$. You could find this out by sketching the expressions given or plugging in numbers to test, as we did with initial expression.

17. C. Factor and simplify: $\frac{ab-b}{2b-4b} < 0 \rightarrow \frac{b(a-1)}{-2b} < 0 \rightarrow \frac{(a-1)}{-2} < 0$. Since the denominator is a positive constant, the numerator must always be positive in order for the fraction to be less than zero (negative). Thus, $a - 1 > 0 \rightarrow a > 1$.

18. A. Let's use substitution to solve this problem. First divide both sides of the expression $4j - 2k = 12$ by 2 and isolate k to get $k = -6 + 2j$. We know that $k < 8$, so we can set the equivalent expression to be less than 8 as well to get $-6 + 2j < 8$. Solving for j leaves us with $j < 7$.

19. B. The ONLY way for the product to be less than *both* its factors is for both its factors to be fractions themselves. To test this theory out, try picking your own arbitrary numbers. If $a = 2$ and $b = 6$, the product ab would be 12, so that doesn't work. If $a = \frac{1}{2}$ and $b = 6$, $ab = 3$, which is less than b but greater than a, so that doesn't work either. Now let $a = \frac{1}{2}$ and $b = \frac{1}{6}$. Their product is $\frac{1}{18}$, which *is* less than both a and b. Thus, it is given that $a > 0$ and $b > 0$ and in order for a and b, to be fractions, $a < 1$ and $b < 1$.

20. A. This question looks overwhelming because of the fractions and variables but it's not that bad. We can assign arbitrary numbers to our variables, which will make it easier. Because the numbers are less than 1 and positive they must be fractions, which means the numerator a must be less than all of the three other variables. Because the fractions all share a numerator, so $b < c < d$ to make the inequality given to us true. Thus, let $a = 2$, $b = 3$, $c = 4$, and $d = 5$. When we take the reciprocals we get $\frac{b}{a} = \frac{5}{2}, \frac{c}{a} = \frac{4}{2}$, and $\frac{d}{a} = \frac{3}{2}$. We know that $1 < \frac{3}{2} < \frac{4}{3} < \frac{5}{2}$, so $1 < \frac{d}{a} < \frac{c}{a} < \frac{b}{a}$.

21. B. This problem is best done by simplifying what you know and then evaluating. We know $a > b + 1$ and $(a + b) < (a^2 - b^2)$

so let's try to simplify that second inequality. We can factor using the difference of squares (see Ch 3): $(a+b)<(a+b)(a-b)$ Now wouldn't it be nice to divide by a+b? I assume it can't be zero, because if it were, then the inequality couldn't be true: both sides of the inequality would equal zero. Now that I've tested for zero I can divide by this inequality by the variable expression a+b with two conditions: 1. Assuming $a+b>0$, I don't flip the sign and get: $\frac{a+b}{a+b}<a-b$ or $1<a-b$. Assuming $a+b<0$ I do flip the sign and get: I $1>a-b$. Going back to what I know, I see a coincidence: $a>b+1$ and my first inequality in these two cases is $1<a-b$ simplifies to $a>1+b$: this is the same inequality! I know the assumption I made to get to this parameter is $a+b>0$, so I also know that this must define the sum of a and b. Thus my answer is B, all positive numbers. Though D overlaps with this answer, it is too restrictive, as there is nothing stopping the sum from being larger than one. Alternatively, come up with arbitrary pairs of numbers and test each of the answers given.

22. **A.** The only thing slightly tricky about this problem is the diction, but the question just asks you to solve the inequality, so combine like terms and simplify: $14x-3\leq 9x+4 \to 5x\leq 7 \to x\leq \frac{7}{5}$.

23. **A.** The amount of money raised is equal to $8x+4y$, where x is the number of adult tickets sold and y is the number of student tickets sold. This sum must be at least \$200 to make a profit, so $8x+4y\geq 200$ is our equal. If we wanted change that into a more recognizable form, our equation would be $y\geq 50-2x$, so we look for a graph with a y-intercept of 50, and an x-intercept of 25 (per this equation) and only answer (A) has that. We do not need to test regions because all of the graphs shade on the same side, but if they didn't we would need to determine which side was shaded as well.

24. **D.** Only (A) and (D) have positive slopes and positive y-intercepts, which are what a and b represent respectively. Because the inequality is "greater than," not "greater than or equal to" we want the graph a dashed, not solid line, which leaves us with graph (D).

25. **D.** The graph has a negative slope and a dotted line, so we know that the a term, which was said to be positive, must have a negative sign in front of it, and the inequality sign must be either "greater than" or "less than," not "equal to." At this point we are deciding between (B) and (D). Now we must test a point to see what direction the sign is facing, plugging in that point's x and y coordinates to see if the inequality is true. In general, if you can, use the origin to test. $0>a(0)+b$ is false, since we are told (and we see on the graph) that b is positive, but $0<a(0)+b$ is true, so our answer is (D).

26. **D.** The question is testing your understanding of both inequalities and graphs. This graph has dashed lines, not solid, so our signs must be either < or >, not ≤ or ≥, narrowing our answer choices down to (B), (D), or (E). Now we must test points, and the easiest point to test is the origin, which is part of the shaded area. Testing answer B: $0>\left(\frac{0}{2}-2\right)^2+6$ is false, so answer (B) is incorrect. Now test answer (D): $0<(0-2)^2+6$ is true and $0>\frac{0}{2}-2$ is true, so (D) is our answer.

27. **B.** Really what this question is asking is: when does $y_1>y_2$, since their y values are being directly compared. When you look at the graph, you see that the graph y_1 is above y_2 when $-1<x<0$ and $x>1$.

28. **C.** The first two inequalities tell you that the solution set is within a certain range of x and y values. What you must look for in the shaded region is within this range (which forms a rectangle) and whose boundaries are the third and fourth inequalities. We can identify which line is which in the graph easily because the third inequality has a positive slope but the fourth has a negative slope. This eliminates (A), because the inequality $y\leq x-2$ is not evaluated, not even incorrectly (you can see that the region is shaded both above and below the line $y=x-2$, which is impossible in graphing an inequality), and (E) because the shaded region in (E) goes outside the box. Now we must test points to determine which of the remaining graphs, (B), (C), or (D), is shaded correctly. When we solve $-7x+50=x-2$, we find that the lines intersect at the point $(6.5, 4.5)$. Thus, it can safely be assumed that the region in answer (B) contains the point $(6.5, 8)$. Testing that point in the third inequality given, $y\leq x-2$, we find that $8\leq 6.5-2$ is false, so the graph must be shaded to

the *right* of this positive sloping line, not the left. Now that we know this, we can also eliminate answer choice (D), since that is also shaded to the left. This leaves (C) as our answer.

29. A. $|x-y|<1$, which means $-1<x-y$ and $x-y<1$, so the *boundary* lines of our inequality are $y=x-1$ and $y=x+1$. Because of this we know that our answer is either (A) or (E), as those are the only two choices that include both of those lines. To know whether we shade inside or outside, we can test the origin: $-1>0$ and $0<1$ are true, so we do shade between the lines, making answer (A) correct.

30. B. Looking at the graph we see that the upward facing parabola is dotted as it bounded the region and the downward facing parabola is solid. This means that in the system of inequalities the expression with a positive leading term is part of a $\leq$ or $\geq$ equation, but the expression with a negative leading term is part of a $<$ or $>$ equation (it doesn't actually matter *what* the expressions are, because we can easily tell them apart by their sign). This eliminates all answer choices except (C), which must be our answer. Note that if it didn't, we would have had to test out the inequalities with points, and in this case we could have used the origin.

CHAPTER

RATIOS, RATES & UNITS

SKILLS TO KNOW

- How to set up ratios from word problems
- How to set up proportions and cross-multiply to solve
- How to convert units
- Dimensional Analysis
- How to solve distance problems
- How to solve work problems
- How to solve wheel distance problems

NOTE: **Chapter 12** in this book, **Direct and Inverse Variation**, encompasses many proportion problems and builds on this chapter.

RATIOS: THE BASICS

While percents and fractions measure part to whole relationships, ratios typically measure part to part. When you see the word RATIO, unless the "whole" or "all" is specifically stated in the sentence, the problem is telling you a relationship between two "parts." For example, the ratio of blue marbles to red marbles is $3:7$ means that for every 3 blue marbles there are 7 red ones.

Beware of FRACTION problems that use the word RATIO.

Tricky problems may use the word ratio to describe a fraction. For example, the ratio of blue marbles to all marbles in the bag is $3:10$ means that for every 3 blue marbles, there are 10 total marbles. In other words, there are 7 non-blue marbles and $\frac{3}{10}$ of all the marbles are blue. When your ratio is "to all," you may have a fraction in disguise.

Most basic ratio problems can be solved in one of two ways:

Method 1: Turn your ratios into fractions
Method 2: Put an x next to it.

For both methods, it's important to understand you'll be dealing with numbers that represent the **ratio** AND numbers that represent the **actual** numbers. Also remember **the whole is always the sum of the parts.**

Let's see an example of these methods in action.

A triangle has interior angles that are in the ratio of $4:5:7$. What is the difference in degrees between the largest and smallest angle measure?

From chapters on triangles and angles and lines (both in Book 2), you should know that triangles have an interior angle sum of 180 degrees. This is the "actual" whole. We also know the ratio of the parts.

Method 1: Turn your ratios into fractions

One way we can solve this problem is to turn all our information into fractions to solve for the actual angle measures. When we know all the ratio "parts" we can also find the ratio "whole" by adding these parts together. 4, 5, and 7 represent the proportional parts. If I add these together, I get the "whole" of the proportion:

$$4+5+7=16$$

We have sixteen parts in total. This is our "whole," or the bottom of each fraction we will create.

Now I can turn each "part" into a fraction. Remember, a fraction is always "part over whole." Each part of the proportion represents a different fraction, i.e.:

$$4\rightarrow\frac{4}{16} \qquad 5\rightarrow\frac{5}{16} \qquad 7\rightarrow\frac{7}{16}$$

I need the difference of the largest angle and the smallest angle. I now know the largest angle is $\frac{7}{16}$ of the total degrees in the triangle, which in math translates to:

$$\frac{7}{16}\times180=78.75$$

I also know the smallest angle is $\frac{4}{16}$ or $\frac{1}{4}$ of the total degrees in the triangle:

$$\frac{4}{16}\times180=45$$

Now I find the difference of the largest and smallest angles: $78.75-45=33.75$ degrees.

Method 2: Put an "x" next to it:

A great trick for solving ratio problems is to know that you can always represent the "actual" amount of something that you only know the ratio part of by putting a variable next to all the "ratio" numbers you know. Ratios are like a "reduced fraction" of actual numbers. To get back to those actual number if we were dealing with reduced fractions, we would multiply by a common ratio. Here we do the same, with that common ratio equal to x: Given a ratio $4:5:7$, I can express the **actual measures** of each angle as $4x$, $5x$, and $7x$.

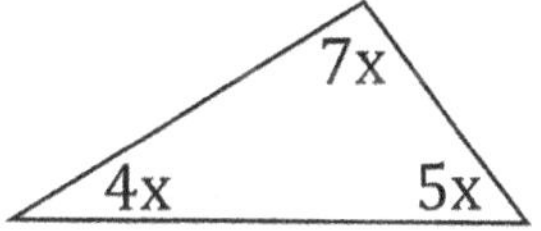

Now that I have an algebraic representation of the actual numbers, I can sum these to be the actual total, 180. Now I just use algebra to simplify:

$$\begin{aligned} 4x+5x+7x&=180 \\ 16x&=180 \\ x&=11.25 \end{aligned}$$

Now be careful, x **<u>is not the answer to the question</u>**! I can solve for $7x-4x$, or $3x$, the difference between the largest and smallest angles, by plugging x into the expression for what I need:

$$3(11.25)=33.75$$

Answer: 33.75.

NOTE: This "put an x next to it" method works very well in similar shape problems as well. (See **Book 2, Chapter 15: Similar Shapes**.)

SOLVING PROPORTIONS

If a proportion is appropriate, it should represent a linear relationship such that as one value increases, so does the other. Proportions can also be used to solve for unit conversions. Be sure you have a proportion situation before assuming you can use a proportion. See the Direct and Inverse Variation chapter for more on determining this.

The easiest way to solve simple proportions is cross-multiplying. To do this, we multiply the numerators of each fraction by the denominator of the other.

CROSS MULTIPLICATION

$$\frac{a}{b} \times \frac{c}{d}$$
$$ad = bc$$

If Maria bought 6 pairs of socks from the department store for 13 dollars, how much did Sheila pay for 9 pairs of the same socks from the same store?

To set up proportions, **create a parallelism with your labels**. For example, flour over cookies equals flour over cookies. Or miles over minutes equals miles over minutes.

Here, we can set up our proportion as dollars over socks equals dollars over socks, keeping Maria on one side and Sheila on the other. We get:

$$\frac{\$13}{6\text{socks}}=\frac{\$x}{9\text{socks}} \quad \frac{\leftarrow\text{dollars}}{\leftarrow\text{socks}}$$

The numerators' units are dollars; the denominators' units are socks. x denotes the number of dollars Sheila paid for 9 pairs of socks.

MISTAKE ALERT: Many students don't keep the order the same in proportions (i.e. they'll set dollars over socks equal to socks over dollars). Always keep your proportions in the same pattern! However, you CAN put whatever label you want in the numerator as long as you are consistent (i.e. you could put socks over dollars for both fractions).

Remember that anything "vertical" in a proportion has something in common (either same person/case or units), as does anything on the same horizontal. Here, our horizontal similarity is socks and dollars. Our vertical similarity is the person involved in the purchase. We can set up proportions in several different ways, but there should always be a similarity on the vertical and horizontal regardless.

Cross-multiplying, we get:

$$6x = 9(13)$$
$$6x = 117$$
$$x = \frac{117}{6} \text{ or } \$19.50$$

Answer: $\$19.50$.

CONVERTING UNITS

One basic way to convert units is to set up a proportion. In a proportion, as described above, the units must be the same along the vertical or along the horizontal of the two fractions.

How many feet there are in 52 inches?

We can solve this by setting up a proportion: feet over inches equals feet over inches. I place the "actuals" on the left side, and the "conversion factor" on the right side (the number of feet I know contain a certain number of inches).

$$\frac{x\,ft}{52\,in} = \frac{1\,ft}{12\,in}$$

Cross-multiplying, we get $12x = 52$. Solving for x, we have the answer: $4\frac{1}{3}$ feet.

That way is fine for simple problems. But there's a better way for most students aiming to score above a 25 on the math: Dimensional Analysis.

DIMENSIONAL ANALYSIS

If you have struggled with unit conversions, ratios, and rates when encountering word problems, dimensional analysis is the awesome sauce you need to crush most any problems with ratios, rates, and/or units. On the new SAT® it is absolutely vital. On the ACT® it's still an important skill to have.

You may have used this process in chemistry class. It's an intuitive way of converting anything using either a **unit conversion factor** or even a **rate**. Let's talk about both of those.

Unit conversion factors:

A unit conversion factor, such as three feet in a yard, is a ratio of two values that are equal to each other. For example, 12 inches are in a foot, and 60 seconds are in a minute (and by the way, you should have all these memorized for the ACT®). If we divide one of these equivalent amounts by the other, we essential create a "1," i.e.:

$$\frac{60\,seconds}{1\,minute}=1 \text{ or } \frac{1\,minute}{60\,seconds}=1$$

Because these values are the same amount of time, if we divide equal amounts, we get "1", right?

Now, let's take that "1" and multiply by it. When we do so, we won't be changing any amounts, but rather, will be changing the labels.

Let's say I have 300 seconds and I want to convert it to minutes. I multiply by the version of "1" above that will help me "cancel" out the units.

$$300\,\cancel{sec}\times\frac{1\,min}{60\,\cancel{sec}}=\frac{300}{60}\,min=\frac{30}{6}\,min=5\text{ minutes}$$

The one with seconds on the bottom so that the seconds will cancel. Just as I can cancel numbers when I multiply fractions by each other (i.e. $\frac{\cancel{2}}{5}\cdot\frac{1}{\cancel{2}}=\frac{1}{5}$), I can cancel my units when I multiply by this version of "1."Now what's great about dimensional analysis is that you can convert several times in a row very easily. Let's say you want to know how many seconds are in a year. Here, I'll multiply by one conversion factor after another to get my units to cancel until I get to seconds. Whatever I want to get rid of I put on the bottom. Whatever I next need I put on top:

$$1\,year\times\frac{365\,days}{1\,year}\times\frac{24\,hours}{1\,day}\times\frac{60\,min}{1\,hour}\times\frac{60\,sec}{1\,min}$$

$$1\,\cancel{year}\times\frac{365\,\cancel{days}}{1\,\cancel{year}}\times\frac{24\,\cancel{hours}}{1\,\cancel{day}}\times\frac{60\,\cancel{min}}{1\,\cancel{hour}}\times\frac{60\,sec}{1\,\cancel{min}}=31{,}356{,}000\text{ seconds}$$

As you can see, writing this out in one line is much easier with this method than using proportions.

Rates

Now here's the kicker: we can use this SAME technique to find values using rates. While units are simply units of measuring the same thing (such as length or temperature), RATES allow us to jump between two totally different ideas, such as miles and hours or gallons and square feet. Still, the method is the same.

TIP: Rates typically use words such as **"per," "for," "every," or "each," which all mean divide.**

For example, if Judy stuffs 4 envelopes **per** hour, I can create a rate that is $\frac{4\,envelopes}{1\,hour}$. Remember, **per** means divide! Now if I know envelopes I can calculate hours or if I know hours, I can calculate envelopes using this method.

If the recipe for chocolate cornbread calls for $3\frac{1}{4}$ cups of milk for 5 total pieces, how many cups does Ralph need to buy if he wants to make exactly 17 pieces?

A. $6\frac{1}{2}$ **B.** $9\frac{3}{4}$ **C.** $10\frac{1}{2}$ **D.** $11\frac{1}{20}$ **E.** 13

The first thing we need to do is set up the ratio. If it takes $3\frac{1}{4}$ cups to make 5 pieces and we are trying to find the number of cups needed, the rate is:

$$\frac{3\frac{1}{4}\,cups}{5\,pieces}$$

Now I ask what I have: pieces. I write that down first. I also think what I need: cups.

I put what I "need" first or on top in my rate, so this rate's orientation above is correct as is (i.e. I don't want its reciprocal).

Now I use dimensional analysis:

$$17\,\cancel{pieces} \times \frac{3\frac{1}{4}\,cups}{5\,\cancel{pieces}} = 11.05\,\text{cups}, \text{ or } 11\frac{1}{20}\text{ cups}$$

What I have	Rate	What I need

Answer: **D**.

At the lumber yard, $10\frac{1}{3}$ feet of wood sells for $\$24.50$. If Sheldon only needs 20 inches of wood for his model train kit, how much will the wood cost in dollars, to the closest dollar?

A. 2 **B.** 3 **C.** 4 **D.** 5 **E.** 6

You may have tried to set up a ratio for this one already, but don't be too hasty. Notice how the units are different for the lengths of wood: one is in feet and the other in inches!

First, we have to convert these numbers to uniform values. Either inches or feet will work, but I will use inches. Set up the conversion equation. I put what I have first (feet) and multiply by the unit conversion factor with what I "need" on top (inches) and what I want to eliminate on the bottom (feet):

$$10\frac{1}{3}\,\cancel{feet} \times \frac{12\,inches}{1\,\cancel{foot}} = 124 \text{ inches}$$

Now that our units are the same, we can set up the equation for the price. We could use dimensional analysis OR a ratio. Let's do a ratio this time. Inches over dollars equals inches over dollars:

$$\frac{124\,inches}{\$24\frac{1}{2}} = \frac{20\,inches}{x\,dollars}$$

Cross multiply to get $124x = 490$. Divide both sides by 124 to arrive at the correct solution, $\frac{490}{124}$. We can use our calculators to help round this to the nearest dollar (per the question), 4.

Answer: **C.**

Ratios in Word Problems

Ratios also often appear in various word problems. We can convert our understanding of ratios into algebra to solve these problems. More problems of this nature can be found in the respective chapters on other topics, such as geometry (Book 2) or general word problems (this book).

In the cube below, if width was multiplied by 3, and both length and height were halved, which of the following would equal the volume of the cube in terms of the original volume, V?

$$V = w \times l \times h$$

A. $1.25V$ **B.** $0.75V$ **C.** $0.6V$ **D.** $0.5V$ **E.** $0.25V$

We can write our new volume's equation out by modifying each of the original dimensions based on the wording of the problem. The new width is $3w$ (three times the old width) while the new length is $0.5l$ (half the old length) and the new height is $0.5h$ (half the old height). Thus:

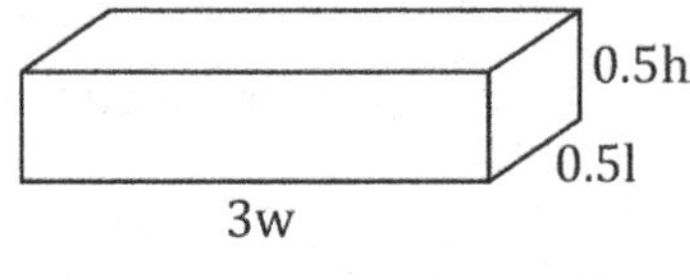

$$V_1 = 3w \times 0.5l \times 0.5h$$

Using the commutative property of multiplication, we rearrange to isolate the constant:

$$V_1 = 0.75(w \times l \times h)$$

Since the original volume, V, was wlh, we know this new volume is 0.75 times that. Substituting in V for wlh, we get $0.75V$.

Answer: **B.**

We also find ratios in speed and distance problems. These can also be approached by memorizing the distance formula. Still, dimensional analysis works, too. **Remember that your rate is an equation in disguise. <u>You can always turn it into a ratio equation or setup a series of elements to cancel out using a rate.</u>**

DISTANCE PROBLEMS

THE DISTANCE FORMULA

$$d = rt$$

Memorize this. When in doubt set up as many equations as necessary using sub-numbers or letters (subscript, i.e. $d_a = r_a t_a$) to denote who/what you're talking about. Then start a chain of substitution until you get to the answer you need.

Nathaniel drove 360 miles in 9 hours of driving time. How much faster would he have to drive than his average speed to cut his driving time by 3 hours?

A. 20 *mph* **B.** 30 *mph* **C.** 40 *mph* **D.** 50 *mph* **E.** 60 *mph*

Use the distance formula (I use subscript to keep each case straight): $d_a = r_a t_a$

For the actual trip (trip a), we let $d_a = 360$ and $t_a = 9hr$, and plug in to the distance formula:

$$360\,mi = r_a(9\,hr)$$

Then solve for r_a:

$$\frac{360\,mi}{9\,hr} = r_a$$

$$40\frac{mi}{hr} = r_a \text{ is his speed for the actual trip}$$

Now make a formula for the hypothetical trip (trip h) and fill in the pieces we know: $d_h = r_h t_h$

We know we want to cut the time by three hours—so $9-3=6$; $t_h = 6$

In terms of the distance, he's trying to take the same trip, so $d_h = d_a = 360$

$$360\,mi = r_h(6\,hr)$$

$$\frac{360\,mi}{6\,hr} = r_h$$

$$60\,mph = r_h$$

Now read the question—we need how much faster this answer is than our speed for the actual trip, or the difference between the hypothetical faster trip and the actual one:

$$r_h - r_a$$

$$60\,mph - 40\,mph = 20\,mph$$

Answer: **A.**

WORK PROBLEMS

Work problems are another kind of rate problem. We can use formulas or dimensional analysis and ratios to solve these too.

WORK FORMULA

$$w = rt$$

Where w is the amount of work (ex. number of pages), r is the rate of work (ex. pages per hour) and t is the time (ex: hours).

COMBINED WORK FORMULA

$$w = (r_1 + r_2)t$$

Where r_1 is the rate of one worker or machine, and r_2 is the rate of another worker or machine working for the same amount of time.

A whatsit machine makes 35 whatsits per minute. A second whatsit machine makes 50 whatsits per minute. The second whatsit machine starts making whatsits 5 minutes after the first whatsit machine starts. Both machines stop 10 minutes after the first machine starts. How many whatsits were produced by the two machines together?

A. 212.5 **B.** 425 **C.** 600 **D.** 675 **E.** 850

If work equals rate times time $w = rt$, and the number of whatsits produced represents the "work", then we need to figure out rate and time to solve for w. We can also think in terms of ratios, or dimensional analysis, described above.

Remember PER means DIVIDE. So 35 whatsits per minute means divide 35 by 1 minute! 50 whatsits per minute means divide 50 by 1 minute! "Per" also indicates a rate.

Thus here are our two rates:

First machine: $\frac{35\ whatsits}{1\ min}$

Second machine: $\frac{50\ whatsits}{1 min}$

From the problem we know the first machine stops after 10 minutes. The second machine works 5 minutes less as it starts 5 minutes later, but ends at the same time. To find the total time that the second machine runs, we subtract:

$$10\ min - 5 min = 5\ min$$

Now we can either analyze the rate and know that multiplying it by minutes will make the minutes cancel, and give us what we need (whatsits) or we can know to multiply rate times time from the formula.

Remember, $w = rt$!

The first machine will produce:

$$\underset{\text{rate}}{\frac{35 whatsits}{1 \cancel{min}}}\underset{\text{(time)}}{(10 \cancel{min})} = \underset{\text{work}}{350\ whatsits}$$

The second machine will produce:

$$\underset{\text{rate}}{\frac{50\ whatsits}{1 \cancel{min}}}\underset{\text{(time)}}{(10-5)\cancel{min}} = \underset{\text{work}}{250\ whatsits}$$

We can then add together the sum of each machine's production:

$$350\ whatsits + 250\ whatsits = 600\ whatsits$$

Answer: **C.**

Notice how the "minutes" on the bottom of the rate cancel with the time in "minutes"—you can always use labels to help you set up rate word problems correctly.

WHEEL PROBLEMS

Sarah rode a bike with wheels 36 inches in diameter. During 4 minutes of her ride, the wheels made 180 revolutions. At what average speed, in *feet per second*, did Sarah travel, rounded to the nearest *foot per second*?

A. 2 **B.** 7 **C.** 15 **D.** 135 **E.** 424

With problems like this, we first figure out what we NEED. **We need FEET PER SECOND. Here we use the fact that the word PER means DIVIDE. So we need $\frac{FEET}{SECONDS}$.**

To find that, we need to find the TOTAL FEET she traveled and the TOTAL SECONDS she traveled.

TIP: Average Rate is **always** TOTAL DISTANCE over TOTAL TIME.

Let's start with the feet. We have a 36" diameter wheel, but that's not the feet we need. We need the distance traveled. When a wheel travels, the distance it completes is equal to the circumference of the wheel times the number of rotations. With each rotation, the entire circumference touches the ground and equals the amount traveled. Imagine if your wheel was made of a piece of string. Cut it and lay it flat—you'll realize that's also the distance your wheel has traveled.

Again we need $C \times (180\, revolutions)$.

To find C (circumference): $C = \pi d$ where d is the diameter. Let's first make our 36" diameter into feet: $36 in = 3 ft$.

Now we'll find the circumference:

$$C = \pi d$$
$$C = \pi(3) = 3\pi$$

This is the number of **feet** we travel **per revolution** (remember PER means DIVIDE!) Take that idea and plug into our equation $C \times (180\,revolutions)$:

$$3\pi \frac{ft}{\cancel{rev}} * 180\,\cancel{rev} \approx 1696\,ft$$

The revolutions "cancel."

Let's go back to our original need: $\frac{FEET}{SECONDS}$. We can now fill in the feet: $\frac{1696\,ft}{SECONDS}$.

To find the seconds, we convert **4** minutes to seconds using **dimensional analysis.**

First come up with the conversion factor: $\frac{60\,sec}{1\,min}$.

If necessary, flip your conversion rate upside down—but make sure the denominator (bottom) has the same label as the number you're multiplying by (here "minutes" is that label, so we want minutes on the bottom so it will cancel):

$$(4\,min)\frac{60\,sec}{1\,min} = 240\,sec$$

Now we complete our original need item by filling in the missing seconds: $\frac{1696\,ft}{SECONDS}$ is $\frac{1696\,ft}{240\,sec} \approx 7$.

Answer: **B.**

TIPS IF YOU'RE STUCK:

1. **Remember that oftentimes you have hidden equal quantities**—for example a round trip flight is the same distance both directions; if two people start at the same time, their time is the same, etc.

2. **Check your units**—sometimes these problems will mix measurements (i.e. inches/feet, seconds/minutes). Use **dimensional analysis!**

3. **Re-read the question**—there may be one more detail you've missed. Rates and ratios sometimes show up in three part questions. Look up if you're lost to see if you might have missed some introductory information.

4. **Seeing the word "distance" doesn't always mean it's a distance/rate problem.** It may be a **Function as a Model** problem—see **Chapter 10** for identifying that question type. If you have a given equation, forget about the distance formula and use what the test gives you. If you have a given chart of information, forget about the formulas and use the given information first! Then turn to memorized formulas if you need more.

1. Doug and Emily are driving cars on a track, and they both do 20 laps around a 1-mile track. Doug drives at an average of 60 miles per hour around the track while Emily drives at an average of 80 miles per hour. When Emily finishes her 20 laps, how many laps does Doug have left?

A. 0
B. 5
C. 10
D. 12
E. 4

2. Bryan and David go running in the park every morning. Bryan starts before David and is 30 meters ahead when David starts. Bryan runs at a rate of 2.5 meters per second, while David runs at a rate of 4.0 meters per second. Which function shows the time t it takes David to catch up to Bryan?

A. $t = 30$
B. $4.0t = 30$
C. $2.5t = 30$
D. $1.5t = 30$
E. $1.5t = 60$

3. Ronny can walk 4 miles in t minutes. At that rate, how long will it take her to walk 11 miles?

A. $\frac{11t}{4}$
B. $t + 7$
C. $\frac{4t}{11}$
D. $\frac{t}{11}$
E. $\frac{4}{t}$

4. Two delivery trucks leave from a warehouse at the same time. The first one moves at 45 miles per hour and drives 1 hour west and 2 hours south. The second one moves at 60 miles per hour and drives 2 hours west and 1 hour south. What expression gives the distance between the two delivery trucks 3 hours after they leave the warehouse?

A. $\sqrt{(45+120)^2+(90+60)^2}$
B. $\sqrt{(45-120)^2-(90-60)^2}$
C. $\sqrt{(45-120)^2+(90-60)^2}$
D. $\sqrt{(45-60)^2+(90-120)^2}$
E. $\sqrt{(45-60)^2+(90-45)^2}$

5. Neela ran uphill to her friend's house and it took her $\frac{t}{6}$ minutes. When she ran downhill back home, she ran 2.5 times as fast as she had uphill. How many minutes did her journey home take?

A. $\frac{5t}{12}$
B. $\frac{t}{15}$
C. $\frac{7}{30t}$
D. $\frac{12}{7t}$
E. $2.5t$

6. A monkey eats 13 bananas in 8 days. At this rate, how many bananas does the monkey eat in $8 + n$ days where n is every additional day?

A. $\frac{13n}{8}$
B. $13 + \frac{n}{8}$
C. $13 + \frac{13n}{8}$
D. $13 + n$
E. $\frac{8+n}{13}$

7. A car's windshield washer fluid reservoir holds 540 ounces of washing fluid, and there is a small leak causing the fluid to leak out at a constant rate of 9 ounces per minute. The car is traveling at 50 miles per hour. If the reservoir starts full, in how many miles will the washer fluid reservoir be empty?

A. 45
B. 50
C. 55
D. 60
E. 65

8. A cylinder in a diesel engine displaces 2×10^3 cubic centimeters and there are 6×10^5 oxygen molecules in the cylinder, what is the average number of oxygen molecules per cubic centimeter?

A. 3×10^3
B. 3×10^1
C. 3×10^0
D. 3×10^2
E. 3

For Questions 9-10, refer to the table below.

Monthly payment per $500 borrowed for various annual rates and numbers of payments			
Annual Interest Rate	Number of monthly payments		
	48	60	72
4%	11.29	9.21	7.82
7%	11.79	9.9	8.52
11%	12.92	10.87	9.52
13%	13.41	11.38	10.04

Paul is planning on purchasing a motorcycle, and he will need to borrow some money. The chart shows different monthly payments based on interest rates and loan terms.

9. Paul finds a motorcycle that costs $10,365. He will have to borrow $7,000. What will his monthly payment be if he borrows the money for 60 months at 11% interest?

A. $152.18
B. $76.09
C. $304.36
D. $159.32
E. $13.28

10. Paul visits another dealership that is offering an end-of-the-year deal with 4% annual interest on loans for 48, 60, or 72 months. Paul can only manage to pay $100 per month. Of the following options, what is the largest loan he can afford with his budget?

A. $4000
B. $5000
C. $6000
D. $7000
E. $8000

11. A wheel of a tricycle is 5 inches in diameter. If it rolls along a line without slipping, how many inches has the wheel traveled after 50 revolutions?

A. 250π
B. 250
C. 500π
D. 500
E. 750π

12. Greg pushes a cart so that one wheel rotates $\frac{2\pi}{3}$ radians. What fraction of the circumference of the wheel has the wheel traveled?

A. $\frac{2\pi}{3}$
B. $\frac{1}{6}$
C. $\frac{1}{3}$
D. $\frac{2}{3}$
E. $\frac{3}{2}$

13. A cable has to be 7 mm thick for every 10 kg it supports. Which of the following expressions gives the thickness of the cable *in* centimeters required to support a weight of 300 stones? (Note: 1 stone ≈ 6.35 kg)

A. $\frac{(7)(6.35)(10)}{30}$

B. $\frac{(7)(6.35)(30)}{10}$

C. $\frac{(7)(6.35)}{(10)(10)(30)}$

D. $\frac{(7)(10)}{(6.35)(30)}$

E. $\frac{7}{(6.35)(30)}$

14. Chrissie is learning how to ride a unicycle with a wheel 30 inches in diameter. She can ride for 2 minutes before falling over. If she only moved in one direction and her wheel made 100 rotations during this time, what was her average speed rounded to the nearest thousandth *in feet per second* over that time interval?

A. 2.083π
B. 3.125π
C. 6.25π
D. 12.5π
E. 25π

15. A recipe for a turkey says to cook at $375°F$ for 35 minutes per 1.5 pounds. How long should a 9-pound turkey be cooked?

A. 3 hours
B. 2 hours 55 minutes
C. 3 hours 15 minutes
D. 3 hours 30 minutes
E. 4 hours

16. A motorcycle has wheels 24 inches in diameter. During 3 minutes, it makes 1620 revolutions. What is the average speed of the motorcycle, to the nearest mph? (Note: 5280 ft=1 mile)

A. 37 mph
B. 38 mph
C. 39 mph
D. 40 mph
E. 41 mph

17. There are 3 pieces of wood whose combined lengths equal to 75 inches. The lengths of the pieces are in the ratio 3:5:7. What is the length, in inches, of the shortest piece?

A. 5 inches
B. 15 inches
C. 20 inches
D. 24 inches
E. 45 inches

18. Carlos, Rebecca, and Kylie all invest in a company and receive 40, 50, and 70 shares respectively. If the entire investment was $40,200, then how much money did Carlos invest in the company?

A. $10,050
B. $251.25
C. $160
D. $12,550
E. $12,562.50

19. The ratio of Roger's age to his son's age is 7:3. If the sum of their ages is 60, then how old is Roger's son?

A. 17 years old
B. 18 years old
C. 19 years old
D. 20 years old
E. 21 years old

20. What value for x makes the proportion below true?

$$\frac{8}{x-10}=\frac{4}{x+2}$$

A. 14
B. 7
C. 0
D. −7
E. −14

21. The local college accepts 2 out of every 11 students. If the school got 77,000 applicants this year, how many students were rejected from the school?

A. 63,000 students
B. 60,000 students
C. 14,000 students
D. 15,400 students
E. 56,000 students

22. Carol is baking cupcakes for a bake sale. She knows that 3 eggs will yield 5 cupcakes. If Carol wants to make 100 cupcakes, how many eggs would she have to buy?

A. 3 dozen
B. 4 dozen
C. 5 dozen
D. 6 dozen
E. 7 dozen

23. A larger rug is 30 feet long and 24 feet wide. What is the area, in square yards, of this rug?

A. $720\,yd^2$
B. $270\,yd^2$
C. $9\,yd^2$
D. $140\,yd^2$
E. $80\,yd^2$

24. James ran at a rate of 7 mph for the first 20 minutes of his run and 10 mph for 10 minutes after. What was the average rate, in miles per hour, that James ran during his 30 minute run?

A. 6 mph
B. 7 mph
C. 8 mph
D. 9 mph
E. 10 mph

25. A rectangular tabletop's length and width is in the ratio of $5:2$. If the area of the tabletop is 90 square feet, then what is the perimeter of the tabletop in yards?

A. 16 yards
B. 36 yards
C. 90 yards
D. 14 yards
E. 9 yards

26. Every summer Andre drives to his cousin's house to visit. If he maintains a steady speed, he can reach his cousin's house in $a+3$ hours. If Andre drives for $b-2$ hours where b is less than $a+5$, what portion of his drive does he have left?

A. $\dfrac{-2}{a+3}$

B. $a-b+5$

C. $b-a+5$

D. $\dfrac{a-b+5}{a+3}$

E. $\dfrac{b-a-5}{a+3}$

27. The ratio of the average weight of an African elephant to the average weight of a Northern giraffe is about $20:3$. The ratio of the average weight of a Northern giraffe to the average weight of a Great White shark is about $13:14$. What is the ratio of the average weight of an African elephant to the average weight of a Great White shark?

A. $20:14$
B. $81:14$
C. $650:81$
D. $130:21$
E. $280:39$

ANSWER KEY

1. B	**2. D**	**3. A**	**4. C**	**5. B**	**6. C**	**7. B**	**8. D**	**9. A**	**10. C**	**11. A**	**12. C**	**13. B**	**14. A**
15. D	**16. C**	**17. B**	**18. A**	**19. B**	**20. E**	**21. A**	**22. C**	**23. E**	**24. C**	**25. D**	**26. D**	**27. D**	

ANSWER EXPLANATIONS

1. **B.** When Emily finishes her 20 laps (20 miles), she would have spent $20\,miles \times \frac{1\,hour}{80\,miles} = 0.25$ hours driving. In that same amount of time, Doug would have driven $0.25\,hours \times \frac{60\,miles}{1\,hour} = 15$ miles (15 laps). So, at that time, Doug will still have $20 - 15 = 5$ laps left to drive.

2. **D.** If we set $t = 0$ to be the time at which David starts to run, then we can write Bryan's distance (B) as the formula $B = 30 + 2.5t$. David's distance (D) would be written as $D = 4t$. We set these distances to be equal and then solve for the time t that makes the equality true. The equality is when $B = D \rightarrow 30 + 2.5t = 4t$. Subtracting $2.5t$ on both sides, we get $30 = 1.5t$.

3. **A.** $11\,miles \times \frac{t\ minutes}{4\,miles} = \frac{11t}{4}$ minutes for Ronny to walk 11 miles.

4. **C.** The first truck's distance traveled is calculated using the formula $distance = rate \times time$. So, the first truck's distance traveled west is $d = 45\frac{mi}{hr} \times 1hr = 45mi$ and his distance traveled south is $d = 45\frac{mi}{hr} \times 2hrs = 90mi$. Likewise, the second truck's distance traveled west is $d = 60\frac{mi}{hr} \times 2hrs = 120mi$ and his distance traveled south is $d = 60\frac{mi}{hr} \times 1hr = 60mi$. This gives us the following diagram.

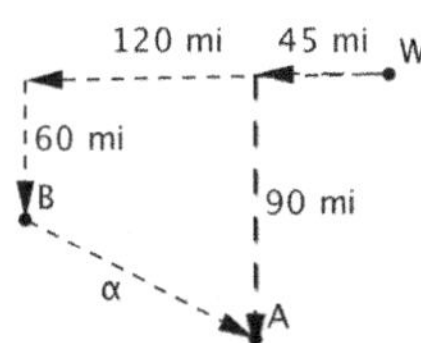

The picture above illustrates the first truck's route starting from the warehouse (point W) to point A and the second truck's route from point W to point B. The distance between their locations after three hours is represented by $\propto$. To find $\propto$, we find the horizontal and vertical differences from point A to point B and use the Pythagorean Theorem to find $\propto$. The horizontal distance is $120\,mi - 45\,mi$ and the vertical distance is $90\,mi - 60\,mi$. So, plugging these values into the Pythagorean Theorem, we get $\propto = \sqrt{(120-45)^2 + (90-60)^2} = \sqrt{(45-120)^2 + (90-60)^2}$.

5. **B.** If Neela takes $\frac{t}{6}$ minutes to run uphill, and she runs 2.5 times as fast as her initial speed on her way home, then we can divide out 2.5 times from her time on the return trip to account for the adjustment to the rate because given the fixed distance, these values vary inversely with each other (as one goes up, the other goes down, and their product is a constant). Thus, it will take her $\frac{t}{6}\left(\frac{1}{2.5}\right) = \frac{t}{15}$ minutes to get home. Though this is the "fast" way, many find it confusing (particularly if you haven't yet done Chapter 12 on direct and inverse variation). To avoid confusion, use the distance formula $d = rt$ or $distance = rate \times time$. For her first uphill trip, we substitute in $\frac{t}{6}$ for the time: $d = r(\frac{t}{6})$. For the return trip, we modify the rate from the first equation by increasing it by 2.5 times to get $d = 2.5r(?)$, where time is the unknown (?) we need. Setting these two equations equal to each other (the distance (d) is the same for both): $r(\frac{t}{6}) = 2.5r(?)$. We

isolate our unknown and see that the r's cancel: $\frac{r(\frac{t}{6})}{2.5r} = (?)$. This simplifies to $(\frac{t}{6})(\frac{1}{2.5})$ or $\frac{t}{15}$.

6. **C.** "Convert" from days to bananas using dimensional analysis and the rate you know:

$$(8+n)\,days \times \frac{13\ bananas}{8\ days} = \frac{(8+n)13}{8} = \frac{13(8)+13n}{8} = 13 + \frac{13n}{8}\,bananas.$$

7. **B.** First calculating the amount of fluid that is lost due to the leakage, we calculate $540-9t$ where t is in minutes. The fluid reservoir will be completely empty when $540-9t=0$. Adding $9t$ to both sides, we get $540=9t$. Dividing both sides by 9, we get $60=t$. So, in one hour, the reservoir will be empty. In one hour, the car will have traveled $1\,hour \times 50\frac{miles}{hour} = 50\,miles$.

8. **D.** "Per" means divide, and an average rate is total of one amount over total of the other, so take all the molecules and divide by cubic centimeters: $\frac{6\times10^5\,molecules}{2\times10^3\,cubic\,centimeters} = 3\times10^2\frac{molecules}{cubic\,centimeters}$.

9. **A.** Looking at the chart, we see that if Paul borrows the money for 60 months at 11% interest, he will have to pay 10.87 every month per \$500 borrowed. Paul is borrowing \$7,000, so his monthly payment can be calculated as $\$7{,}000 \times \frac{\$10.87}{\$500} = \$152.18$ per month.

10. **C.** We are looking for the maximum loan amount Paul can afford, so we want to look at the minimum rate of monthly payments. In this case, the minimum monthly payment amount offered with 4% annual interest for 48, 60, or 72 months is \$7.82 (the 72 month loan).We want to find the maximum loan amount x that would yield a monthly payment of less than \$100. This can be calculated by $\frac{\$7.82}{\$500}\times x \le 100$. $x \le 100 \times \frac{\$500}{\$7.82} \rightarrow x \le \$6393.86$. So, the maximum loan amount he can afford of the available answers is \$6000.

11. **A.** The circumference of the circle is calculated by the formula $C=\pi d$. Plugging in $d=5$, we get $C=5\pi$. So, for every revolution, the wheel travels 5π inches. After 50 revolutions, the wheel would have traveled $5\pi \times 50 = 250\pi$ inches.

12. **C.** One complete circle is 2π radians, so $\frac{2\pi}{3}$ radians is $\left(\frac{1}{3}\right)2\pi$ or $\frac{1}{3}$ of the circle. See **Circles** and **Trigonometry** in Book 2 for more on circumference and radians.

13. **B.** We are given the rate of mm per kg (remember "for every" denotes a rate): $\frac{7\,mm}{10\,kg}$. We also know that 1 stone is 6.35 kg, so we write that as a conversion factor: $\frac{1\ stone}{6.35\ \text{kg}}$. Now we can set up a line of dimensional analysis to convert what we have to start (300 stones) into centimeters. Remember each rate and conversion factor can also be flipped to their reciprocals if necessary. Whatever we need to cancel, we'll place on the bottom of the subsequent rate or conversion factor in the line. We start with stones and so need the stones to cancel, so we flip the stone/kg conversion factor upside down and multiply. Then we need the kg to cancel, so we multiply by mm/kg. Finally, we need cm but have mm, so we multiply by the conversion factor of $\frac{1cm}{10mm}$ (there are 10 mm in a cm, a fact you must memorize). At the end we cancel as necessary to get an answer that is among the choices: $300\ stones \times \frac{6.35kg}{1\ stone} \times \frac{7mm}{10kg} \times \frac{1cm}{10mm} = \frac{300(6.35)(7)}{(10)(10)}cm = \frac{30(6.35)(7)}{(10)}cm$

14. **A.** Average speed is equal to the distance traveled divided by the time taken. In this case, the distance taken is equal to the number of revolutions times the circumference of the circle. The circumference of the circle is the diameter times π or

30π inches. The total distance covered is thus 100 rotations of this: $100(30\pi)$ inches or 3000π inches. The time taken was 2 minutes. We now can form a rate by dividing our distance by time, $\frac{3000\pi \text{ inches}}{2 \text{ minutes}}$. However, we are looking for the speed in **feet per second.** Thus, we convert the inches to feet (there are 12 inches in a foot) and the minutes to seconds (there are 60 seconds in a minute) using dimensional analysis, orienting the conversion factors so the labels cancel: $\frac{3000\pi \text{ inches}}{2 \text{ minutes}} \times \frac{1 \text{ foot}}{12 \text{ inches}} \times \frac{1 \text{ minute}}{60 \text{ seconds}} = \frac{3000\pi \text{ feet}}{2(12)(60) \text{ seconds}}$. Using a calculator, we can simplify this to 2.083π.

15. D. The turkey is cooked at a rate of $\frac{1.5\, pounds}{35\, minutes}$. We can use this rate to "convert" our given amount, 9 pounds, into its equivalent cook time using dimensional analysis. We'll need to flip this rate upside down so that the pounds cancel (remember you can always flip a rate or conversion factor upside down and multiply by the reciprocal if necessary). $9 \text{ pounds} \times \frac{35 \text{ minutes}}{1.5 \text{ pounds}} = 210 \text{ minutes}$. We can use dimensional analysis again to convert minutes to hours: $210 \text{ minutes} \times \frac{1 \text{ hour}}{60 \text{ minutes}} = 3.5 \text{ hours}$, or 3 hours and 30 minutes (or divide 210 by 60 to find the time in hours).

16. C. Here we use the fact that the word PER means DIVIDE. So we need $\frac{Miles}{Hour}$. Let's first convert the diameter from inches to miles. Find the unit equivalents and put the labels in the right places to make the labels cancel: $24\, inches \times \frac{1\, foot}{12\, inches} \times \frac{1\, mile}{5280\, feet} = \frac{1}{2640}\, miles$. Now I need to turn this diameter into the miles traveled. To do so, multiply by π to get the circumference $C = d\pi = \frac{1}{2640}\, mi \times \pi$ and then multiply the circumference times the number of revolutions $\frac{1}{2640} \times \pi \frac{mi}{rev} \times 1620\, rev \approx 1.93\, mi$. Now we can find our number of hours, converting from 3 minutes: $3min\left(\frac{1hr}{60min}\right) = \frac{1}{20}hr$. Finally we put these two pieces back into our original rate in miles per hour (miles divided by hours): $\frac{1.93mi}{\frac{1}{20}hr} = 1.93 \times 20\, mph = 38.6\, mph$.

17. B. First, we set up an equation in which we use the ratio to find out how the 75 inches is divided amongst the three pieces: $3x + 5x + 7x = 75 \rightarrow 15x = 75 \rightarrow x = 5$. Now we plug in x to the smallest wood piece to find out its length: $3(5) = 15\, inches$.

18. A. If the entire investment was worth $\$40{,}200$, then we can use algebra to figure out how much each share costs: $40x + 50x + 70x = 40{,}200 \rightarrow 160x = 40{,}200 \rightarrow x = 251.25$. If each share costs $\$251.25$, and Carlos has 40 shares he would have invested this much in the company: $40(251.25) = \$10{,}050$.

19. B. If the sum of their ages is 60, we can set up a simple equation to find out their individual ages: $7x + 3x = 60 \rightarrow 10x = 60 \rightarrow x = 6$. Now we plug that into the ratio for the son's age: $3(6) = 18\, years\, old$.

20. E. We cross multiply: $8(x+2) = 4(x-10)$. Now we can solve for x: $8x + 16 = 4x - 40 \rightarrow 4x = -56 \rightarrow x = -14$.

21. A If 2 out of every 11 students are accepted, that means 9 out of 11 get rejected. We can use this new ratio to find out how many students were rejected this year: $\frac{9}{11} = \frac{x}{77{,}000} \rightarrow 693{,}000 = 11x \rightarrow x = 63{,}000$.

22. C. Let's set up a proportion, cross multiply, and solve for how many eggs Carol will need: $\frac{3}{5} = \frac{x}{100} \rightarrow 300 = 5x \rightarrow x = 60$ Now we divide the number of eggs to find out how many dozen we will need: $60/12 = 5$ dozen eggs. (Remember 12 are in a dozen).

23. E. The ratio of feet to yards is $3:1$. The ratio of square feet to square yards is the square of the single dimension ratio, or $9:1$. The rugs area in square feet is $30\times 24 = 720\,ft^2$. So we set up a proportion for square yards, and cross multiply to solve: $\frac{1}{9}=\frac{x}{720}\rightarrow 720=9x\rightarrow x=80\,yd^2$.

24. C. First, let's find out the total distance James ran. Don't forget to convert minutes to hours! We can do this by setting up proportions (miles/hours):

$$\text{First 20 minutes}\left(\frac{1}{3}\text{of hour}\right):\frac{7\,miles}{1\,hour}=\frac{x\,miles}{\frac{1}{3}hour}\rightarrow x=2\frac{1}{3}miles$$

$$\text{Last 10 minutes}\left(\frac{1}{6}\text{of hour}\right):\frac{10\,miles}{1\,hour}=\frac{x\,miles}{\frac{1}{6}hour}\rightarrow x=1\frac{2}{3}miles$$

If we add up the distances, we find that James ran 4 miles on his entire 30 minute ($\frac{1}{2}$ hour) run. Remember: $\frac{total\ distance}{total\ time}=average\ rate$. So: $\frac{4\,miles}{\frac{1}{2}hour}=8\,mph$.

25. D. First let's find out the dimensions of the table in feet using algebra. We "put an x" next to each of our ratio pieces to represent the actual length (5x) and actual width (2x). Now we create an equation for the area and solve for x: $(5x)(2x)=90\rightarrow 10x^2=90\rightarrow x^2=9\rightarrow x=3$. If $x=3$, we can plug into the expressions 5x and 2x to find the dimensions of the tabletop, 15 feet by 6 feet. Divide each by 3 feet to find yards: $15\,\cancel{feet}\times\frac{1\,yard}{3\,\cancel{feet}}=5\,yards$; $6\,\cancel{feet}\times\frac{1\,yard}{3\,\cancel{feet}}=2\,yards$. In yards, that is 5 yards by 2 yards.

The perimeter than would be twice the length plus twice the width: $2(5)+2(2)=14\,yards$.

26. D. The fraction of the drive time remaining is the drive time left divided by the total drive time. The drive left is the total drive time, $(a+3)$, minus the time he has already driven, $b-2$: $(a+3)-(b-2)$. We then divide this by the total drive time, which is $(a+3)$, so our formula is $\frac{(a+3)-(b-2)}{a+3}$. Simplifying this yields $\frac{a-b+5}{a+3}$.

27. D. Let e equal the average weight of an African elephant, g equal the average weight of a Northern giraffe and s equal the average weight of a Great White shark. If the ratio of the average weight of an African elephant to the average weight of a Northern giraffe is about 20:3, then $\frac{a}{g}=\frac{20}{3}$ and if the ratio of the average weight of a Northern giraffe to the average weight of a Great White shark is about 13:14, then $\frac{g}{s}=\frac{13}{14}$. We can isolate g in our first equation and substitute to solve for the ratio of e to s:

$$\frac{a}{g}=\frac{20}{3}\rightarrow 3a=20g\rightarrow\frac{3a}{20}=g$$

We can plug this into our second equation to find that:

$$\frac{g}{s}=\frac{13}{14}\rightarrow g=\frac{13s}{14}\rightarrow\frac{3a}{20}=\frac{13s}{14}\rightarrow 14(3a)=20(13s)\rightarrow 14(3)\left(\frac{a}{s}\right)=20(13)\rightarrow\frac{a}{s}=\frac{20(13)}{14(3)}\rightarrow\frac{a}{s}=\frac{260}{42}\rightarrow\frac{a}{s}=\frac{130}{21}$$

Therefore, the ratio of the average weight of an African elephant to the average weight of a Great White shark is 130:21.

CHAPTER

10 FUNCTION AS A MODEL

SKILLS TO KNOW

- How to understand function notation
- How to plug in values and solve for other values when given equations
- How to understand given equations—what means what, and how do the parts work together?

NOTE: For more on function notation check out **Chapter 16: Functions**, in this book. More Function as a Model questions also appear in **Chapter 17: Quadratics and Polynomials**.

Oftentimes on the ACT® you'll see **word problems that include a formula.** We call these problems "Function as a Model." The equation models some particular circumstance; most often, you're given some values that you can plug into that equation, and then you're asked to solve for something given the information. At other times, you'll be asked to interpret these functions and what particular variables mean.

A few things to remember when working on these types of problems:

FUNCTION NOTATION

Function notation confuses many students. Don't let that be you!

A microscope company's profit, P dollars, when m microscopes are made and sold can be modeled by $P(m) = m^2 - 440m - 30{,}000$. What is the least number of microscopes the company must make and sell in order for the company to not lose money on this production run?

Function notation ($P(m)$, $F(x)$, etc.) means one value—it's just like a "y"!
$P(m)$ is a single value—the PROFIT—it doesn't mean multiply P times some number m! It's just a fancy way of writing what would be y in an equation you would graph. You can even scribble out the entire ugly function notation and write "y" to keep things straight. Same goes for times when a problem uses a letter such as P—without the function notation—if it's isolated on one side of the equation, treat it like a y!

Function notation is interchangeable with a single letter!
$F(x)$ can also be written as F, $C(n)$ can also be written as C, etc. As you see in the problem above P dollars is the same as $P(m)$. This is confusing, but it's a fact you need to be aware of. A problem can ask for the value of $P(m)$ or P but it's asking for the same thing.

PLUG IN

One of the most common tactics when you see these given equation problems is to plug in! You have to solve and simplify, and as formulas have multiple variables you'll typically need to eliminate one or more by figuring out what number to plug in. **First plug in any given values—or do any simple calculations necessary to come up with a value to plug in.** These problems often have given values (the initial investment was $300, she swam for two weeks, etc.). Always plug in given values, or figure out how to use those values to find the value you need.

TIPS: **When in doubt, try zero!** If you're short on what number to plug in, think about whether there's a "zero" inherent in the question—are you trying to make a profit? Show a loss? Both of those happen when some value crosses zero. Is there a variable that is the "number of years" an investment has grown—think about it—if your investment just started, that's zero years or investment periods. In other words, if you think there's nothing to plug in, zero may help you out—experiment, play around with the numbers until the problem clicks.

Make up numbers! Even if you have five equations in the answer choices, plugging in can help you understand how the numbers work together—make the problem real and understandable. You can even make up other numbers (try using 1, 2, etc.) to understand how a function works.

The number, N, of students at Fitzgerald High School who will catch a cold through week t of school is modeled by the function $N(t)=\frac{600t^2-450}{t^2+5}$. If there are 3000 students in the school and the semester is 15 weeks, according to the model approximately what percent of students will catch the flu by the end of the semester?

To solve, we simply identify what "t" is—15 weeks of school—and plug that number in. Then we can find the number of students who caught the flu.

$$N(15)=\frac{600(15)^2-450}{(15)^2+5}$$

$$\frac{600(225)-450}{225+5}$$

$$\frac{135000-450}{230}$$

$$\frac{134550}{230}=585$$

Now we need to turn this number into a percent. Remember percent is part over whole. The whole is the total number of students in the school, 3000.

$$\frac{585}{3000}=0.195=19.5\%$$

Answer: 20% (notice the question says "approximately").

A colony of bacteria grows exponentially, as described by the equation $y = y_0(3)^t$, where t represents the number of days, y is the number of bacteria, and y_0 is the original population when $t=0$. When the colony is initially placed in a petri dish, there are 10 bacteria. According to this formula, how many cells will be in the group at the end of the week?

Now we must identify what t is (number of days) and what the original population is (10). To find the number of days—use your brain—how many days are in a week? 7! So $t=7$.

$$t=7; y_0=10$$

$$y=y_0(3)^t$$

$$y=(10)(3)^7$$

$$y=10(2187)=21870$$

Answer: 21870.

TIP: **Always double check your logic!** What makes these problems tough is that they're integrated with real world problems. Even if something is modeled by an equation, if that something is a physical distance, it can't be negative. If it's a number of items sold, it also can't be negative—make sure your answer at the end makes sense. Quadratic models, in particular, may introduce extraneous solutions. Ignore any negative solutions that don't make sense.

A microscope company's profit, P dollars, when m microscopes are made and sold can be modeled by $P(m)=m^2-440m-30{,}000$. What is the least number of microscopes the company must make and sell in order for the company to not lose money on this production run?

Let's take the problem above—we want to make a profit—how do you make a profit? By getting more than "0" for the letter P—because P is profit! How do we get P above zero? Well, imagining $P=y$ and $m=x$, this is an upwards facing parabola, so it's going to sink down and then rise back up. What we want to know is when P rises above 0—so to solve for that I set P equal to 0 and solve:

$$0=m^2-440m-30{,}000$$

This is a basic quadratic equation—we can solve by factoring or the quadratic equation. I'll factor, but you can always use the quadratic equation (you can even program your calculator to do the quadratic equation for you), or solve by graphing.

$$0=(m-500)(m+60)$$

$$m=500 \text{ or } -60$$

Now here's the tough part—because this is a word problem, you're not actually looking for -60, even though it's the smaller answer of the two. You have to realize that you can't make a negative number of microscopes. Thus the answer is 500. If you make between $0-499$ microscopes, you'll lose money, because at 500, the company neither loses nor earns money.

Answer: 500.

TIP: **When in doubt, you're looking for "zeros," points of intersection, and vertexes.** At the heart of all these problems are the same ideas that you're solving for in regular problems involving quadratics, exponential functions, and linear functions. If you're having trouble figuring out what the problem is looking for, ask yourself if any of these ideas (zeros, points of intersection, vertexes) will push you forward.

UNDERSTANDING EQUATIONS

Finally, you'll confront questions that test your ability to understand which numbers mean what in a given equation. You'll need to be able to decipher where each variable and constant came from and what they mean. Many of the same tips that we just mentioned will come into play in these types of problems. One of the best things you can do on these is to start plugging in values and understand the equation by working with it.

A paper airplane is thrown from the second story balcony of a building. The flight of a paper airplane can be modeled by the equation $h(t)=-\frac{t^2}{5}+t+10$, where t is the time in seconds after the paper airplane has been launched and h is the height of the paper airplane in feet. According to this equation, which of the following statements is true about the paper airplane?

- **A.** After 10 seconds, the paper airplane reached its maximum height.
- **B.** After 5 seconds, the paper airplane reached its maximum height.
- **C.** After 11.25 seconds the paper airplane reaches its maximum height.
- **D.** After 10 seconds the paper airplane hits the ground.
- **E.** After 2.5 seconds the paper airplane hits the ground.

Notice that ALL the choices give us a "seconds" to deal with—we can solve this out by backsolving and plugging in the seconds for the value t. Let's start with 10 (answers A & D) because it's in two different answer choices AND in the problem.

$$h(t)=-\frac{t^2}{5}+t+10$$

$$h(10)=-\frac{10^2}{5}+10+10$$

$$h(10)=-\frac{100}{5}+20$$

$$h(10)=-20+20=0$$

The height (h) is zero when we're at 10 seconds. That's answer choice D and we're done. If we weren't so lucky, we could continue to plug in, look for other intercepts (plug in zero), or look for the vertex value to help us narrow the field. Remember, the greatest height is often the vertex, the ground or landing is a zero, etc. Thus, after 10 seconds, the paper airplane hits the ground.

Answer: **D.**

NOTE: Find more problems like this in **Quadratics & Polynomials** (Ch. 17).

1. In the desert biome, the temperature can range greatly from day to night. A typical range in Celsius would be $-18° \le C \le 20°$. Given the formula $F = \frac{9}{5} \times C + 32$, where is the temperature in degrees Fahrenheit and is the temperature in degrees Celsius, what would this biome's temperature range be in Fahrenheit?

A. $1.4° \le F \le 72°$
B. $22° \le F \le 43.11°$
C. $64.4° \le F \le 68°$
D. $-0.4° \le F \le 68°$
E. $22° \le F \le 68°$

2. Candice deposits $2,000 at 5% annual interest compounded yearly at a bank which use the formula $A = P(1+r)^n$, where A is the current value; P is the amount deposited initially; r is the rate of interest for one compounding period, expressed as a decimal; and n is the number of periods, to determine the current value of the account. Which of the following would be the approximate value of the account after 10 years?

A. $35
B. $1228
C. $3258
D. $21000
E. $115330

3. The number of clients Candy Cane Cable acquired in each year can be modeled by the function $C(t) = 800t + 250$, where $t = 0$ corresponds to the year 1999. Using the given model, how many clients would you expect Candy Cane Cable to acquire in the year 2007?

A. 1605850
B. 64250
C. 7450
D. 5850
E. 6650

4. The time, t seconds, required for a simple pendulum x feet long to make 1 complete swing can be modeled by $t = 4\pi\sqrt{\frac{x}{12}}$. How many seconds will it take for a simple pendulum that is 3 feet long to make one complete swing?

A. Between 7 and 8 seconds
B. Between 6 and 7 seconds
C. Between 2 and 3 seconds
D. Between 1 and 2 seconds
E. Less than 1 second

5. For a certain species of frog, the optimal temperature range for the tadpoles in degrees Celsius is $20° \le C \le 35°$. The conversion from Celsius to Fahrenheit is $F = \frac{9}{5}C + 32$. What is the optimal temperature range for the tadpoles in degrees Fahrenheit?

A. $68° \le F \le 95°$
B. $68° \le F \le 51.44°$
C. $43.11° \le F \le 95°$
D. $43.11° \le F \le 51.44°$
E. $36° \le F \le 63°$

6. Jose invested $1400 in a high-yield savings account. In 7 months, his investment had earned $62 interest. To the nearest tenth, what is Jose's annual interest rate? (Use $I = Prt$, where I is the amount of interest earned, P is the initial investment, r is the interest rate, and t is the time, in years.)

A. 0.6%
B. 7.6%
C. 14.3%
D. 8.9%
E. 10.6%

7. $y = a(1-r)^x$ is an equation that models decay. a is the initial amount before decay, r is the rate of decay, and x is the number of time intervals that have passed. A certain element decays by 12% every hour. If there are 500 grams of this element at 5:00 AM, which expression shows how many grams will be left at 11:00 AM the same day?

A. $500(1+.12)^6$
B. $500(1-12)^6$
C. $500(12)^6$
D. $500(.12)^6$
E. $500(1-.12)^6$

8. The length L, in centimeters, of a simple pendulum is given by the equation $L = \frac{gT^2}{4\pi^2}$, where g is the acceleration of gravity in meters per second squared and T is the period of the pendulum. What is the length of the pendulum, knowing that the acceleration due to gravity is $9.8m/s^2$ and the period of the pendulum is $3.5s$?

A. 2.63
B. 3.04
C. 3.63
D. 4
E. 5.21

9. A ball is kicked into the air from ground. Its height, h feet above ground, t seconds after it is launched, is given by $h = -16t^2 + 64t$. During the ball's descent, at what value of t is the ball 48 feet off of the ground?

A. 3
B. 1
C. 2
D. 4
E. 5

10. A microwave company's profit, P dollars, when m microwaves are made and sold can be modeled by $P = m^2 - 400m - 120{,}000$. What is the least number of microwaves the company must make and sell in order for the company to not lose money on this production run?

A. 400
B. 500
C. 550
D. 600
E. 650

11. The amount, in kilograms, of usable wood left in a wood-burning furnace starting with 25 kilograms of wood can be approximated by the equation $y = 25 - 2.5h$, where h is the number of hours the stove has been burning for $0 \le h \le 10$, and y is the number of kilograms of wood remaining. According to this equation, which of the following statements is true about the stove?

A. After burning for 1 hour, the furnace still has 22.5 kilograms of wood left.
B. After burning for 1 hour, the furnace has used 22.5 kilograms of wood.
C. After burning for 2.5 hours, the furnace has used 2.5 kilograms of wood.
D. After burning for 9 hours, the furnace still has 22.5 kilograms of wood.
E. After burning for 0.5 hours, the furnace has burned 5 kilograms of wood.

12. Temperatures measured in degrees Fahrenheit (F) are related to temperatures measured in degrees Celsius (C) by the formula $F = \frac{9}{5}C + 32$. There is 1 value of x for which x degrees Fahrenheit equals $x - 30$ degrees Celsius. What is that value?

A. −77.5
B. −2.5
C. 5.75
D. 27.5
E. 77.5

13. A triangle has an angle measuring 60° and the leg opposite that measures 30 inches. One of the legs adjacent to the 60° angle has a changing length corresponding to $f(t) = 80 + 2t - 2t^2$ inches, starting from $t = 0$ where t is the time in seconds. What is the time, t, at which the resulting triangle will have an area that is $\frac{1}{2}$ that of the original triangle?

A. 4 seconds
B. 5 seconds
C. 6 seconds
D. 7 seconds
E. 8 seconds

14. The distance, d in meters, that an accelerating object travels is given by $d = \frac{1}{2}at^2$, where a is the acceleration rate in meters per second per second and t is the time in seconds. A motorbike accelerates from a stop with an acceleration of 40 meters per second squared over a distance of 60 meters. About how many seconds did the motorbike travel?

A. Between 4 and 5
B. Between 2 and 3
C. 1
D. Between 1 and 2
E. 2

15. A children's wind-up car toy travels a distance, d inches, as modeled by the function $d = 12 + 8t$ where t is the number of seconds after the car has been wound up and released. If the car is released at 12 feet, to the nearest second, how many seconds would it take to reach 21 feet?

A. 10
B. 11
C. 12
D. 13
E. 14

ANSWER KEY

1. D **2. C** **3. E** **4. B** **5. A** **6. B** **7. E** **8. B** **9. A** **10. D** **11. A** **12. D** **13. B** **14. D** **15. C**

ANSWER EXPLANATIONS

1. **D.** Converting the lower bound of the range from Celsius to Fahrenheit, we get $F=\frac{9}{5}(-18)+32\rightarrow-\frac{162}{5}+32=$ $=-32.4+32\rightarrow-0.4$ Converting the upper bound of the range from Celsius to Fahrenheit, we get $F=\frac{9}{5}(20)+32\rightarrow\frac{180}{5}+32=36+32\rightarrow68$. So, the range of temperature in Fahrenheit is $-0.4\leq F\leq68$.

2. **C.** We are given that $P=2000,\ r=0.05,$ and $n=10$. We know n is ten because interest compounds annually and the number of periods is thus equal to the number of years. Plugging these values into the equation $A=P(1+r)^n$, we solve for A. $A=2000(1+0.05)^{10}\rightarrow2000(1.05)^{10}\rightarrow3257.79\approx3258$.

3. **E.** Since $t=0$ corresponds to the year 1999, the year 2007 corresponds to when $t=2007-1999=8$. So, plugging in $t=8$, we get $C(8)=800(8)+250\rightarrow6400+250=6650$.

4. **B.** Plugging in $x=3$, we get $t=4\pi\sqrt{\frac{3}{12}}=4\pi\sqrt{\frac{1}{4}}=4\pi\left(\frac{1}{2}\right)=2\pi$. The value of 2π is ≈6.28 which is between 6 and 7 seconds.

5. **A.** Converting the lower bound of the range from Celsius to Fahrenheit, we get $F=\frac{9}{5}(20)+32\rightarrow\frac{180}{5}+32\rightarrow36+32=68$.

   Converting the upper bound of the range from Celsius to Fahrenheit, we get $F=\frac{9}{5}(35)+32\rightarrow\frac{315}{5}+32\rightarrow63+32=95$.

   So, the range of temperature in Fahrenheit is $68\leq F\leq95$.

6. **B.** We first need to know that 7 months is equal to $\frac{7}{12}$ years. Plugging in $I=62,\ P=1400,$ and $t=\frac{7}{12}$ into $I=Prt$, we get $62=1400r\left(\frac{7}{12}\right)$. Dividing both sides by $(1400)\left(\frac{7}{12}\right)$, we get $r=0.0759$. So, the interest rate r is approximately 7.6%.

7. **E.** $r=0.12,\ a=500,$ and $x=6$. Plugging these values into the given formula, we get $y=500(1-0.12)^6$.

8. **B.** This problem is as simple as plugging numbers into their respective variables and putting the expression into your calculator. Plugging the values given for period and acceleration due to gravity, we get $L=\frac{9.8\frac{m}{s^2}\times(3.5s)^2}{4\pi^2}\approx3.04$.

9. **A.** The question gives us the value of t and wants us to find the value of h that makes the equation true. So, plugging in $h=48$, we get $-16x^2+64x=-48$. Moving all the values to the right side of the equation, we get $0=-16x^2+64x-48$. Factoring out, 16 we get $16(-x^2+4x-3)$. Now, factoring the polynomial inside the parenthesis, we get $16(-x+1)(x-3)=0$. This gives us two t values of 1 and 3, but the problem asks for the time at which the ball has a height of 48 feet AND is descending. Because the ball is kicked from rest, rises, and then falls, the time closer to 0 $(t=1)$ will be when the ball is rising, so the time further from 0 $(t=3)$ will be when the ball is falling and is the correct answer.

10. **D.** For the company to not lose money, the profit P must be zero or positive. Setting $P=0$ and solving for m would give us the least amount of microwaves the company must sell for the company to not lose money. We have $0=m^2-400m-120{,}000$. To solve for n, we must find two numbers that add up to equal -400 and multiply to be $-120{,}000$. The numbers -600 and 200 satisfy these conditions, so we can factor the equation to be $(m-600)(m+200)=0$. This means $m-600=0\rightarrow m=600$ or $m+200=0\rightarrow m=-200$. The positive value for m is our answer because the company cannot sell a negative number of microwaves.

11. A. The equation $y = 25 - 2.5h$ has a y-intercept of 25 and a slope of -2.5. This means that for every hour (h), the amount of wood (y) is decreasing by 2.5. So, running through the answer choices, we see that for answer choice A, after burning for 1 hour, the furnace still has 22.5 kilograms of wood left. This is true because the amount of wood left after 1 hour is equivalent to the equation evaluated at $h = 1$. This gives us $y = 25 - 2.5 \times 1 = 22.5$. So there are 22.5 kilograms of wood left after 1 hour.

12. D. If x degrees Fahrenheit equals $x - 30$ degrees Celsius, we can plug these values into our formula as $x = \frac{9}{5}(x-30)+32$. Distributing the $\frac{9}{5}$ gets us $x = \frac{9}{5}x - 54 + 32$. This simplifies to $\frac{4}{5}x = 22$. Multiplying each side by $\frac{5}{4}$ isolates the x to get $x = \frac{55}{2} = 27.5$.

13. B. The area of a triangle is equal to any of the bases times the corresponding height (the height is the length of the altitude drawn from the point opposite the base). Since the point opposite the leg with a changing length does not move, neither does the altitude, so the height is constant. Therefore, if the base doubles its area or halves its length, the area will also double or halve, respectively. At $t = 0$, the length is 80. To find when the area is $\frac{1}{2}$ that of the original triangle, we need to find the time when the length of the leg is $\frac{1}{2} \times 80 = 40$. Set up our equation as $80 + 2t - 2t^2 = 40$ and rearrange to $2t^2 - 2t - 40 = 0$. This simplifies to $t^2 - t - 20 = 0$. This factors into $(t+4)(t-5) = 0$. The solutions are $t = -4$ and $t = 5$. t can only be a positive value, so the solution is 5 seconds.

14. D. Plugging in $d = 60$ and $a = 40$, we have $60 = \frac{1}{2}(40)t^2$. Dividing by 20 on both sides, we get $3 = t^2$ or $t = \sqrt{3}$. Since $\sqrt{1} < \sqrt{3} < \sqrt{4}, \sqrt{3}$ is between 1 and 2.

15. C. Calculate the distance the car travels by subtraction: 21-12=9 feet. The car must travel 9 feet, which is $9 \times 12 = 108$ inches. So, we plug in 108 for the distance d and solve for the time t. We get $108 = 12 + 8t \rightarrow 96 = 8t \rightarrow t = 12$.

CHAPTER

11 WORD PROBLEMS

SKILLS TO KNOW

- How to avoid careless errors
- How to identify key words that are equivalent to math symbols (translating from English to math!)
- Using made up or concrete numbers to reason the relationship between variables
- How to approach logic word problems
- How to approach Venn Diagram problems
- How to set up and solve linear equations, quadratic and other algebraic word problems
- How to approach multi-step word problems

NOTE: Word problems occur in nearly every subtopic of math. You'll find many in this chapter, but if you're looking for more, be sure to see our chapters on:

Ratios, Rates, & Units (this book, Chapter 9)
Direct & Inverse Variation (this book, Chapter 12)
Data Analysis (3-part problems) **(Book 2, Chapter 8)**
Fractions (Book 2, Chapter 4)
Percents (Book 2, Chapter 5)
Averages (Book 2, Chapter 7)
Properties of Numbers (Book 2, Chapter 2)

AVOIDING CARELESS ERRORS

Word problems require you to weed out unnecessary information and convert words to numerical expressions. The difficult part is decoding the problem.

Most word problems can be solved in three steps:

STEP 1: Read a question and translate words to numbers and symbols.
STEP 2: Turn it into a solvable equation or system of equations.
STEP 3: Ensure you've answered the right question.

In general, I see more careless errors on word problems than on any other type of math question on the ACT®. **Always re-read the question before you put your final answer.** This is a habit you must practice. Almost half of missed word problems are the result of good math and poor attention to detail.

You can also rework a word problem once you have an answer choice by reading along again with the word problem and plugging in the answer you got. In other words, **double check your answers if you have time.**

Finally, whenever **I have extra time** at the end of a test, if I've already reviewed anything I'm unsure of, **word problems are the first type of question I double check**.

TRANSLATING ENGLISH TO MATH

One skill you'll need for these questions is how to translate key terms that correlate with mathematical symbols.

Sometimes this relationship is pretty obvious:

The sum of a number and seven is nine. What is four times the number?

SUM means add $(+)$.
A NUMBER means use a variable (i.e. x or n).
IS means equals $(=)$.
TIMES means multiply $(\times)$.

First we formulate an equation and an expression:

$$n+7=9 \qquad ?=4n$$
$$n=2$$

We can solve the equation to get $n=2$, and then plug 2 in the expression:

$$4n=?$$
$$4(2)=8$$

That was easy—but sometimes things get more confusing.

Let's first review a load of English words and what they translate to in math.

Word	Meaning	Notes	Expression
OF	Multiply	Generally used with percents or fractions	Half **of** the toys $=0.5 \times t=0.5t$
LESS THAN*	Subtract	Be sure to put the number **before** the phrase "less than" **after** the subtraction sign. I like to think of less than like a fishing pole—you must "throw the line out" to the far side of the equation.	The apples weigh 2 lbs. **less than** the pears. $a=p-2$
MORE / -ER THAN*	Add	Order doesn't matter with this one as it does with "less than"	Jon is 5 inches **taller than** Sue. $j=5+s$ or $j=s+5$
PRODUCT	Multiply		The **product** of 5 and x is t. $5 \times x=5x=t$
QUOTIENT	Divide	Be sure to divide the first number after this word by the second word after it.	The **quotient** of 10 and b is c. $10 \div b=c$ or $\frac{10}{b}=c$
WHAT (NUMBER/ FRACTION)	Create a variable	Be sure to make up a new variable	**What number** is five less than 7? $x=7-5$
DIFFERENCE	Subtract	Replace the "and" with a subtraction sign.	The **difference** between x and y is 5. $x-y=5$
IS / EQUALS	Equals	Other words may also mean equals (are, equivalent to, etc.)	What fraction **is** half of $\frac{3}{4}$? $n=\left(\frac{1}{2}\right)\left(\frac{3}{4}\right)$
THE DISTANCE BETWEEN	Absolute value of the difference	These are called "open sentence" equations and inequalities	The distance between a number and five is less than .25. $\|x-5\|<0.25$

*These represent the MOST COMMON elements students mess up. If you ever are confused, make up REAL NUMBERS to remember what these words men. For example, if apples weigh 2 lbs. LESS than pears, I could have 3 lbs. apples, 5 lbs. pears. I can then see $\text{apples}=\text{pears}-2$ because $3=5-2$. I use real numbers to help prevent confusion and ensure the right step.

Which of the following inequalities is equivalent to the expression below?

11 is greater than or equal to 22 less than the product of n and $\frac{1}{3}$

A. $11>22-\frac{1}{3}n$ **B.** $11\geq 22+\frac{1}{3n}$ **C.** $-11\leq -\frac{1}{3}n$

D. $11\geq \frac{1}{3}n-22$ **E.** $11\leq \frac{1}{3}n+22$

MISTAKE ALERT: Many students get into trouble when translating "LESS THAN." For example: "5 less than x" is NOT "$5-x$" but actually "$x-5$"! As I mentioned earlier, "less than" is like a fishing pole—you have to "throw the line out" to the far side of the equation!

11	is greater than or equal to	22 less than	the product of n and $\frac{1}{3}$
↓	↓		↓
11	$\geq$		$\frac{1}{3}\times n$ -22

Our first number is 11, and it says that 11 **is greater than or equal to**. This means the equation must be $11\geq$ so far. Reading on, we see 22 **less than**, which means "____-22." The blank is "the **product** of n and $\frac{1}{3}$," so the second part of the equation is $\left(n\times\frac{1}{3}\right)-22$. (Note: it's always a good option to use parentheses when you plug in elements.) Joining the two parts together, we get $11\geq\left(n\times\frac{1}{3}\right)-22$. This expression is equivalent to $11\geq\frac{1}{3}n-22$.

Answer: **D**.

MAKING IT CONCRETE

A technique I often use to help solve word problems is to make the ideas concrete. What that often entails is making up some numbers so that I understand how the numbers given should work together and so that I can set up an equation. This technique works when I see variables in answer choices and need to know how to set up an equation, as well as when I simply want to solve a word problem and need to set up an algebraic equation to solve for my unknowns.

For a fund-raiser, a club decides to make layered hot cocoa jars and to deliver them the week before winter break begins. They estimate it will cost $\$7$ for ingredients per cocoa jar and $\$65$ to order all the mason jars and decorations. The club decides to sell each jar for $\$10$. Assuming they have no other expenses, which of the following equations represents the profit in P that they will make by selling j jars of hot cocoa?

A. $P=10j-65$ **B.** $P=7j+65$ **C.** $P=3j-65$

D. $P=-17j+65$ **E.** $P=17j-65$

For this question, we want to know how much money they will make overall, including costs. We know that each jar j costs $\$7$, there is a fixed cost of $\$65$ for the other materials, and the complete jars sell for $\$10$ each. But where do we put the numbers? One way to know is to make up a number of jars and then figure out how you would set up the calculation. Let's pretend we had 10 jars. If we are paying $\$7$ for each jar, that would be $\$7\times10$ or $\$70$ that we'd pay for the ingredients. Then if we sold each one for $\$10$ that would be $\$100$ we'd make. That leaves a profit of $\$30$—however, we still have to pay $\$65$ in fees, so we'd be down $\$35$ and our P would be -35. Now I can plug in that number 10 into j and try to figure out which choice gives me $-\$35$. Choice (C) makes the most sense—we're profiting $\$3$ for each of the 10 jars but then subtract $\$65$ to get a loss of $\$35$. I can plug in and back solve and know that it works.

We can also think of this problem algebraically—the overall cost for the jars is $7j+65$. But we also need to account for the profit. We know that each jar will make them $\$10$, so that translates to $10j$ dollars they make before subtracting costs. As for profit P, it will be the amount of money they make minus the cost. This means that $P=10j-(7j+65)$. By distributing the negative sign and simplifying, we get $P=3j-65$.

If the algebra method confuses you, try making up numbers to help!

Answer: **C.**

LOGIC QUESTIONS

Some other word problems are called "logic problems." These problems don't show up as often.

The basic logic "equation" is an "if—then" statement. We call this statement the **"positive"** or "Modus Ponens" if you're taking a logic class:

IF A...THEN B

You should memorize the fact that when you have an "if—then" statement as above, the following statement is ALWAYS true (called the **"contrapositive"** or "Modus Tollens"):

IF NOT B...THEN NOT A

However, you cannot assume that:

IF B...THEN A

or

IF NOT A...THEN NOT B

I know that sounds confusing, so here's my favorite example for logic problems to help make sense of the above.

Let's let "A" equal **it is snowing** and "B" equal **it is below 32 degrees outside*.**

A ┐ **B** ┐

GIVEN: IF **it is snowing**...THEN **it is below 32 degrees outside.**

By extension we can assume the following:

NOT B ┐ **NOT A** ┐

CORRECT: IF **it is NOT below 32 degrees outside**...THEN **it cannot be snowing.**

*To all the pedantic ones out there: I realize the true statement would more accurately read "at or below" this temperature at the altitude of snowflake formation but let's not get crazy technical. This is just a logic lesson.

Logic will tell you this is true, and it makes sense. You can't have snow if it's not freezing out. At the same time we can't assume the following:

INCORRECT: IF **it is below 32 degrees outside** THEN **it is snowing.**

Just because it's really cold doesn't mean snow is in the air, right?

INCORRECT: IF **it is NOT snowing** THEN **it is NOT below 32 degrees outside.**

Yet another statement that simply can't be assumed—just because it's not snowing doesn't mean the weather isn't cold.

If the idea If $A \rightarrow B =$ If NOT $B \rightarrow A$ confuses you, replace your statement with the temperature/snow example above, substituting these phrases for whatever phrases you're analyzing.

"If Hannah is home, then her shoes are in the hallway."

If the previous sentence is true, then which of the following sentences MUST also be true?

A. If Hannah's shoes are not in the hallway, then she is not home.
B. If Hannah is not home, then her shoes are not in the hallway.
C. If Hannah's shoes are not in the hallway, then she is home.
D. If Hannah's shoes are in the hallway, then she is home.
E. If Hannah is not home, then her shoes are in the hallway.

TIP: For "If-then" statements, the only other statement that must be true is the **contrapositive** of that statement. So to anticipate the answer, flip the two phrases' order and add a "not."

For this problem, the contrapositive would be that "if Hannah's shoes are **not** in the hallway," then "Hannah is **not** home." If this confuses me, I can remember the pattern with my earlier snow example, and replace "Hannah is home" with "it is snowing" and "her shoes are in the hallway" with "it is freezing outside" and then think logically about what else is necessarily true.

Answer: **A.**

VENN DIAGRAMS

Another type of word problem you'll occasionally need to solve is the Venn Diagram problem. These problems ask you to divide elements into groups and deduce information based on groups that overlap.

Sam asks 50 students questions about traveling.

Question	Yes	No
1. Have you ever traveled on a train or airplane?	35	15
2. If you answered Yes to Question 1, did you travel on an airplane?	29	6
3. If you answered Yes to Question 1, did you travel on a train?	18	17

How many students have been on both a train and an airplane?

A. 35 **B.** 37 **C.** 2 **D.** 12 **E.** 13

One fail-safe and intuitive way to solve a problem like this is to sort through the information in a Venn Diagram, or diagram of overlapping loops (see example below). You could then label each region a, b, and c and create equations:

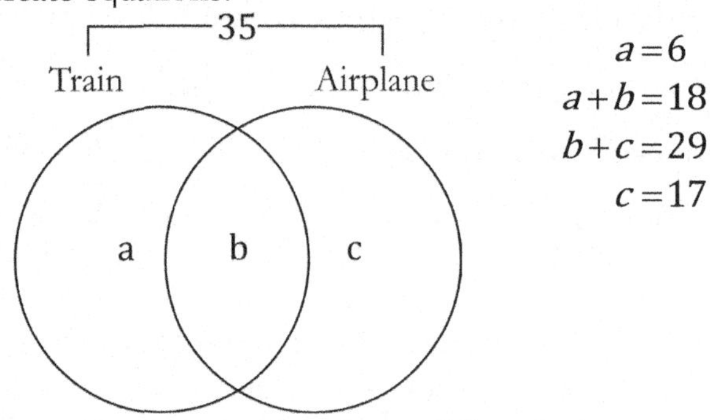

$$a=6$$
$$a+b=18$$
$$b+c=29$$
$$c=17$$

We want to solve for b. Let's take the equation $b+c=29$ and plug in c to solve for b:

$$b+17=29$$
$$b=12$$

There are many other ways we could have solved. For example, we could have used the $a+b=18$ equation and substituted for a. I know some students memorize a formula that askes them to add the two full circle amounts together and subtract from the total (i.e. $35-(18+29)$), but this formula only works in certain cases; drawing it out works for ALL cases and helps prevent careless errors.

Answer: **D.**

ALGEBRAIC WORD PROBLEMS: WHAT YOU HAVE, KNOW, AND NEED

The most traditional word problems require you to set up a system of equations or a linear equation and solve for an unknown variable. Often these word problems involve rates (also known as slope) and may also involve an initial value (typically the y-intercept of the corresponding graph of the function). For some problems, you may even have more variables or more unique situations. In any case, the technique is the same: create variables for your unknowns, set up equations, and solve. If you struggle with these questions, I also encourage you to check out this book's chapters on Systems of Equations and Ratios, Rates & Units—both of those chapters cover similar problems.

Isme, Una, and Elliot are selling one type of candy bar for their school fundraiser. If Isme sells 50 bars, Una sells 80 bars and Elliot sells 110 bars, and the gross sales total of all three students' sales is $\$300$, what is the value of Elliot's gross sales?

Remember "gross" means the total profit before any deductions, so we won't need to worry about how much profit students made or what the wholesale cost is, etc.

For this problem, we can first try to figure out the price per bar, and then multiply that price times the number that Elliot sold to find out how much he sold.

If this is confusing, start with what you:

- **<u>Have</u>** (given elements in the problem)
- **<u>Know</u>** (any formulas or facts) and
- **<u>Need</u>** (what the question asks you to solve for)

Write each of these down on your paper to help anchor your problem solving process.

NEED: The total dollar amount Elliot sold.

Think—if Elliot sold 10 bars for $\$2$ each he would have sold $\$20$. How did I get that $\$20$? I multiplied the number of bars times the cost per bar. So that's what I write down:

$$NEED = Cost\ per\ bar \times Number\ of\ Bars\ Elliot\ Sold$$

Don't be afraid to make up numbers as I did above to make headway in a problem, and don't be afraid to write things out in English before filling in the numbers you know. I **know** that he sold 110 bars, so I **need**:

$$NEED = Cost\ per\ bar \times 110$$

So how can I find the "cost per bar"? We'll assume all the bars sell for the same price.

Think about what you **KNOW**: The word PER means DIVIDE! So "cost PER bar" translates to TOTAL COST divided by TOTAL BARS.

We call this the unit rate (cost per bar). I want to divide all the money made by all the bars sold, so I take $\$300$ and divide by the sum of all the bars sold:

$$\frac{Total\ Sales\ of\ All\ Bars}{Total\ Number\ of\ Bars} = \frac{\$300}{\#Una\ Sold + \#Isme\ Sold + \#Elliot\ Sold}$$

Now I plug in what I **HAVE** (the given numbers): The bottom of this fraction is the sum of the three kids' bars, and the top is the total sales they had. The top number is a given number ($\$300$) and the bottom I can easily calculate:

$$\frac{\$300}{50+80+110} = \frac{\$300}{240} = \$1.25$$

I go back to my NEED equation:

$$NEED = Cost\ per\ bar \times (110)$$

And plug in:

$$\$1.25(110) = \$137.50$$

Answer: $\$137.50$

Remember when you are trying to solve word problems, if you feel lost:
1. Make up numbers here or there to understand relationships.
2. Remember what you HAVE, KNOW, and NEED.
3. Write out equations using English to help you know where to put what.
4. Remember PER means DIVIDE.

A fabric store sells cotton and velvet fabrics. Myra pays $\$50$ for 3 yards of cotton and 2 yards of velvet. Jaime pays $\$22$ for 1 yard of cotton and 1 yard of velvet. How much does 1 yard of velvet cost?

For this problem let's make up some variables and use algebra. Let's start with what we NEED to figure out the first variable to make up (this is always a good starting place). We need the cost per one yard of velvet. Let's call that v. We likely also want the cost per one yard of cotton—as that would be the corresponding variable for our other fabric, cotton. Let's call that c.

Let's say cotton costs $\$15/yard$: $c=\$15$. We would figure out how much someone paid for two yards by multiplying $2(\$15)$ or $2c$, so that's how we set up the equation. Again I just make up numbers for a moment to make sure I'm putting everything in the right place. We'd do the same for velvet. $2v$ would be the total cost someone spent on velvet if they purchased two yards. So let's create an equation for Myra:

$$\$50=3c+2v$$

Jaime's equation is even easier:

$$\$22=c+v$$

Now let's figure out what we need. We need to solve for v. Because I want to KEEP v, I'll **isolate c to eliminate** it. (remember, isolate whatever variable you want to get rid of and then substitute.) I try to remember the phrase "**isolate to eliminate**" to help me more efficiently work word problems. You could also isolate v, but then you'd need to plug in at the end of the question to solve for v. Let's work off the bottom equation—it has no coefficients so will be easy to use substitution with.

$$\$22-v=c+v-v$$
$$\$22-v=c$$

Now we plug in:

$$\$50=3(\$22-v)+2v$$
$$\$50=\$66-3v+2v$$
$$\$50-\$66=-v$$
$$-\$16=-v$$
$$\$16=v$$

Because we need the cost of velvet, and we solved for v, we are done.

Remember to always **RE-READ** the question before you put your answer to avoid careless mistakes on problems like this one!

A fitness apparel store sells t-shirts for $\$32$ each. At this price, 50 t-shirts are sold per week. For every $\$1$ decrease in the price, the store will sell five more t-shirts per week. If one week the store wants to maximize revenue, and prices its t-shirts accordingly, how many t-shirts will it sell that week? (Note: revenue is the total dollar amount of retail sales before any costs are deducted and not including any taxes).

This one is tricky. Let's start with what we need.

We need to maximize revenue. How do we calculate revenue? Well, for the first week it's $\$32$ times 50—that would be the cost per shirt times the number of shirts.

$$(\textit{cost per shirt})(\textit{\# of shirts})=(32)(50)$$

Let's say we decrease the cost per shirt by $\$1$, what happens? The number of shirts goes up by 5:

$$(\$32-1)(50+5)$$

$$\underset{(COST)}{\nearrow} \qquad \underset{(\#\,OF\,SHIRTS)}{\nwarrow}$$

Let's say we decrease the cost of the shirt by 2:

$$(\$32-2)(50+5(2))$$

Let's say we decrease it by 3:

$$(\$32-3)(50+5(3))$$

I'm starting to see a pattern here, and now I can see where my variable will go, i.e. it will be the decrease in cost in the shirt. I will let the cost decrease be $\$x$ and the sales increase is going to be $5x$.

$$(\$32-x)(50+5(x))$$

As you can see, I used baby steps to work out the problem one example at a time until the relationship "clicked." When you're not sure, just dive in! Then look for patterns and try to figure out your variables. In this case, and in all cases when we have "maximums" with a single equation, we're dealing with a quadratic or polynomial. Inevitably figuring those out is often best done by writing down one example at a time, stepping forward, and looking at what you have. The variable I made up represents "dollars off" not "number of shirts" (what I need)—but once I figure out the dollars off, however, I can figure out the number of shirts by plugging x into the second expression in the product above: $(50+5(x))$. Remember this expression is the number of shirts.

Now, let's find the maximum of this function:

$$f(x)=(\$32-x)(50+5(x))$$

Remember, "maximum" always refers to the highest "y-value" of a function. In a quadratic equation, this is the y-value of the vertex. Here we don't need the maximum itself, we need a piece of information given that we are at the max y-value. Bottom line is, we need to find the vertex.

To find the vertex, we could plug this function into our calculator (not a bad idea—and not much more time consuming than my method below), or we could waste time expanding this and then trying to figure out the vertex. Or we could waste even more time completing the square after we expand. But I want the fastest path to the answer.

Whenever you have a quadratic *in factored form*, the fastest path to the vertex or maximum is to average the zeros to find the x value of the vertex, as that is where the maximum will be. The x-value of the vertex is always the average of the two zeros,* which can be easily found by using the Zero Product Property:

**Zeroes are the x-values when* $y=0$ *(x-intercepts of the graph).*

$$(\$32-x)=0 \text{ or } 50+5(x)=0$$

The first equation gives us:

$$x=32$$

The second:

$$50=-5(x)$$
$$x=-10$$

Now remember because parabolas have symmetry, the x-value of the vertex will be the average of these two:

$$\frac{32-10}{2}=11$$

Thus, the maximum will occur when we reduce the price of the t-shirts by $\$11$. But we're not done yet! Oftentimes, problems want you to find the maximum, or y-value if we plug in 11. But here we need something else.

Re-read the question! The question asks for the number of shirts sold—we plug $x=11$ into the expression that describes the number of shirts sold:

$$50+5x$$
$$50+55=110$$

Answer: 110

Our answer is 110 shirts will sell if the sales are at a maximum.

MULTI-STEP WORD PROBLEMS

These are word problems that do not fit nicely into a few algebraic formulas or are better solved by simply churning through one step at a time. Oftentimes with harder multi-step problems, it can be difficult to see how to get to the end of the problem.

The best approach for these is first to not panic! And second, to simply digest the information one little piece at a time. Write down notes AS YOU READ to help you!

The pep club is organizing a bake sale and is planning to purchase baking supplies. Each dozen cookies require 1.5 cups of flour and $7/8$ cup of sugar. Originally, the club planned to purchase enough flour and sugar to make exactly 200 cookies, adjusting the recipe scale as necessary, but found that ingredients were cheaper when purchased in specific sized quantities: 14-cup bags of flour and 8-cup bags of sugar. How many extra whole cookies was the pep club able to bake assuming it purchased at least enough flour and sugar to bake 200 cookies?

For this problem we want to take things one at a time.

NEED: First we need to figure out how much flour and sugar we need to make 200 cookies.

KNOW: We know ratios according to each dozen cookies, so let's figure out how many dozen cookies are in 200 cookies.

Step 1: Find out how many dozen cookies are in 200 cookies.

Divide 200 by 12: $16\frac{2}{3}$ dozen cookies

Now we need $16\frac{2}{3}$ times each portion of baking product necessary for a dozen cookies.

Step 2: Find out how much total flour and how much total sugar is necessary to make $16\frac{2}{3}$ dozen cookies.

Multiply $16\frac{2}{3}$ times the amount of flour needed per dozen.

$$\left(16\tfrac{2}{3}\ \cancel{dozen}\right)\left(1.5\frac{cups}{\cancel{dozen}}\right) \text{ (see how the "dozen" label cancels?) } = 25 \text{ cups of flour}$$

Multiply $16\frac{2}{3}$ times the amount of sugar needed per dozen.

$$\left(16\tfrac{2}{3}\ \cancel{dozen}\right)\left(\tfrac{7}{8}\frac{cups}{\cancel{dozen}}\right) \text{ (see how the "dozen" label cancels?) } = 14.58333 \text{ or } 14\tfrac{7}{12} \text{ cups of sugar}$$

Step 3: Figure out how much flour / sugar would be purchased.

To cover 25 cups of flour, we'd need 2 bags that have 14 cups each (28 cups total).
To cover 14.583 cups of sugar, we'd need 2 bags of sugar with 8 cups each, totaling 16 cups sugar.

Step 4: Figure out how much excess flour sugar exists.

28 cups flour minus 25 cups flour would give us 3 **leftover cups of flour.**
16 cups sugar minus 14.58333 would give us $1\frac{5}{12}$ **or 1.416667 cups of sugar.**

Step 5: Calculate how many dozen cookies you can make with the leftovers.

Flour: This one is easy. 3 cups can make two dozen cookies as each dozen requires 1.5 cups. But flour may not be our "limiting factor."

Sugar: This one is harder. How did we do the flour one? We divided 3 cups by 1.5 cups for the dozen. Let's do the same with the $1\frac{5}{12}$—let's divide that by $\frac{7}{8}$: 1.619. We can make 1.619 dozen cookies with the sugar we have left. Because this number is smaller than the flour number, this is our limiting factor that will determine how many cookies we can make.

Step 6: Calculate the extra cookies.

Let's multiply that by 12 now to turn this "dozen" number from our limiting factor of sugar into number of cookies:

$$1.619 \times 12 = 19.42 \text{ cookies.}$$

We won't be baking partial cookies, so that means we can make 19 **extra cookies.**

As you can see, we need to just take this one step at a time and apply all the ideas we used in other types of word problems.

1. The smaller of two numbers is 12 less than one third of the larger number. Three times the larger number plus twice the smaller is 152. Which equation below could be used to solve for the smaller number, s?

A. $2s^2+3$
B. $9(s+12)+2s=152$
C. $5s+12=152$
D. $3s+12=152$
E. $9(s-12)+2s=152$

2. Each side of a square is s inches long. If the length is doubled and the width is increased by 3, what is the area of the rectangle that is formed?

A. $2s^2+3$
B. $(2s+3)^2$
C. $2s^2+6$
D. $2s^2+6s$
E. $6s^2$

3. A rectangle has length $n+2$ and width $n-3$. If a square with side $n-1$ is removed from the rectangle, what is the area?

A. $n-5$
B. $-2n-7$
C. $n-7$
D. n^2-2n-4
E. $(n-1)^2$

4. Ricky is buying pieces of colored cloth to build a costume. For the costume, he needs 7 10-inch pieces of red cloth, 4 14-inch pieces of blue cloth, and 3 19-inch pieces of black cloth. The cloth costs $0.83 per *yard* regardless of color. If Ricky has no coins, how many dollars will he need to bring to pay?

A. $3
B. $4
C. $5
D. $6
E. $7

5. A clothing store surveyed 125 of its customers about what articles of clothing they had bought from the store. Of the 125 customers, 104 had bought T-shirts, 74 had bought jeans, and 63 had bought both. Some customers bought neither. Of the 125 customers, how many had bought T-shirts, jeans, or both?

A. 178
B. 115
C. 62
D. 52
E. 10

6. Jemaine is a fruit farmer and he sells apples for $0.17 per pound. He recently sold a shipment of apples for $300. On average there are 3 apples per pound. Which value is closest to the number of apples in the shipment?

A. 300
B. 5,300
C. 6,000
D. 50,000
E. 3,000

7. Vijay has 89 marbles that are blue, green, or white. He spreads them out on the floor and sees that there are three times as many blue marbles as white marbles and 16 fewer green marbles than white marbles. What is the difference between the amount of blue marbles and green marbles than Vijay has?

A. 21
B. 5
C. 16
D. 58
E. 42

8. At a racetrack, motorcyclists start with 50 points. Each racer gains 15 points for every lap of the track in 1 minute or under and loses 10 points for every lap over 1 minute. Marco finishes 12 more laps in 1 minute or under than he does in more than 1 minute. He ends with 320 points. How many laps did Marco finish in over 1 minute?

A. 12
B. 14
C. 16
D. 18
E. 20

9. A schoolteacher rewards good students with gummy bears. He has an initial amount x of gummy bears. He gives $\frac{1}{4}$ of the gummy bears to the first student. Then he gives $\frac{1}{4}$ of the remaining gummy bears to the next student. He repeats this process one more time. If the teacher only gave out whole gummy bears, what is the minimum number of gummy bears he could have started with?

A. 4
B. 16
C. 64
D. 124
E. 256

10. If 7 times a number q is subtracted from 28, the result is negative. Which of the following gives the possible value(s) for q?

A. 0 only
B. 4 only
C. 12 only
D. All $q < 4$
E. All $q > 4$

11. Which of the following actions will produce the smallest result when replacing the blank in the expression 23________$(-\frac{3}{25})$?

A. multiplied by
B. divided by
C. minus
D. plus
E. to the power of

12. If the statement "If it is Thursday, then the café is closed" were true, which of the following statements would also have to be true?

A. "If the café is closed, then it is Thursday."
B. "If the café is not closed, then it is Thursday."
C. "If the café is not closed, then it is not Thursday."
D. "If it is not Thursday, then the café is closed."
E. "If it is not Thursday, then the café is not closed."

13. Adam told Candice that if he spent $70 from his savings account, his saving account would have at least $\frac{4}{5}$ as much in it as it has now without spending. From Adam's statement, Candice can deduce that the *least* amount of money that Adam could have in his savings account now is:

A. $350
B. $87.50
C. $56
D. $280
E. $375

14. Mr. Lee is a teacher whose salary is $25,650 for the school year, which has 195 days. In Mr. Lee's school district, substitute teachers are paid $85 per day. If Mr. Lee takes two days off without pay and a substitute teacher is paid to teach Mr. Lee's classes, how much less does the school district pay in salary by paying a substitute teacher instead of paying Mr. Lee for those days?

A. $29.45
B. $46.54
C. $93.08
D. $107.30
E. $9,075.00

15. Consider the 3 statements below to be true.
- No mammals are squirrels.
- Some mammals are bats.
- Some bats are squirrels.

Which of the following statements *must* be true?

A. All squirrels are bats.
B. All bats are mammals.
C. Some squirrels are not bats.
D. Some bats are not mammals.
E. Some mammals are squirrels.

16. Two scientists are running an experiment to produce a solid that can be used for further experimentation. The formula says that of the total solid formed, S, the usable amount, U, can be calculated by subtracting three-fourths the amount of Nitrogen, N, used in the experiment from 25% of the total solid formed. What equation gives the amount of usable solid?

A. $U=\frac{1}{4}S-\frac{3}{4}N$

B. $U=\frac{3}{4}N-\frac{1}{4}S$

C. $U=\frac{3}{4}S-\frac{1}{4}N$

D. $U=\frac{1}{4}N-\frac{3}{4}S$

E. $U=\frac{1}{4}S+\frac{3}{4}N$

17. For every 3 cents the price of a pound of bananas increases, a farmer sells 120 fewer pounds of bananas. If the farmer normally sells 5000 pounds of bananas, what expression models the number of pounds of bananas he sells if the price increases by an amount of c cents?

A. $5000-120c$
B. $5000-40c$
C. $5000-c$
D. $5000+120c$
E. $5000+40c$

18. What rational number is one third of the way from $\frac{2}{5}$ to $\frac{6}{7}$?

A. $\frac{58}{105}$

B. $\frac{116}{105}$

C. $\frac{2}{3}$

D. $\frac{16}{115}$

E. $\frac{3}{5}$

19. 300 households are surveyed about their news channel preferences. 147 answered that they watch Channel 7 news daily, and 130 answered that they watch Channel 42 news daily. What is the minimum number of households that watch both Channel 7 and Channel 42 news?

A. 17
B. 23
C. 0
D. 147
E. 130

ANSWER KEY

1. B **2. D** **3. C** **4. C** **5. B** **6. B** **7. D** **8. D** **9. C** **10. E** **11. B** **12. C** **13. A** **14. C**
15. D **16. A** **17. B** **18. A** **19. C**

ANSWER EXPLANATIONS

1. **B.** Let s represent the small number and let l represent the larger number. We can make these two equations from the question: $s=\frac{1}{3}l-12$ and $3l+2s=152$. Taking the first equation and multiplying both sides by 3, we get $3s=l-36$. Adding 36 on both sides, we get $3s+36=l$. Now, substituting in $l=3s+36$ into the equation $3l+2s=152$ we get $3(3s+36)+2s=152$. Factoring out a 3 from the expression in the parenthesis, we get $9(s+12)+2s=152$.

2. **D.** The new length is $2s$ and the new width is $s+3$, so the new area is $A=width\times length \rightarrow 2s(s+3)$. Distributing the $2s$ we get $A=2s^2+6s$.

3. **C.** The area of the original rectangle is $(n+2)(n-3)=n^2+2n-3n-6=n^2-n-6$. The area of the square is $(n-1)(n-1)=n^2-n-n+1=n^2-2n+1$. So, the area remaining after subtracting the square from the rectangle is $n^2-n-6-(n^2-2n+1)=-n-6+2n-1=n-7$.

4. **C.** One yard is equal to 3 feet, which is equal to 36 inches, so Ricky will have to pay \$0.83 for every 36 inches. He needs a total of $7(10)+4(14)+3(19)=183$ inches of cloth. That means he must pay $183\ \cancel{inches}\times\frac{\$0.83}{36\ \cancel{inches}}=\$4.22$. Rounding this up, because Ricky has no coins but must have at least this amount, we get \$5.

5. **B.** If 63 customers had bought both and 74 had bought jeans, then $74-63=11$ had bought only jeans. If 104 had bought shirts, then $104-63=41$ had bought only shirts. This counts for a total of $63+11+41=115$ customers who bought a shirt, jeans, or both. To understand this question better, draw a venn diagram and make up variables for each segment. Then solve your equations with substitution.

6. **B.** Convert using dimensional analysis, lining up labels to get them to cancel. $\$300\times\frac{1\ \cancel{pound}}{\$0.17}\times\frac{3\ apples}{1\ \cancel{pound}}=5294.12$ apples. The closest answer choice value is 5300 apples. This is a Ratios, Rates & Units word problem. See Chapter 9 for more problems like this and problem solving tips.

7. **D.** Let b= the number of blue marbles, g = the number of green marbles, and w = the number of white marbles. There are a total of 89 marbles, so $b+g+w=89$. There are 3 times as many blue marbles as white marbles, so $b=3w$. There are 16 fewer green marbles than white marbles, so $g=w-16$. If you made an error setting these up, remember to make up numbers as you go to better understand relationships. Substituting $b=3w$ and $g=w-16$ in the equation $b+g+w=89$, we get $3w+w-16+w=89$. Simplifying and solving for w we get $5w-16=89\rightarrow 5w=105\rightarrow w=21$. Plugging in this value for w we can solve for the values of b and g. $b=3w=3(21)=63$. $g=w-16\rightarrow 21-16=5$. So, the difference between the amount of blue marbles and of green marbles is $b-g=63-5\rightarrow 58$.

8. **D.** Let m =the number of laps he finishes in more than 1 minute. Then the number of laps he finished in less than or equal to 1 minute is 12 more than m and can be represented as $m+12$. Then, the number of points he gets from laps finished over 1 minute is equal to $-10m$ and the number of points he gets from laps finished under 1 minute is equal to $15(12+m)$. The total points he gets for all his laps is then $15(12+m)-10m$. Every racer starts with 50 points and we are given that he ends with 320 points, so $50+15(12+m)-10m=320$. Simplifying and solving for m, we get $50+15(12+m)-10m=320\rightarrow 50+180+15m-10m=320\rightarrow 230+5m=320\rightarrow 5m=90\rightarrow m=18$.

9. **C.** Starting with x gummy bears, the teacher gives out $\frac{1}{4}x$ to the first student, which means that he is left with $x-\frac{1}{4}x=\frac{3}{4}x$ gummy bears. Of these $\frac{3}{4}x$ gummy bears, he gives away $\frac{1}{4}\times\frac{3}{4}x=\frac{3}{16}x$ gummy bears to the second student, so he is left with $\frac{3}{4}x-\frac{3}{16}x\rightarrow\frac{12}{16}x-\frac{3}{16}x\rightarrow\frac{9}{16}x$. Of these $\frac{9}{16}x$ gummy bears, he gives away $=\frac{1}{4}x\times\frac{9}{16}x\rightarrow\frac{9}{64}x$ of these gummy bears to the last student. He is then left with $\frac{9}{16}x-\frac{9}{64}x\rightarrow\frac{36}{64}x-\frac{9}{64}x=\frac{27}{64}x$ gummy bears. The smallest number x that will make $\frac{27}{64}x$ a whole number is $x=64$.

10. **E.** Writing out the given information into an equation, we have $28-7q<0$. We add the q term to the other side and divide by $7: 28<7q\rightarrow 4<q$, which is the same as $q>4$.

11. **B.** We can eliminate answer choices C, D, and E because these all give us positive values when the operation is performed. Only choices A and B yield negative results. $23\times(-\frac{3}{25})=-\frac{23\times3}{25}\approx-3$ while $\frac{23}{-\frac{3}{25}}=23\times(-\frac{25}{3})\approx-125$. So, we can see that by roughly estimating 23 as 25, we can already see that the result reached by dividing is much smaller than the result reached by multiplying.

12. **C.** The statement "If it is Thursday, then the Café is closed" is in the form "If p then q" where p= the event it is Thursday and q = the event the café is closed. If the statement is true, then the contrapositive "If not q then not p" is also true. This is the statement "If the café is not closed, then it is not Thursday."

13. **A.** If we assume that his savings account has $\frac{4}{5}$ as much as what he has now after spending $\$70$, then we can calculate the amount of money in his savings account now as $\frac{4}{5}s=s-70$ so $s=7\times50=350$. This is the least amount of money he could have in his savings account to spend $\$70$ and $\frac{4}{5}$ or more of his savings remain. If he had more money saved, he would save a greater portion of his savings, but if he had less than $\$350$, $\$70$ would be more than $\frac{1}{5}$ of his savings, and this he would have less than $\frac{4}{5}$ remaining.

14. **C.** Mr. Lee earns $\$25{,}650$ over 195 days, which means the school district pays him $\frac{25{,}650}{195}=\$131.54$ per day. If Mr. Lee takes two days off without pay and the school district pays substitute $\$85$ a day for two days to cover Mr. Lee, then the school district saves $\$131.54\times2-\$85\times2=\$93.08$ for those two days.

15. **D.** These types of logic puzzles can be difficult, so we should try to draw a Venn diagram. First, draw a circle that represents squirrels. Our first statement is that "all mammals are not squirrels". Draw two separate circles to represent squirrels and mammals. The next statement, "some mammals are bats" means that we can intersect a circle called bats with mammals. The next statement, "some bats are squirrels" means we can also intersect the bats circle with squirrels. Note that even though there may be space in our diagram where the bats circle doesn't intersect with mammals or squirrels, this is not necessarily true of bats themselves. We do not know if there are bats that are neither mammals nor squirrels. We do know, from looking at the diagram, that there are some bats that are not mammals (those that are in squirrels) and vice versa. Thus, (D) is true.

16. **A.** 25% of the total solid formed is equal to $\frac{1}{4}S$. Subtracting $\frac{3}{4}$ of the amount of Nitrogen N from this, we get $\frac{1}{4}S-\frac{3}{4}N$. So, $U=\frac{1}{4}S-\frac{3}{4}N$.

17. B. For every 3 cents the price increases, the farmer sells 120 fewer pounds of bananas. So, if the price increases by c cents, the farmer will sell $\frac{c}{3}$ times 120 fewer pounds. If you have trouble seeing this, make up some numbers. A 6 cent increase would mean 3 cents twice or 240 fewer pounds of bananas, or 2(120). Where did the two come from? 6 divided by 3. So divide c by three. So, with every price increase of c cents, he will sell $\frac{c}{3}(120)=40c$ fewer pounds than his usual 5000 pounds. This means he will sell $5000-40c$ pounds of bananas if he increases the price by c cents.

18. A. We first write $\frac{2}{5}$ and $\frac{6}{7}$ with the same denominator. Making the denominator the greatest common multiple $5(7)=35$ we get $\frac{2}{5}\times\frac{7}{7}=\frac{14}{35}$ and $\frac{6}{7}\times\frac{5}{5}=\frac{30}{35}$. The difference between these two fractions can then be easily calculated as $\frac{30}{35}-\frac{14}{35}=\frac{16}{35}$. One third of this difference is $\frac{1}{3}\times\frac{16}{35}=\frac{16}{105}$. So, the number that is one third of the way from $\frac{2}{5}$ to $\frac{6}{7}$ is $\frac{2}{5}+\frac{16}{105}\rightarrow\frac{14}{35}+\frac{16}{105}\rightarrow\frac{42}{105}+\frac{16}{105}=\frac{58}{105}$.

19. C. For this problem, draw a Venn Diagram:

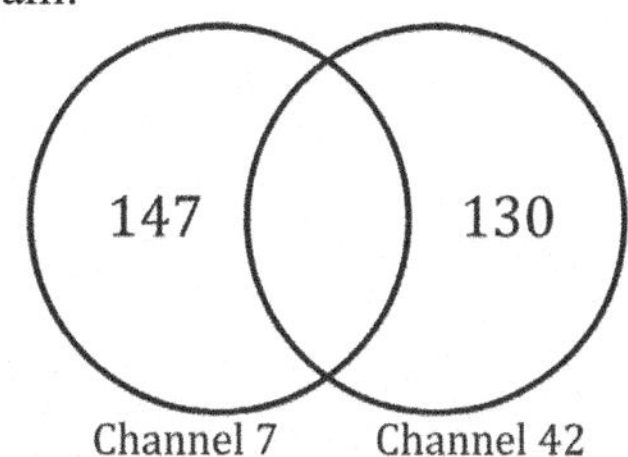

Neither: 300 - (147+130) = 23 people

We may quickly notice it is quite possible to have zero people overlap, given that there is a total of 300 viewers—as we can place both numbers in the diagram and not total 300—but rather only 277 ! In fact, we can calculate the minimum number that watch neither to be 23 people! And that's the minimum number of people who watch neither! Because I can easily find a scenario with no overlap, a minimum of 0 people must overlap. If you blindly added the smaller values together and subtracted from 300 to get 23 as your answer, you could do well to think more about what's going on before applying a formula. Always make sure the problem makes sense. Be wary of formulas and draw out your ideas to be certain you know what is happening. If a problem near the end of the test feels too easy, you may have missed something. Use logic!

CHAPTER

12 DIRECT AND INVERSE VARIATION

SKILLS TO KNOW

- Definition of direct and inverse relationships
- How to identify examples of each
- How to set up a direct or inverse relationship from word problems

NOTE: Direct and Inverse Variation problems are a slightly more complex style of speed/rate/proportion problems. This chapter directly builds on knowledge presented in **Chapter 9: Ratios, Rates & Units** in this book.

DEFINITIONS

DIRECT VARIATION

A direct variation relationship can be summarized by the equation $y = kx$, where k is some constant. Notice that as x increases, y increases (and vice versa). Sometimes we say that "x varies directly as y," or "y varies directly as x." A direct variation equation can be graphed as a line that goes through the origin $(0,0)$. Also, $y = kx$ is a slope-intercept form of a line with $b = 0$; $y = kx + 0$ looks a lot like $y = mx + b$.

This kind of a relationship is also the same as a standard proportion. For instance, if 2 cups of flour make 24 cookies, and we wanted to know how many cookies 3 cups could make, we could also use direct variation to solve. This kind of a relationship can be represented by $y = kx$. Anytime a problem could be solved using direct proportions, the problem involves direct variation.

TIP: When a problem indicates a **direct variation relationship**, place one variable on the left side of the equation, one on the right side, and then place a **constant of variation** (k) on the right side of the equation: $y = kx$.

INVERSE (OR INDIRECT) VARIATION

An inverse relationship can be summarized by the equation $y = \frac{k}{x}$ or $xy = k$, where k is some constant. As x increases, y decreases, since x is in the denominator as opposed to the numerator. Sometimes we say that "x varies inversely as y," or "y varies inversely as x."

Inverse proportions are the same as inverse variations. Problems that involve inverse variation involve one value increasing while the other decreases. For example, the number of hours it takes to do a task and the number of people helping at a constant rate would be inversely related: the more people who help, the less time it takes. As one goes up, the other goes down.

TIP: When a problem indicates an **inverse variation relationship**, place one variable on the left side of the equation, one on the right side in a **denominator** (lower half) of a fraction, and a **constant of variation** (k) on the right side of the equation in the **numerator** of the fraction: $y = \frac{k}{x}$.

CONCEPTS TO KNOW

- When an increase in one variable means an increase in the other (e.g. number of items sold and revenue), and if one variable equals zero, the other is also zero, there is potentially a direct relationship between the two variables
- When an increase in one variable means a decrease in another (e.g. speed and time that it takes to travel a certain distance), there is potentially an inverse relationship between the two variables

JOINT VARIATION

Sometimes you may also see the phrase "x varies jointly as y and z," which simply means that x varies directly as y, directly as z, and $yz = kx$ or $k = \frac{x}{yz}$.

This concept has not appeared on any ACT® tests I have seen, but is the type of information I might expect the ACT® to introduce as a "curve ball" to make a question particularly challenging.

WORD PROBLEM TRANSLATION

To solve word problems that involve inverse or direct variation, you must become familiar with key words and phrases that indicate the two types of relationships, and how to create an equation to guide you through the rest of the problem.

Always be on the lookout for words or phrases like "is proportional to," "varies with," or "directly related to." You must not only recognize these phrases but also understand whether they indicate a direct or indirect relationship. Sometimes you must use common sense to determine what kind of relationship is being described. (e.g. the relationship between hours worked and pay is obviously a direct one: as the numbers of hours worked goes up, so does the pay). Even words like "for each/every" and "per" indicate rates, and sometimes these words are clues you may have a direct or inverse relationship. Unlike word problems best modeled by slope-intercept form linear equations, however, problems that involve direct or inverse variation involve pure proportions: they do not have one time "set up fees," "initial costs," or other one time offsets.

Let's work through some simple example problems to become familiar with this process.

Jimmy sells each box of fruit for n dollars. Establish an expression that demonstrates the relationship between R, Jimmy's revenue, and f, the number of boxes of fruit that Jimmy sells.

Here, we are not given an explicit relationship, but we can use common sense to determine that the more Jimmy sells, the more money he will make; there is a direct relationship between the number of boxes sold and revenue, and the constant is n: the price of a box of fruit. If you're not sure, make up some numbers and play around with them until you see the relationship. For example, if Jimmy sells each box of fruit for \$5 (n), and he sells 3 boxes (f), then he would make \$15 in revenue (R).

Thus R is equal to the product of n and f. The situation is modeled by $R = nf$.

x varies directly with the product of a and b and indirectly with the quotient of y and z. Which equation correctly models this relationship?

Let's take this one part at a time. We are told explicitly that x varies directly with ab, which means that ab will be in the numerator of the right side of the equation. In other works, $x = k(ab)$ (just as $y = kx$ according to our model for direct variation). The next part is a little more tricky. We are also told explicitly that x varies indirectly with $\frac{y}{z}$. The word "indirectly" tells us that this expression must go in the denominator of the right side of the equation. So:

$$x = \frac{k(ab)}{\frac{y}{z}}$$

Just as our model equation tells us $y = \frac{k}{x}$. When we divide by a fraction, we flip the fraction and multiply, so:

$$x = \frac{k(ab)}{1} \times \frac{z}{y}$$

$$x = \frac{kabz}{y}$$

The time that it takes a car to travel a certain distance, d, is given by T and the car's speed is given by s. Write an expression that expresses T in terms of d and s.

Once again, we are going to use common sense to determine the relationship between variables. There is a direct relationship between distance and time, since more distance means more time, and an indirect relationship between speed and time, since more speed means less time.

These relationships are modeled by: $T = \frac{d}{s}$

Are there other ways to think about a problem like this? Yes. You could memorize the distance equation. Alternatively, you could observe or imagine the units and use dimensional analysis detective work (i.e. if a typical speed is miles PER hour and PER means divide then we can write: s miles per hour $= \frac{miles}{hours} = \frac{d}{T}$, where $d =$ a distance in miles and $T =$ a time in hours). At this point you could do further algebraic manipulation to get the solution.

FINDING THE CONSTANT OF PROPORTIONALITY, "K"

Sometimes, direct and inverse variation problems are multi-step processes. For more involved variation questions, first use given information to come up with a standard equation and its constants (such as k). After devising a general formula that models the situation, plug in new, given values into this formula to find an answer.

The perceived intensity of light (lux) produced by a light source varies inversely with the square of the distance, d, from the source, and directly with the square root of the power output, w of the source. If the perceived intensity of light produced 3 feet from a light source outputting 400 watts is 169 lux, which of the following is closest to the perceived intensity of light produced 5 feet from the same source running at 900 watts?

A. 76 lux **B.** 91 lux **C.** 451 lux **D.** 456 lux **E.** 2737 lux

STEP 1: WRITE OUT THE EQUATION

The first step in the problem is to translate the word problem into an equation that accurately describes the relationship between our three variables (perceived intensity, distance, and power). Since perceived intensity is directly related to the square root of power, we can place $\sqrt{w}$ in our numerator along with our constant of proportionality k. And since we are told that the intensity varies inversely with the square of distance, we can place d^2 in the denominator, leaving us with:

$$I = \frac{k\sqrt{w}}{d^2}$$

STEP 2: PLUG IN CASE #1 AND SOLVE FOR K

Now to find the constant of proportionality k, we can plug in the first set of values we were given:

$$169 = \frac{k\sqrt{400}}{3^2}$$

$$k = 76.05$$

STEP 3: CREATE THE GENERAL FORMULA USING K

Now that we have all the information we need, we can plug in our value of k into our equation from Step 1 to create a general formula.

$$I = \frac{76.05\sqrt{w}}{d^2}$$

STEP 4: SOLVE FOR WHAT YOU NEED

Using the above formula from step 3, plug in the final case described by the question and solve for what the question requires of you. We need the perceived intensity, I, which we've already isolated in step 3 on the left of the equation. Plugging in 5 feet, the distance from the source, d, and 900 watts, the power output, w, into the equation we get:

$$I = \frac{76.05\sqrt{900}}{5^2} \approx 91 \text{ lux}$$

Answer: **B.**

The rate of energy consumption of an electric circuit varies proportionally with the square of its voltage, V, and inversely with its resistance, R. If the constant of proportionality is 1, which of the following expressions represents the rate of energy consumption of a circuit?

A. V^2R **B.** V^2R+1 **C.** $\frac{V^2}{R}$ **D.** $\frac{V^2}{R}+1$ **E.** $\frac{V^2}{R^2}$

There are three things that you must understand in this problem: what it means to vary proportionally, what it means to vary inversely, and how to implement a constant of proportionality.

Let's start by assigning a variable, C, to the idea of the "rate of energy consumption." The problem does not name this variable, but it asks for an expression equal to it. I like to have a variable, though, as having one makes the problem easier to set up.

Now, we know that C is proportionally variable with V^2, or in other words, varies directly as V^2. Thus, per the pattern we discussed earlier, we can write on variable on the left side, one on the right, and then place "k" next to the element on the right:

$$C=V^2k$$

Now we know that C varies inversely with its resistance, R. So R must go in the denominator on the opposite side of the equation as C.

$$C=\frac{V^2k}{R}$$

The constant of proportionality refers to the term (here k) that gives the specific amount by which the voltage varies due to changes in voltage or resistance.

Here, the constant of proportionality equal to 1, or $k=1$. Substituting 1 in for k:

$$C=\frac{V^2k}{R}$$

$$C=\frac{V^2(1)}{R}$$

$$C=\frac{V^2}{R}$$

Because we are asked for C, or "the rate of energy consumption" in terms of these variable, the answer is simply: $\frac{V^2}{R}$

Answer: **C.**

1. Which of the following expressions correctly translates the statement: a varies inversely as the quotient of b and c^2, and directly as d^3?

A. $a = \frac{kd^3c^2}{b}$

B. $a = \frac{kb}{d^3c^2}$

C. $a = \frac{kd^3}{bc^2}$

D. $a = \frac{kd^3b}{c^2}$

E. $a = \frac{k}{bc^2d^3}$

2. Which of the following equations correctly shows the statement: p varies directly as the square root of q and inversely as the product of r and s?

A. $p = \frac{kr\sqrt{q}}{s}$

B. $p = \frac{krs}{\sqrt{q}}$

C. $p = \frac{ks\sqrt{q}}{r}$

D. $p = \frac{k\sqrt{q}}{rs}$

E. $p = krs\sqrt{q}$

3. In physics, Power is the amount of work done divided by the time it takes, or the rate of work. Power (P) varies directly as the product of Force (F) and Distance (D) and inversely as Time (T). Given these parameters, which of the following expressions could correctly show the formula for Power?

A. $P = \frac{FT}{D}$

B. $P = \frac{FD}{T}$

C. $P = \frac{DT}{F}$

D. $P = \frac{T}{FD}$

E. $P = FDT$

4. A certain recipe produces 7 cups of tea and requires $2\frac{1}{4}$ cups of water. If all ingredients are increased proportionally, how many cups of water are required for the recipe to yield $45\frac{1}{2}$ cups of tea?

A. $5\frac{3}{8}$

B. 8

C. $8\frac{1}{2}$

D. $14\frac{5}{8}$

E. $20\frac{1}{2}$

5. Given that $x = 5$ when $y = 16$ for the proportion $\frac{2}{x} = \frac{y}{k}$, what is y when $x = 3$?

A. $\frac{3}{40}$

B. $\frac{1}{4}$

C. 16

D. $26\frac{2}{3}$

E. $40\frac{1}{5}$

6. At a certain college, the acceptance rate of admitted students varies indirectly with the number of applicants. When 5,000 students applied, the rate was 37%. What is the acceptance rate next year, rounded to the nearest tenth, if 8,300 students decide to apply?

A. 8.3%

B. 18.5%

C. 22.3%

D. 44.6%

E. 61.4%

7. The temperature of a gas T varies directly with the square root of the pressure, P. The current pressure of the gas is 15 atm. What is the pressure required to have double the temperature as it currently has?

A. $\frac{2}{\sqrt{15}}$

B. $\sqrt{15}$

C. $4\sqrt{15}$

D. 30

E. 60

8. The dollars d that Katie earns at a certain company is directly proportional to the hours h that she works and inversely proportional to the cube of vacation days v. With k as the constant of proportionality, which equation correctly represents Katie's earnings?

A. $\frac{k}{hv^3}$

B. $\frac{kh}{v^3}$

C. $\frac{kv^3}{h}$

D. $\frac{hv^3}{k}$

E. khv^3

9. The amount of time t needed to build a company's hotels is directly proportional with the number of bricks b used and inversely proportional to the number of workers w employed. The current hotel building system employs 850 workers and 1275 bricks. Without changing the amount of bricks b used, what is the minimum amount of workers w necessary in order to halve time t?

A. 425

B. 637

C. 666

D. 1700

E. 1912

10. Ethan discovers that his weight on Earth and Mars are directly proportional. Ethan weighs 324 lbs. on Earth and 122.4 lbs. on the planet of Mars. If the constant of proportionality remains the same, and his brother is 98 lbs. on Earth, how much does his brother weigh on Mars, to the nearest whole number?

A. 4

B. 19

C. 37

D. 259

E. 405

11. For all $n > 1$, what happens to the value of $\frac{n^4 - n^2}{n^2} + 1$ as n increases?

A. It increases proportionally to n^2

B. It decreases proportionally to n^2

C. It increases proportionally to n^3

D. It decreases proportionally to n^3

E. It remains constant.

12. For the equation $E = \frac{h}{n}$, h is a proportionality constant. When $n = 14$, $E = 20$. So, if $n = 7$, what is the corresponding value of E?

A. 40

B. 0.1

C. 10

D. 0.025

E. 0.25

13. Consider the exponential equation $y = Bx^z$, where B and x are positive real constants and z is a negative real number. The value of y increases as the value of z decreases if and only if which of the following statements about x is true?

A. $-5 < x$

B. $0 < x$

C. $0 \le x \le 1$

D. $0 < x < 1$

E. $1 < x < 2$

14. A driving instructor charges $43 per lesson, plus a fee for the use of his car. The charge for the use of the car varies directly with the cube root of the time spent driving the car. If a driving lesson with 64 minutes of driving time costs $55, how much does a lesson with 27 minutes of driving time cost?

A. $52
B. $46
C. $27
D. $58
E. $43

ANSWER KEY

1. A 2. D 3. B 4. D 5. D 6. C 7. E 8. B 9. D 10. C 11. A 12. A 13. D 14. A

ANSWER EXPLANATIONS

1. **A.** The quotient of b and c^2 is $\frac{b}{c^2}$. So if a varies inversely with $\frac{b}{c^2}$ and directly with d^3, then $a = \frac{kd^3}{\frac{b}{c^2}} = \frac{kd^3c^2}{b}$.

2. **D.** The square root of q is $\sqrt{q}$ and the product of r and s is rs. So, if p varies directly as $\sqrt{q}$ and inversely as rs, then $p = \frac{k\sqrt{q}}{rs}$.

3. **B.** The product of Force (F) and Distance (D) is FD. So, if P varies directly as FD and inversely as T, then $P = \frac{kFD}{T}$.

   None of the answer choices include a k, but that's ok. It just means k=1. Our answer is thus B.

4. **D.** Since the tea and water have a directly proportional relationship, let ***k*** be the constant of proportionality, ***t*** equal the number of cups of tea, and ***w*** equal the number of cups of water such that $t = kw$. Plug in the first set of numbers given, $7 = \frac{9}{4}k$ and solve to find that $k = \frac{28}{9}$. Now set up the equation again, but plug in the ***k*** we just found, 45.5 for t and solve for the unknown amount of water, w. The equation $\frac{91}{2} = \frac{28}{9}w$ yields $w = \frac{112}{14}$, which is $14\frac{5}{8}$.

5. **D.** First, plug in the given values for x and y in the equation to solve for k. Once we find that $k = 40$, we can use the second set of given variables in order to isolate y. When we solve the equation $\frac{2}{3} = \frac{y}{40}$, we find that $y = 26\frac{2}{3}$.

6. **C.** Because the relationship between the acceptance rate and number of applicants is indirect, we can use the formula $y = \frac{k}{x}$, where y is the percent acceptance rate, x is the number of applicants, and k is the constant of proportionality. When we plug in the given values of x and y, $37 = \frac{k}{5000}$, we can find that $k = 185{,}000$. Then plug in the values of k and the new number of applicants into the same formula to find that the acceptance rate is approximately 22.3%.

7. **E.** First, we must determine the formula that corresponds to the information. Since this is direct variation, the formula is $T = k\sqrt{P}$. Since we are looking to double the temperature, T does not have to be a precise number, so for our purposes let $T = 1$. When we plug in $P = 15atm$ and $T = 1$, we find $k = \frac{1}{\sqrt{15}}$. With this value of k and a new equation, we now double our original temperature and set $T = 2$, plug in, and solve for P. Once we plug variables T and k into the equation, $2 = \left(\frac{1}{\sqrt{15}}\right)\left(\sqrt{P}\right) \rightarrow 2\sqrt{15} = \sqrt{P} \rightarrow 60 = P$, we find that $P = 60$. Alternatively, plug in $P = 15atm$ in the first equation $T = k\sqrt{15}$. Now form a second equation to solve for a second pressure with the same k and twice the temperature $2T = k\sqrt{P_2}$. Now substitute in the value of T from the first equation into the second: $2(k\sqrt{15}) = k\sqrt{P_2} \rightarrow 2(\sqrt{15}) = \sqrt{P_2} \rightarrow (2(\sqrt{15}))^2 = (\sqrt{P_2})^2 \rightarrow 4(15) = P_2 \rightarrow 60 = P_2$.

8. **B.** Since the problem tells us hours h is directly proportional to dollars d, we know that $d = kh$. If this is inversely proportional to v^3, then we can find that $d = \frac{kh}{v^3}$.

9. D. We can summarize the given relationships with the equation $t=\frac{kb}{w}$. Since the problem is not concerned with the actual value of time, but rather a change in t, we don't need to plug in any specific values. Halving the time means that we need to get to $\frac{1}{2}t$, which is the same as halving the left side of the equation above. Given that the constant of proportionality, k, and the number of bricks, b, does not change, the number of workers must double in order to make the right side of the equation halve as well: $850\times2=1700$.

10. C. Since the relationship between weight on Earth and weight on Mars are directly proportional, we can set up the equation $324=122.4k$ to model the relationship and solve to find that $k\approx2.647$. Then use this value of k to solve for the weight w of Ethan's brother. Solving $98=(2.647)(w)$, we find the brother's approximate weight to be 37 lbs. on Mars.

11. A. We can simplify our expression to find that $\frac{(n^4-n^2)}{n^2}+1\rightarrow\frac{n^4-n^2}{n^2}+\frac{n^2}{n^2}\rightarrow\frac{n^4}{n^2}\rightarrow n^2$. So, as n increases, the expression increases proportionally to n^2. That proportion happens to be 1:1 as the value is itself n^2. Remember that a number squared increases so long as that number is greater than one. Squaring smaller numbers can actually make their value decrease or switch sign, but we don't have to worry about negatives or fractions here.

12. A. Plugging in $n=14$ and $E=20$, we get $20=\frac{h}{14}\rightarrow h=20(14)=280$. Now, plugging in $n=7$ and $h=280$, we get $E=\frac{280}{7}=40$.

13. D. The value of y increases while the power of z decreases only when x is between 0 and 1. Positive fractions become smaller when multiplied together, so exponents applied to positive fractions are inversely related to the value of the expression. If you need to, try plugging in values from the ranges in the answers to better understand this relationship. The value of x is within the bounds $0<x<1$. Note that x cannot equal 0 or 1 because $0^z=0$ and $1^z=1$ for any negative real z.

14. A. If t represents the time spent driving the car, $f=$ the car use fee, and $C=$ the total car for the lesson, then the charge for car use can be represented as $f=C-43$ and $f=k\sqrt[3]{t}$ where k is a constant. We know that when $t=64$, $C=55$. So, $f=C-43\rightarrow55-4\rightarrow12$. We plug in $f=12$ and $t=64$ in the equation $f=k\sqrt[3]{t}$ to solve for the constant k. We get $12=k\left(\sqrt[3]{64}\right)\rightarrow12=4k\rightarrow k=3$. So, the relationship between f and t is $f=3\sqrt[3]{t}$. Plugging in $t=27$, we can solve for the car use cost $f=3\sqrt[3]{27}\rightarrow3(3)\rightarrow9$. So, the total cost for the 27-minute lesson is $C=43+f\rightarrow43+9\rightarrow52$.

CHAPTER

13 COORDINATE GEOMETRY

SKILLS TO KNOW

- Translating between physical distance and coordinate points
- Non-linear equations in coordinate geometry
- Finding minimums and maximums in a system
- Representing concepts in graphs
- 3D Graphing Basics

NOTE: Coordinate Geometry intersects multiple areas of knowledge. Most problems here overlap heavily with another chapter in these books. Specifically, many rely on geometry knowledge (book 2). Furthermore, several specific cases of coordinate geometry are covered in other chapters in this book:

Chapter 5: Intercepts and Slopes
Chapter 6: Distance and Midpoint
Chapter 20: Conics (Circles & Ellipses)
Chapter 22: Graph Behavior
Chapter 23: Translations and Reflections

I've placed this chapter at the transition point of this book as it covers basics from Algebra 1, though its applications span Algebra 2, Geometry, and beyond. Don't be fooled, this isn't all "easy" Algebra. **Expect to reference other chapters when reviewing this material.**

FINDING PROPERTIES OF LINE SEGMENTS FROM COORDINATES

Oftentimes on the ACT®, you'll be asked to jump between the world of coordinates and the universe of "actual lengths." You'll need to know how to easily convert information from coordinate points into physical distances.

Finding Lengths of Horizontal/Vertical Lines

To find these lengths when you have vertical or horizontal lines, you'll subtract the y-value from another y-value or an x-value from another x-value and find the absolute value of the difference. For example, in the picture below, to find the length of the line defining the right side of the trapezoid with endpoints $(5,3)$ and $(5,-2)$, we calculate $|3--2|$ or $|-2-3|$, which equals 5.

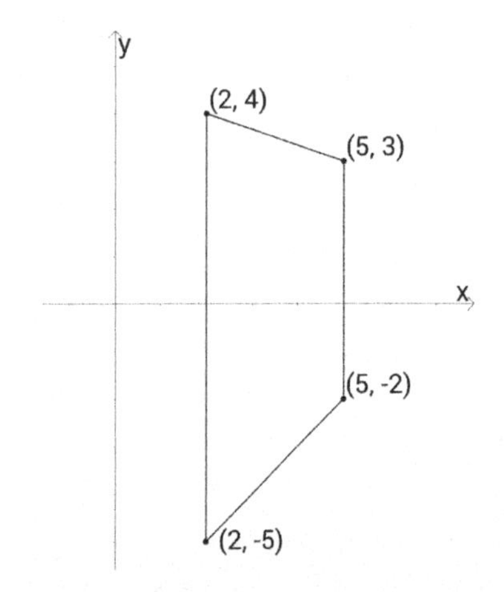

Because we are using absolute value, the order we subtract in does not matter. Remember, absolute value signs simply mean remove the negative sign if there is one. (If you need more review on absolute value, see **Chapter 14** in this book).

We can do the same to find the distance between a point and a line or two parallel lines: simply subtract the x-values or y-values and take the absolute value. This often comes into play when finding the height of a shape in a coordinate plane.

What is the area, in square units, of the trapezoid graphed below?

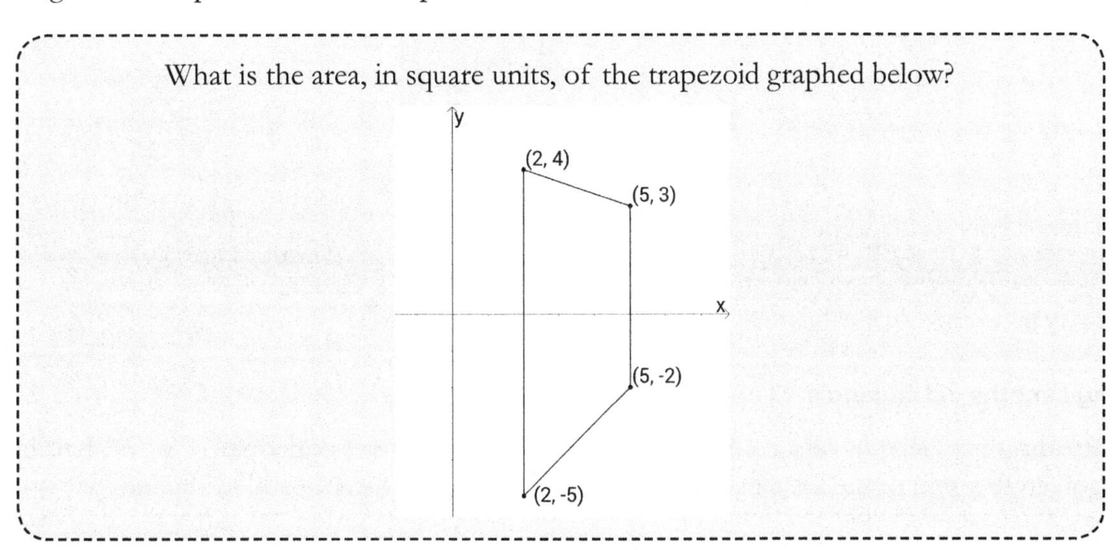

The area of a trapezoid is one-half the sum of the bases times the height. All of these lengths are vertical or horizontal distances, so we can calculate each amount by taking the absolute value of the difference of the respective x or y values. Here, the bases are our vertical lines and the "height" is the horizontal distance between the parallel vertical lines.

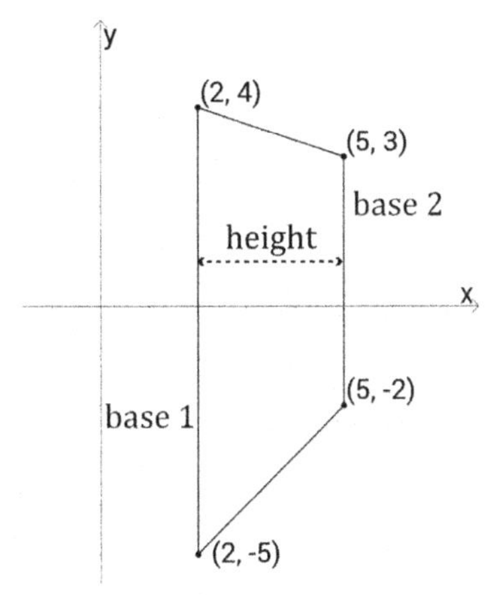

The length of the bases are equal to the absolute value of the difference of the y-values of the endpoints. For the left most vertical line:

$$|4-(-5)|=|9|=9$$

And for the right most vertical line:

$$|3-(-2)|=|5|=5$$

Now that we've converted these "coordinates" into a physical distances, we use a similar process to find the height.

The height of the trapezoid equals the x-distance between the parallel bases, found by subtracting the x-values of these lines (5 for the right line, 2 for the left line) and taking the absolute value:

$$|5-2|=|3|=3$$

We plug in these values into the formula for the area of a trapezoid*:

$$\frac{b_1+b_2}{2}h=\frac{9+5}{2}(3)$$
$$=\frac{14}{2}(3)$$
$$=7(3)$$
$$=21$$

*It's best if you memorize formulas like this one. We cover these in Book 2 (Geometry section) and also have a list of formulas in our online resources: SupertutorTV.com/bookowners. You could also solve this problem by breaking up the shape into rectangles and triangles.

Answer: 21.

Finding Lengths of Diagonals

Calculate diagonals' lengths on a coordinate plane (i.e. lines that aren't horizontal or vertical) using the Pythagorean theorem or the Distance Formula. (See Chapter 6: Distance and Midpoint.)

Find the perimeter of triangle ABC:

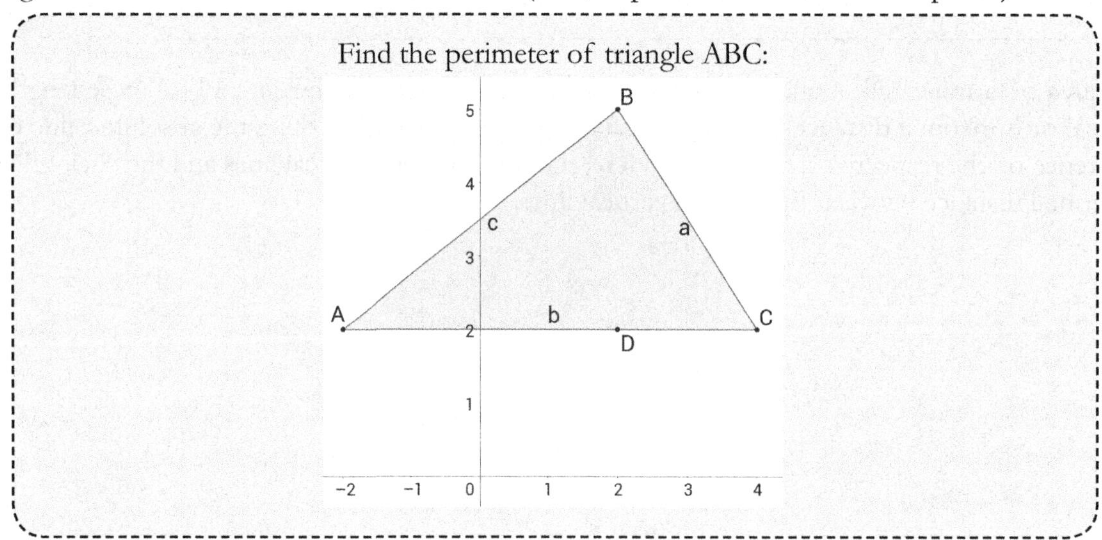

First, we'll find the coordinates: $A(-2,2)$, $B(2,5)$, and $C(4,2)$. I draw a vertical line to create two right triangles and eyeball the physical lengths of the sides of right triangles formed.

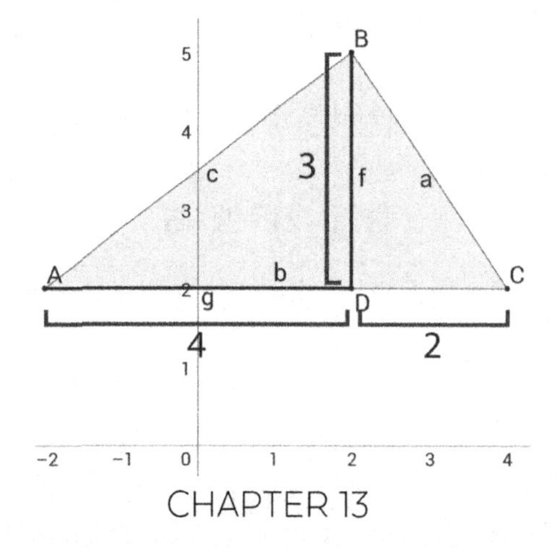

Now we need to find the length of each side. Let's start with side c or AB: I can use Pythagorean triples to find this using my sketch above and lengths g and f. From -2 to 2 (horizontal length g) is 4 units, from 2 to 5 (vertical length, f) is 3 units. This is a 3-4-5 triangle so:

Length $c = 5$ units

(If you don't have that Pythagorean Triple memorized, you could also run the Pythagorean Theorem on these numbers and find $\sqrt{4^2+3^2} = \sqrt{(16+9)} = \sqrt{25} = 5$. For more on Pythagorean Triples and Pythagorean Theorem, see the Triangles chapter in Book 2.)

Now let's find the other two lengths:

Length b (AC): From -2 to 4 is 6 units, and this is horizontal, so no further math is necessary.

Length $b = 6$ units

Length a (BC): from 2 to 4 is 2 units (horizontal length from D to C), and from 2 to 5 is 3 (vertical length, f). With the Pythagorean Theorem: $a^2+b^2=c^2$ or $\sqrt{a^2+b^2}=c$:

$$\sqrt{2^2+3^2} = \sqrt{(4+9)} = \sqrt{13}$$

Length $a = \sqrt{13}$ units

To find the perimeter, we add the three lengths:

$$5+6+\sqrt{13} = 11+\sqrt{13}$$

Answer: $11+\sqrt{13}$.

NON-LINEAR EQUATIONS IN COORDINATE GEOMETRY

Other coordinate geometry problems synthesize skills from Conic Sections (see Chapter 20 in this book), Quadratic Equations (see Chapter 16 in this book), or other non-linear equations. The principle is the same, however. We want to turn what could be an equation or coordinates into physical distances or vice versa. (**NOTE:** next example is tough; recommended for high achieving students scoring 32+)

In the standard (x,y) coordinate plane below, 1 side of a rectangle is on the x-axis, and the vertices of the opposite side of the rectangle are on the graph of the parabola $y=\frac{x^2}{2}-6$. Let k represent any value of x such that $-2\sqrt{3} < x < 2\sqrt{3}$. One potential value of k is depicted in the picture below. Which of the following is an expression in terms of k for the area, in square coordinate units, of any such rectangle?

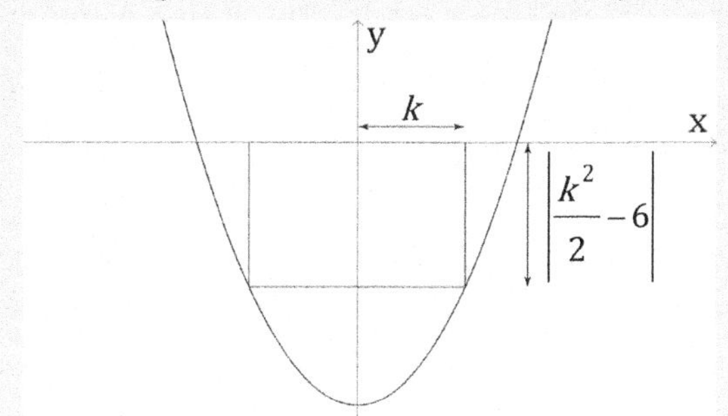

A. k^3-12k **B.** $-9k$ **C.** $\left|\frac{k^3}{2}-6k\right|$ **D.** $-\left|k^3\right|+12\left|k\right|$ **E.** $k^2+4k-12$

This question is pretty confusing. I suggest unpacking the question until you understand it. You might not need all the explanation here that follows, but I try to be complete because it may take some students longer than others for the information to "click."

Let's focus on the idea of the rectangle. We have a rectangle with one side on the x-axis and the other side parallel and touching the parabola. To find the area of this rectangle we need the physical distance of each side of this rectangle.

First let's consider the vertical side of the rectangle. From the picture, we know that it could be expressed as $\left|\frac{k^2}{2}-6\right|$. But why? Where did that come from?

From our work previously in this section, we know this vertical rectangle height is the absolute value of the difference between the y-value of the corner that sits on the x-axis (that y-value is clearly zero) and the y-value of the lower right corner of the rectangle (a solution to the equation $y=\frac{x^2}{2}-6$).

Thus that lower corner will be $\frac{x^2}{2}-6$, and we need to subtract zero from that, and take the absolute value to get the length, i.e. $\left|\frac{x^2}{2}-6-0\right|$. Subtracting zero has no effect, so this equals $\left|\frac{x^2}{2}-6\right|$. That looks just like the label on the picture, right? Remember k is "any value of x ..." that fulfills a particular parameter, so k is just the same as x: $k=x$.

Thus $\left|\frac{k^2}{2}-6\right|$ is our height, or y-value: the vertical physical distance of the rectangle height for any given value k.

The way the question words this whole situation is mind numbing, yes, but let's keep on.

From the problem, we know k is a number that equals x so long as $-2\sqrt{3}<x<2\sqrt{3}$. But why? This looks like the zeros of the equation, and to test my theory, I can plug these end points into the parabola equation.

First I'll plug in $2\sqrt{3}$:

$$y=\frac{x^2}{2}-6$$

$$y=\frac{\left(2\sqrt{3}\right)^2}{2}-6$$

$$y=\frac{4\times3}{2}-6$$

$$y=\frac{12}{2}-6$$

$$y=6-6$$

$$y=0$$

Yes—this is a **zero** of the equation (the spot where the parabola crosses the x-axis). If I were to plug in $-2\sqrt{3}$, the solution would also be zero for y (because this term is squared the sign really doesn't matter). The spread of x values is thus the domain of where you could draw a rectangle that has one side between the zeros and the other side touching the parabola in two places below the x-axis. That's all. If I pick an x-value outside of this domain, I wouldn't be below the axis anymore; I would be above it. This parameter keeps our rectangle within the boundaries of the "u" part of the parabola below the x-axis.

Our horizontal distance is already given for "one" value of k on the picture—and it appears that it spans half the width of the rectangle. To get the full width in this case we would need $2k$. In terms of physical distance, this makes sense. The value k only spans from the line of symmetry ($x=0$) to the x coordinate, but we need the whole distance across the parabola, not just the distance from the midline. But let's be careful. Again, this is one possibility according to the question. If k were say 2, $2k$ would be 4, right? That works. But what if k were negative? What if k were -2? If so, we'd need -2 times 2, which would be -4, but then we need to make that a physical distance. The physical distance is the absolute value, which is 4. Because our k can be anything in this range, $-2\sqrt{3} < x < 2\sqrt{3}$, we need a solution that accounts for the possibility that k is negative.

Thus the width is not simply $2k$, but rather $|2k|$.

Now we multiply our width times our height:

$$|2\boldsymbol{k}| \times \left|\frac{\boldsymbol{k}^2}{2} - 6\right|$$

Here's where many students might get stuck. None of our answers are exactly this.

We can quickly eliminate, however, any answers without a k^3 term (see bolded letters above), as we know, even with absolute value, that there is a third power of k somewhere in the answer. That leaves A, C, and D:

A. $k^3 - 12k$

C. $\left|\frac{k^3}{2} - 6k\right|$

D. $-|k^3| + 12|k|$

You might be tempted to drop the absolute value. If you do this without thinking, and simplify $(2k)\times\left(\frac{k^2}{2}-6\right)$, you'll get A.

But (A) doesn't work because the vertical value $\frac{k^2}{2} - 6$ will never be positive: you cannot ignore the absolute value. Remember back to the idea that this rectangle sits below the x-axis. That means we need this absolute value sign or we'll get a negative value for "y" when we calculate this part of the equation. If this length is negative and the value of k is positive, our area would be negative, and that is not possible.

(C) forgets to account for the idea that k is not the full length of the rectangle, and thus is not our answer.

(D) works. You can figure this out by plugging in numbers into both what you got and this, or by going back to the logic of why A doesn't work: **every y-value that corresponds to the lower corners of the rectangle is always negative, but our physical distance is always positive.** If you realize this fact, you'll know that we can simply multiply $\frac{k^2}{2}-6$ **by negative one** to get the physical distance of the rectangle height. That gives us a vertical height of:

$$\left(\frac{k^2}{2}-6\right)(-1)=\left(-\frac{k^2}{2}+6\right)$$

Now we multiply this new expression by our length $|2k|$ or $2|k|$:

$$(2|k|)\left(-\frac{k^2}{2}+6\right)=\left(-\frac{|k|2k^2}{2}+6\times 2|k|\right)$$
$$=\left(-|k|k^2+12|k|\right)$$
$$=-\left|k^3\right|+12|k|$$

Answer: **D.**

FINDING MINIMUMS/MAXIMUMS IN A SYSTEM

Within a system or a graph, you can find the minimum or maximum by plugging in all the intersection points between any two or more of the inequalities in the system into the function you are trying to maximize. **One of these intersection points will always contain the solution.** This is a concept you simply must memorize.

Consider the set of all points that satisfy all 3 of the conditions below:

$$\begin{cases} 0<y<6 \\ y<\frac{x}{2}+3 \\ y<-3x+15 \end{cases}$$

y=4
B(2,4)
C(11/3, 4)
$y=\frac{x}{2}+3$
y=-3x+15
A(-6,0)
D(5,0)

The graph of this set is trapezoid $ABCD$ and its interior, which is shown shaded in the standard (x,y) coordinate plane above. Let this set be the domain of the function $P(x,y)=4y-x$.

What is the minimum value of $P(x,y)$ when x and y satisfy the conditions given?

A. –24 **B.** 6 **C.** –30 **D.** 14 **E.** –5

We want to test all the intersection points of two or more of the inequalities, in other words the "corners" of the bounded region, to see which ordered pair creates the minimum value of $P(x,y)$.

$$P(-6,0)=4(0)-(-6)=6$$

$$P(2,4)=4(4)-2=14$$

$$P\left(\frac{11}{3},4\right)=4(4)-\frac{11}{3}=\frac{37}{3}$$

$$P(5,0)=4(0)-5=-5$$

Because -5 is the smallest value, the answer is **E**.

3-DIMENSIONAL COORDINATES

3D Coordinate Geometry on the ACT® is rare. Save this section until you're scoring above a 32 on the math section.

For 3D coordinate planes, there's an additional axis, known as the z-axis.

Points in three dimensions are given in (x,y,z) form, where x and y function in the same way as in 2 dimensions, and z simply adds a third "depth" dimension. When any value in a coordinate of this form has a zero, it is an intercept of the plane formed by the other two axes.

As shown in the (x,y,z) coordinate space below, a cube has vertices at $(0,3,0)$ and $(0,0,3)$. What are the coordinates of vertex A?

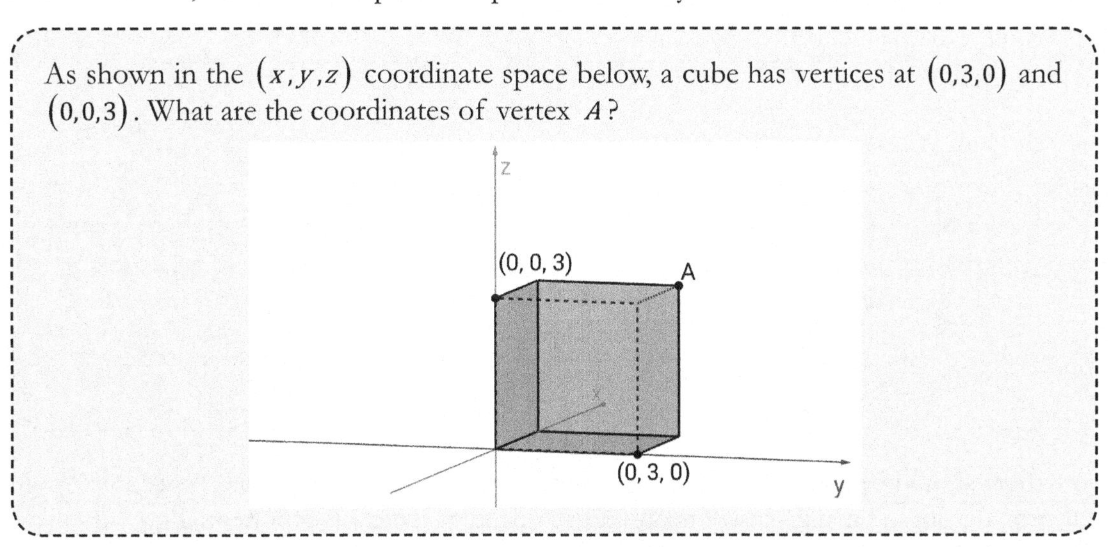

From the picture, we can see that point A shares the same y-coordinate with the point $(0,3,0)$ and lies on the plane formed by the right side of the cube, so its y value is also 3. Similarly, as it also intercepts the top of the cube with z-value of 3, as we see from the coordinate $(0,0,3)$, so it has a z value of 3. Since a cube's sides are of equal length, it extends in the x direction for 3 as well, so its coordinates are $(3,3,3)$. We could also do each coordinate one at a time, visualizing moving back 3 on the x-axis, up 3 on the z-axis or plane, and over 3 on the y-axis or plane. It's important to read the axes: here z and y are up and to the right, while x is deep. You may have thought x looked negative, but we'll assume no negative options exist.

Answer: $(3,3,3)$.

REPRESENTING CONCEPTS IN GRAPHS

These questions usually don't give exact numbers, and if they do, the numbers probably don't matter much. They generally ask for overall trends and want to test if you know what the graph is supposed to look like.

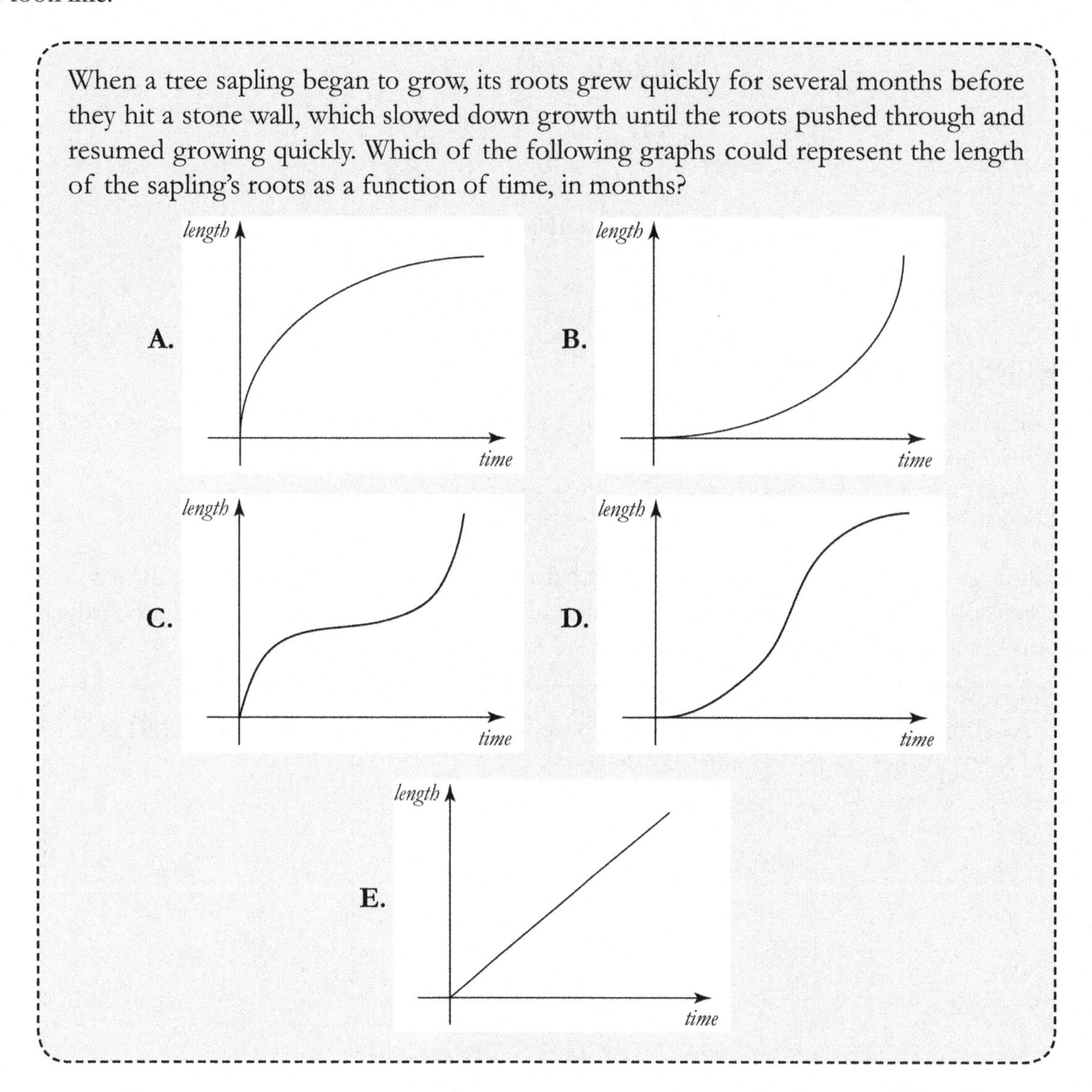

When a tree sapling began to grow, its roots grew quickly for several months before they hit a stone wall, which slowed down growth until the roots pushed through and resumed growing quickly. Which of the following graphs could represent the length of the sapling's roots as a function of time, in months?

We know there should be some funky looking "slow down" point, and another "speed up" point, so E, linear growth, doesn't make sense, and A and B, different types of exponential growth also don't make sense.

We're left with C and D. What we want is high growth in a short span of time. Because time is on the bottom, we want a steep line at the start and at the end. A steep line would minimize time (x-axis) and maximize growth (y-axis). Likewise, we want a flatter line in the middle. A flatter line would extend time (x-axis) with little growth (y-axis). Thus the best choice is C.

Be sure to check your x-axis and y-axis labels. This logic works only because time is on the x-axis and height is on the y-axis. Had these axes labels been flipped, D could have been correct. D has a funky change in rate, but it is growing rapidly in the middle, not slowing down.

Answer: **C.**

NOTE: **Most problems in this set rely heavily on knowledge from other chapters in this book and Book 2 of this series. Relevant chapter names are denoted in parentheses after each problem.**

1. The figure below shows a cube composed of 64 smaller identical cubes in the standard (x,y,z) coordinate system. The vertex A has coordinates $(0,4,0)$. The cube has vertices at the origin, on the positive x-axis, and on the positive z-axis, respectively. What are the coordinates of vertex B on the positive z-axis? (Solids, Book 2)

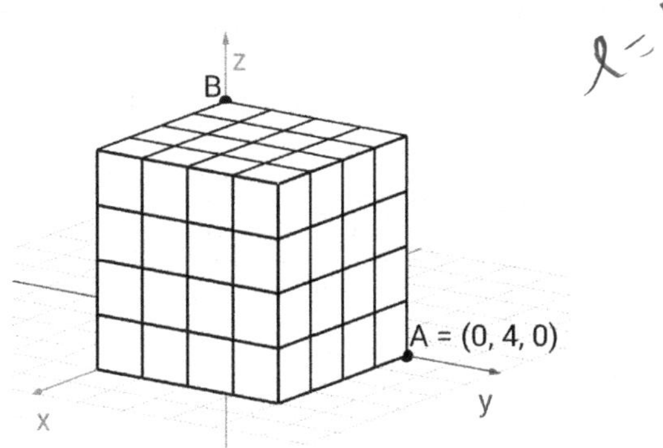

A. (4,0,4)
B. (4,0,0)
C. (0,0,4)
D. (4,4,0)
E. (4,4,4)

2. The angle formed by the positive x-axis and the line $y=-\sqrt{3}x$ is denoted by ϕ in the figure below. What is the value of the angle ϕ? (Triangles, Book 2)

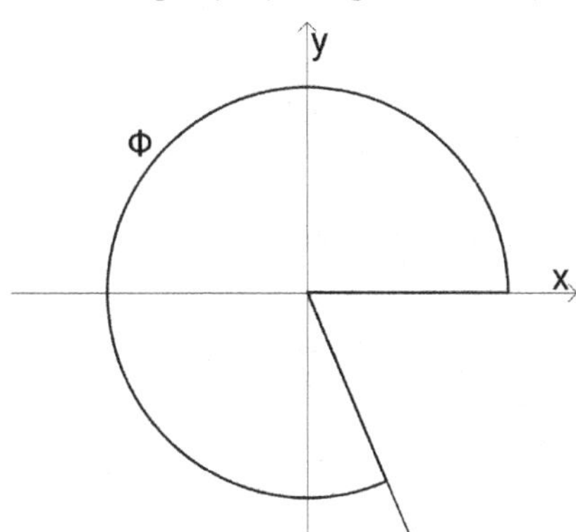

A. −30°
B. 270°
C. 300°
D. 330°
E. Cannot be determined from the information provided

3. The triangle $\triangle ABC$ is shown below with labeled vertices. What is the area of $\triangle ABC$? (Triangles, Book 2)

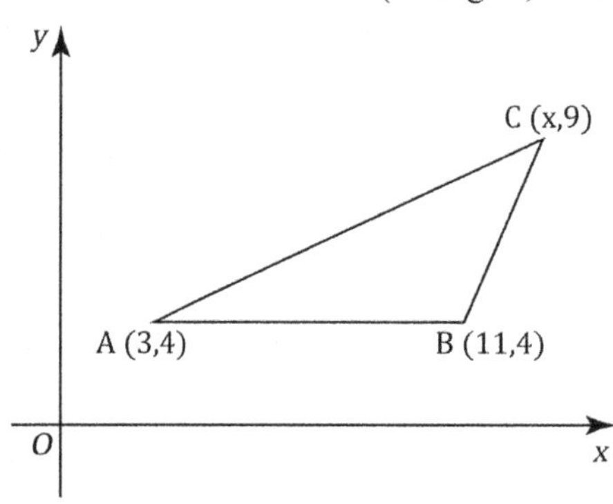

A. 20
B. $20\sqrt{2}$
C. 40
D. $40\sqrt{2}$
E. Cannot be determined from the information provided

4. In the figure below, a circle with center $(-6,-4)$ is shown. What is the equation of a line that is tangent to the circle at point $(0,-8)$? (Conics, Book 1; Circles, Book 2)

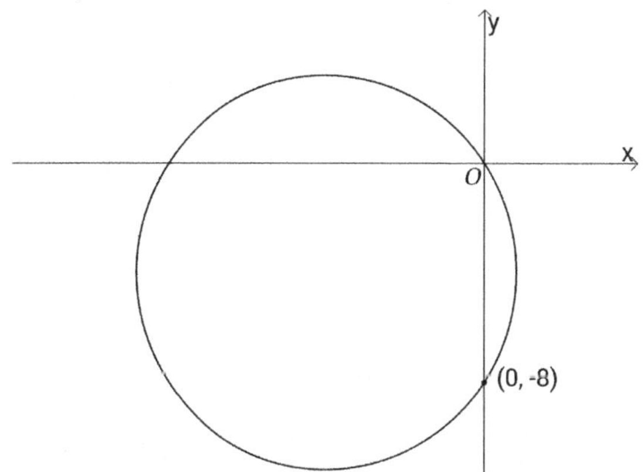

A. $y=\frac{3}{2}x-8$
B. $y=-\frac{3}{2}x-8$
C. $y=-\frac{3}{2}x+8$
D. $y=\frac{3}{2}x+8$
E. $y=\frac{2}{3}x-8$

5. The graphs of the functions $y = x - 2$ and $y = (x-2)^3$ are shown below. The two graphs intersect at points $(-1,-1)$, $(2,0)$, and $(3,1)$. Which values of x satisfy the inequality $x - 2 < (x-2)^3$? (Inequalities Advanced, Book 1)

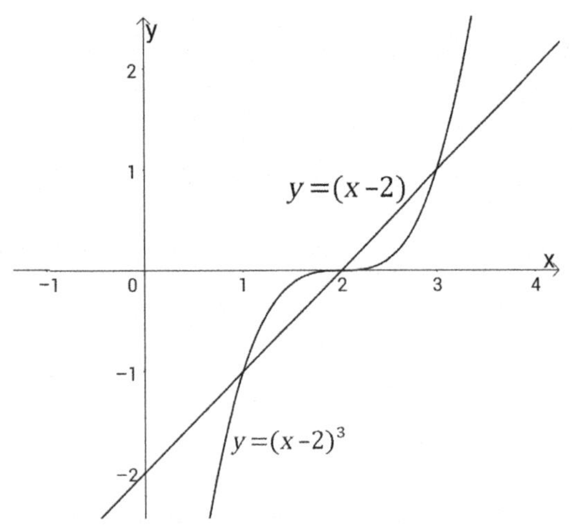

A. No real values
B. $x < 2$ and $x > 3$
C. $x < 1$ and $2 < x < 3$
D. $1 < x < 2$ and $x > 3$
E. $x > 1$ and $x > 3$

6. The shape below is formed by combining a rectangle and two semicircles. According to the given vertices, what is the total area of the shape? (Circles, Book 2)

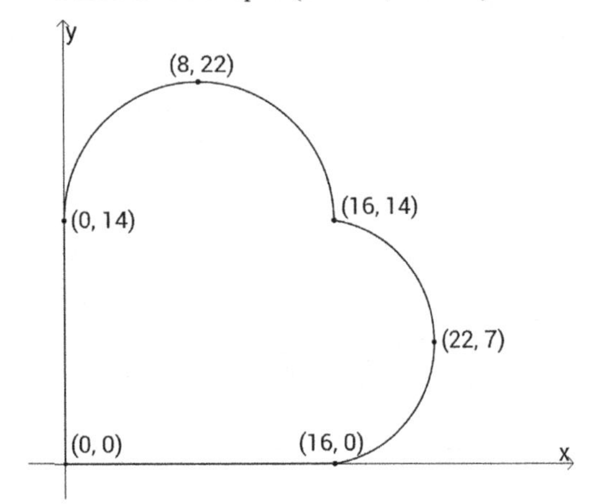

A. $224 + 49\pi$

B. $224 + \frac{113}{2}\pi$

C. $224 + 64\pi$

D. $224 + \frac{113}{4}\pi$

E. $224 + 113\pi$

7. In the figure below, the points F and G are shown at coordinates (m,f) and (m,g) respectively. F' is the point generated by rotating F 90° counterclockwise about G. What are the coordinates of F'? (Translations & Reflections, Book 1)

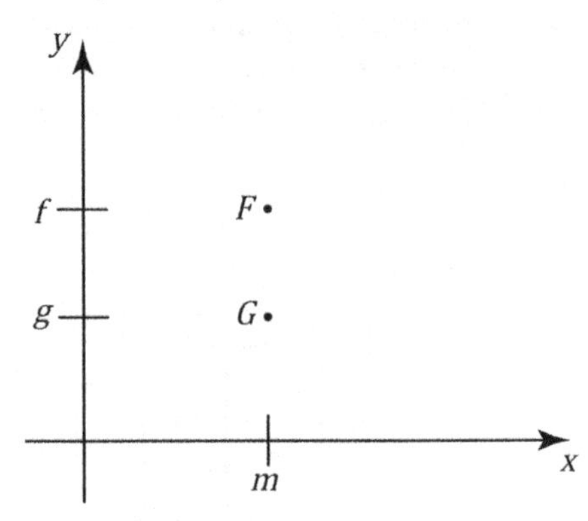

A. $(m+f-g, g)$
B. $(m-f+g, g)$
C. $(m+f-g, f-g)$
D. $(m-f+g, f-g)$
E. $(m-g, g)$

8. The figure below depicts a triangle formed by the labeled vertices. If each vertex point of the triangle is multiplied by $\frac{3}{2}$, what would be the area of the resulting triangle? (Triangles, Book 2)

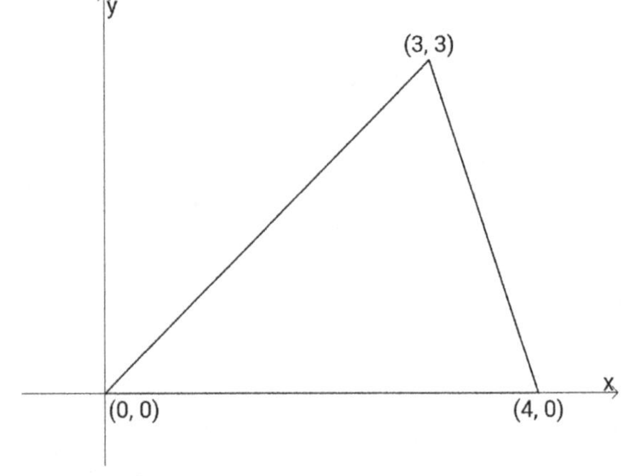

A. 9
B. 13.5
C. 15
D. 18
E. 27

9. The triangle below has vertices $A(-5,5)$, $B(10,5)$, and $C(11,2)$. Point C can be moved to any point on a certain line, and the area of the triangle will remain the same. What is the slope of this line? (Triangles, Book 2; Intercepts and Slopes, Book 1)

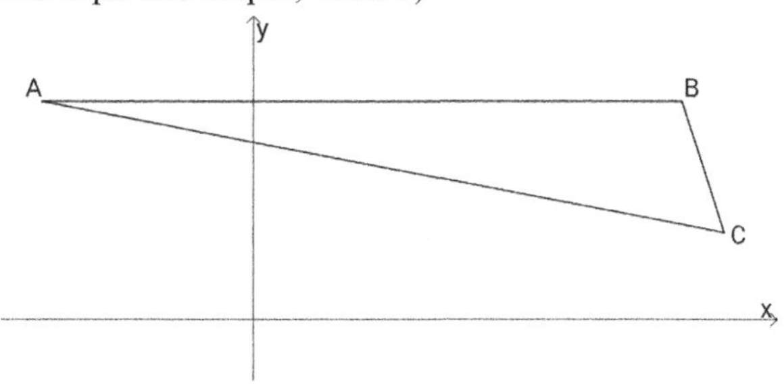

A. -3

B. $-\frac{3}{16}$

C. 0

D. $\frac{3}{16}$

E. 3

For questions 10-12:

Consider the set of all points (x, y) that satisfy all 3 of the conditions below:

I. $y \geq x$

II. $y \geq -2x$

III. $y \leq -\frac{1}{2}x + 3$

The graph of this set is $\triangle ABC$ and its interior, which is shown in the standard (x, y) coordinate plane below. Let this set be the domain of the function $S(x, y) = 3x + y$. (Functions, Book 1)

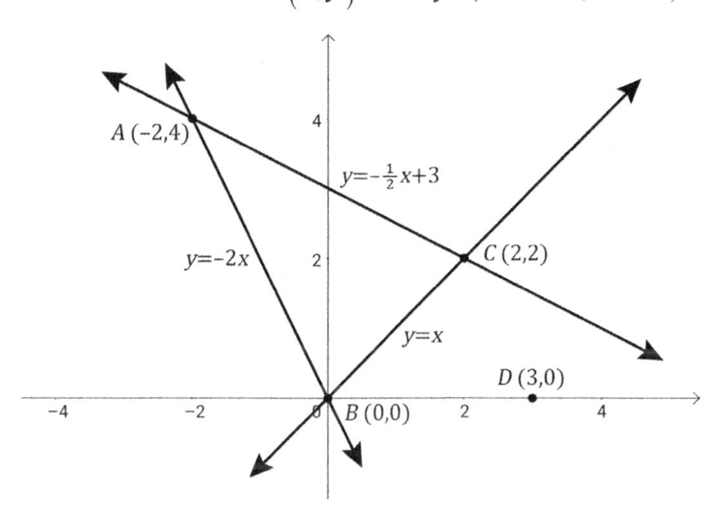

10. What is the minimum value of $S(x, y)$ when x and y satisfy the 3 conditions given? (Functions, Book 1)

A. -5

B. -3

C. -2

D. 0

E. 8

11. What is the area of $\triangle ABC$? (Triangles, Book 2)

A. $2\sqrt{2}$

B. $3\sqrt{2}$

C. 6

D. 8

E. 12

12. What is the cos of $\angle ABD$? (SOHCAHTOA, Trigonometry, Triangles Book 2)

A. $-\frac{\sqrt{5}}{5}$

B. $\frac{\sqrt{5}}{5}$

C. $-\frac{2\sqrt{5}}{5}$

D. $\sqrt{5}$

E. Cannot determine from the information given.

For questions 13-15:

Consider the rational function $f(x)=\dfrac{x^2-16}{x-3}$, whose graph is shown in the standard (x,y) coordinate plane below.

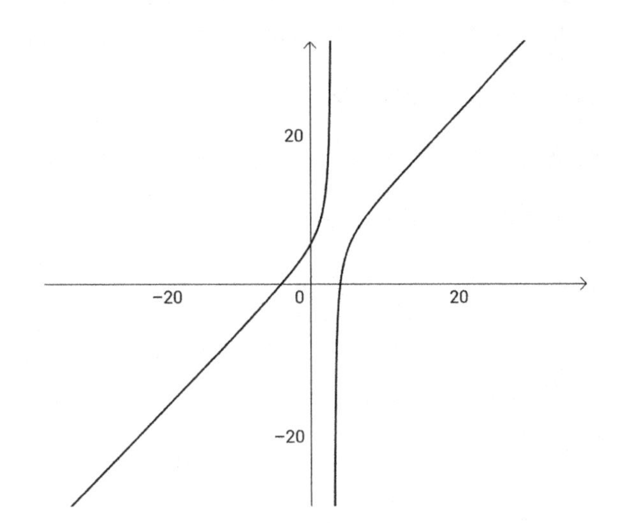

13. What is the value of $f(x)$ at $x=5$? (Functions, Book 1)

A. 3
B. 4.5
C. 5
D. 9
E. 20

14. What is the range of $f(x)$? (Functions, Graph Behavior, Book 1)

A. All real values except 3
B. All real values except ± 4
C. All real values except ± 4 and 3
D. All real values except $\dfrac{4}{3}$
E. All real values

15. How many horizontal and/or vertical asymptotes does the graph of $f(x)$ have? (Graph Behavior, Book 1)

A. 0
B. 1
C. 2
D. 3
E. 4

16. The table below gives the price of riding a vehicle through a safari park per passenger. The price of the trip depends on the number of passengers.

# of Passengers	Price
$0<x<3$	\$11.60
$3\le x<6$	\$16.30
$6\le x\le 10$	\$20.05

Which of the following graphs best represents this information? (Data Analysis, Book 2)

A.

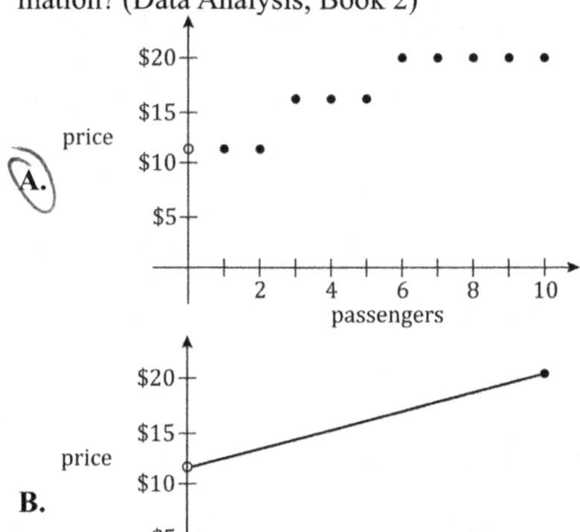

B.

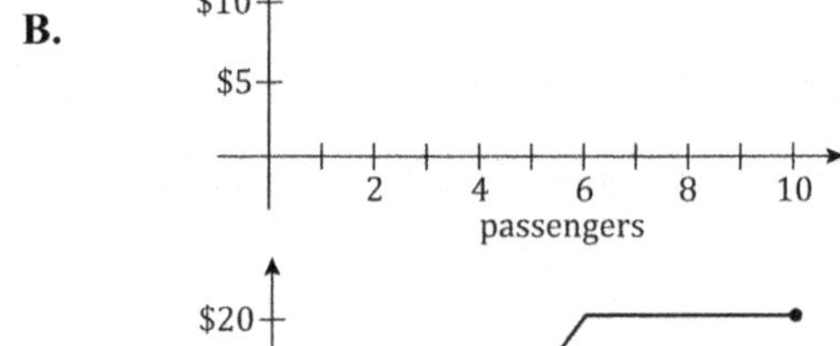

C.

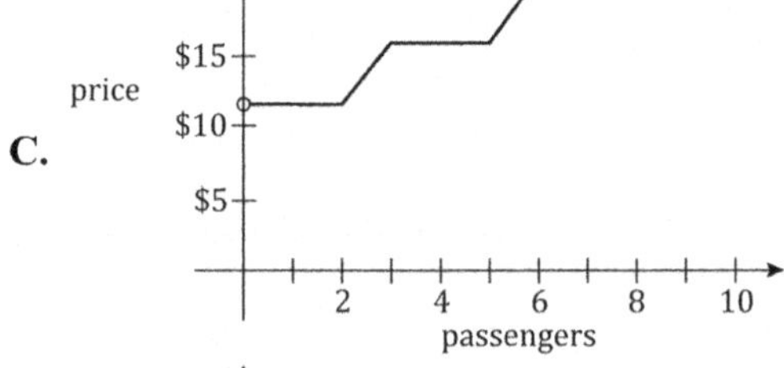

D.

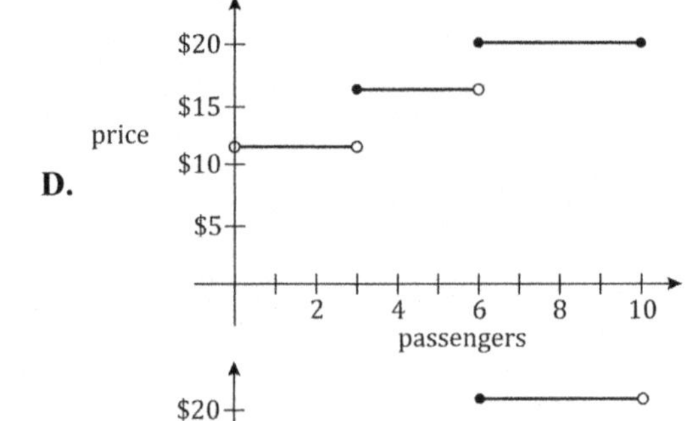

E.

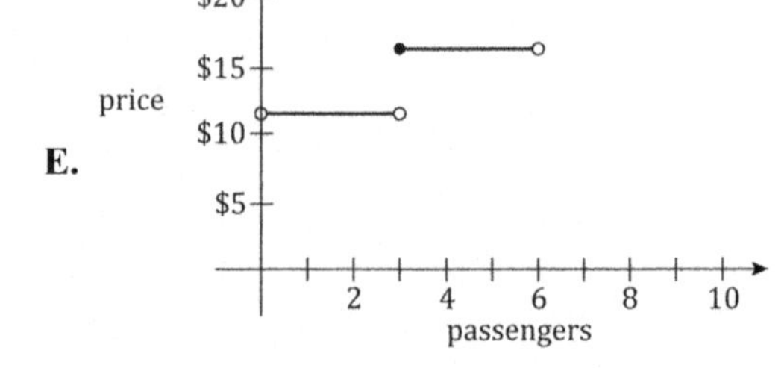

17. Tina sells t-shirts. Her total setup cost, including the cost of the plain t-shirts, is $50 and she charges $5 for each finished t-shirt. Her profit is found by subtracting her expenses from her income. Which of the following graphs represents how much money she needs to recoup her costs as a function of the number of shirts she sells? (Word Problems, Book 1)

A.

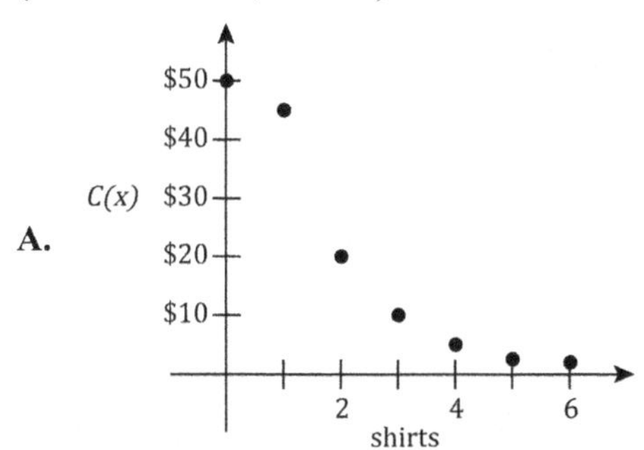

B.

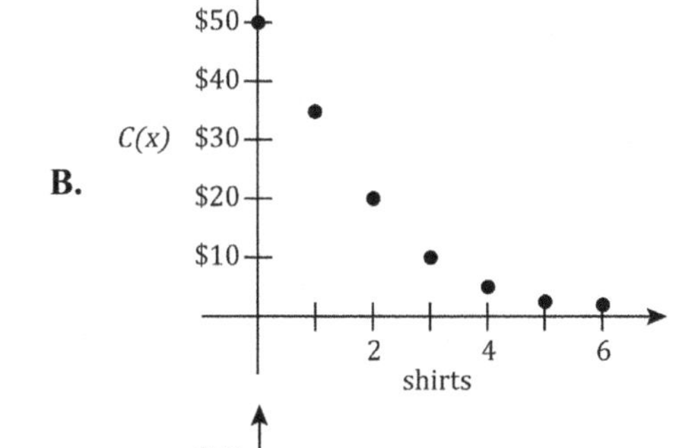

C.

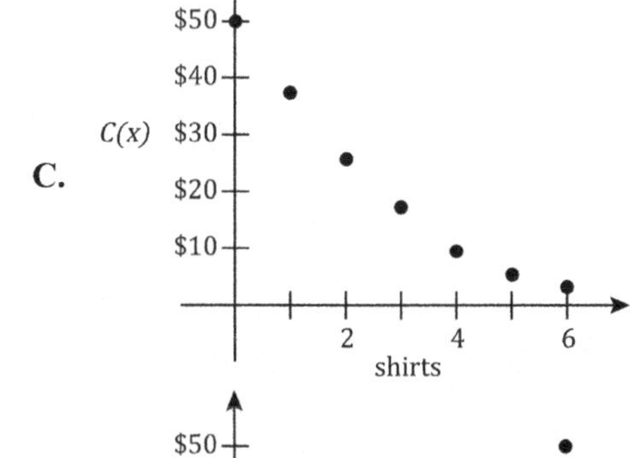

D.

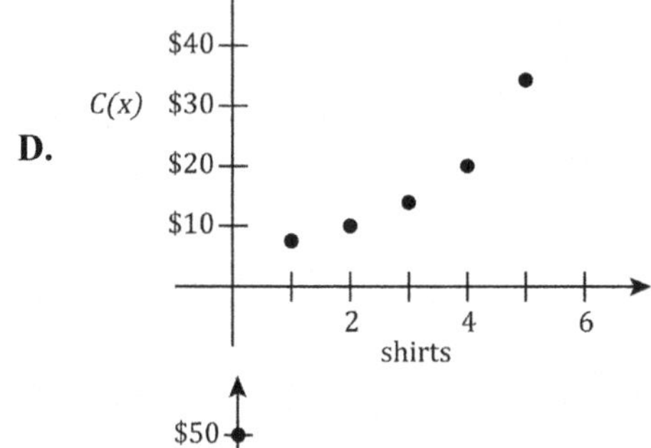

E.

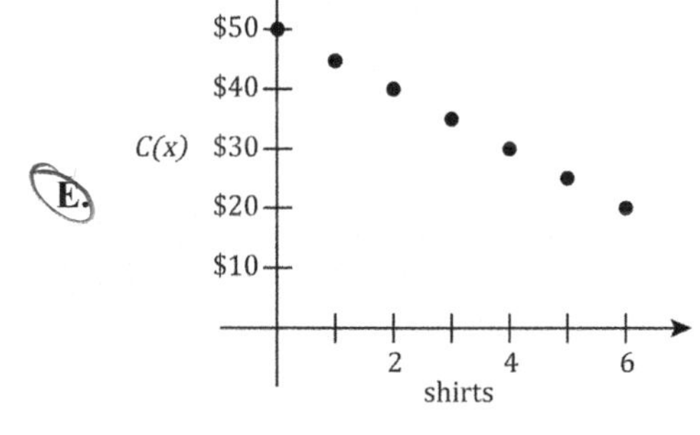

18. An angle in standard position in the standard (x,y) coordinate plane has its vertex at the origin and its initial side on the positive x-axis. If the angle measures $-2493°$ in standard position, it has the same terminal side as all of the following angles except: (Trigonometry, Book 2)

A. $-2133°$
B. $-1413°$
C. $-1053°$
D. $-513°$
E. $27°$

19. In the standard (x,y) coordinate plane, the equation $x^2 + y^2 = 169$ represents a circle. At what two points does the circle cross the x-axis? (Conics, Book 1)

A. $(13,0);(-13,0)$
B. $(0,13);(0,-13)$
C. $(5,12);(-5,-12)$
D. $(12,0);(5,0)$
E. $(16,0);(9,0)$

20. A circle in the standard (x,y) coordinate plane is centered at $(5,-2)$ and passes through $(2,8)$. What is the area, in square coordinate units, of the circle? (Circles, Book 2, Distance & Midpoint, Book 1)

A. 109π
B. $\sqrt{109}\pi$
C. 109
D. 11881π
E. 11881

21. A circle in the standard (x,y) plane is tangent to the y-axis and has its center at $(4,7)$. Where is the point $(3,1)$? (Conics, Book 1; Circles, Book 2)

A. Inside the circle
B. On the circle
C. Outside the circle
D. On the y-axis
E. Cannot be determined from this information

22. There is a right triangle in the standard (x,y) coordinate plane. If its vertices are at $(-4,0)$, $(3,0)$, and $(-4,10)$, what is the length of the hypotenuse in coordinate units? (Triangles, Book 2)

A. $-\sqrt{149}$
B. 5
C. 12
D. 17
E. $\sqrt{149}$

23. In the standard (x,y) coordinate plane, ΔPQR is isosceles. PQ is congruent to QR. Vertex P is at $(-5,4)$ and vertex R is at $(1,4)$. What is the x-coordinate of Q? (Triangles, Book 2)

A. -5
B. -2
C. 1
D. 3
E. 4

24. The sides of a square are 5 inches long. One vertex of the square is at $(3,4)$. Which of the following cannot be a vertex of the square? (Polygons, Book 2)

A. $(8,4)$
B. $(3,9)$
C. $(-2,4)$
D. $(3,1)$
E. $(-1,1)$

25. Points $X(4,1), Y(7,1)$, and $Z(4,6)$ are in the standard (x,y) coordinate plane. If $XYZW$ is a rectangle, what is the length, in coordinate units, of $\overline{XW}$? (Polygons, Book 2)

A. 3
B. 4
C. 5
D. $\sqrt{34}$
E. Impossible to determine from the given information.

26. What is the perimeter of a quadrilateral in the (x,y) coordinate plane with vertices at $(5,5), (6,8), (9,9)$, and $(8,6)$? (Polygons, Book 2)

A. $4\sqrt{10}$
B. 8
C. $2\sqrt{10}+6$
D. $2\sqrt{10}+8$
E. $2\sqrt{10}+10$

27. Let $6x-8y=11$ be an equation of line M in the standard (x,y) coordinate plane. Line N has a slope that is half of the slope of line M and a y-intercept that is 4 more than the y-intercept of line M. Which of the following equations is line N? (Intercepts & Slopes, Book 1)

A. $y=\frac{3}{4}x-\frac{11}{8}$
B. $y=\frac{3}{8}x-\frac{15}{8}$
C. $y=\frac{3}{8}x+\frac{43}{8}$
D. $y=\frac{3x+21}{8}$
E. $y=\frac{3x-15}{8}$

28. A plane contains 14 horizontal and 14 vertical lines that divide the plane into multiple disjoint regions. How many of these regions have a finite, nonzero area?

A. 28
B. 169
C. 196
D. 210
E. 225

29. All of the following create a unique plane in 3-dimensional Euclidean space except:

A. 2 lines that intersect at only 1 point
B. 3 collinear points
C. 2 parallel lines that do not touch
D. 3 different non-collinear points
E. 3 vertices of a scalene triangle

30. Plane D contains line segment K, which measures 5 centimeters. How many lines in D are perpendicular to K?

A. 2
B. 3
C. 4
D. 5
E. Infinitely many

31. A basketball is dribbled in a court. Assume that every time the player dribbles, the basketball bounces and then returns to its former height to be dribbled again. Among the following graphs, which one best represents the relationship between the height, in inches, of the ball and the time, in second, when the ball leaves the player's hand until it returns?

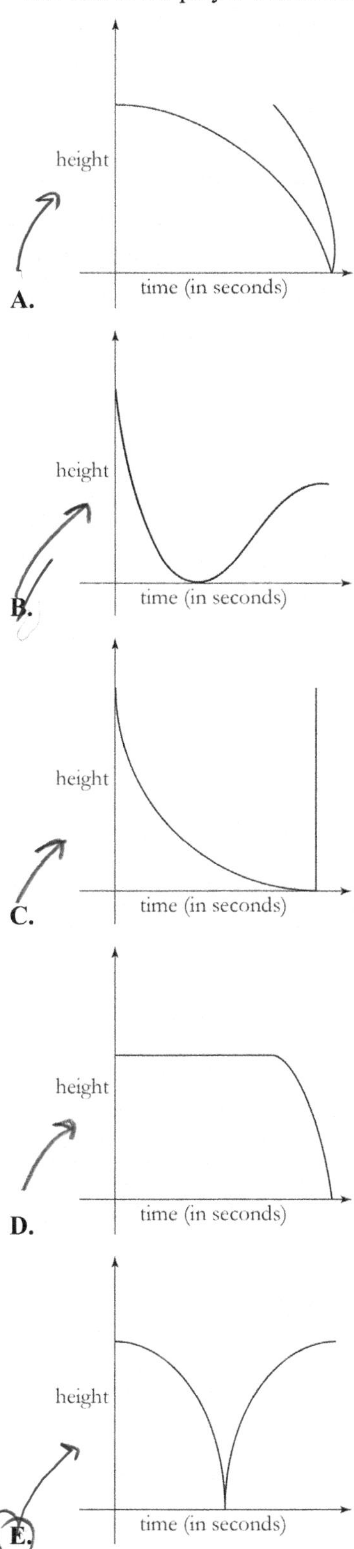

ANSWER KEY

1. C	**2. C**	**3. A**	**4. A**	**5. D**	**6. B**	**7. B**	**8. B**	**9. C**	**10. C**	**11. C**	**12. A**	**13. B**	**14. E**
15. B	**16. A**	**17. E**	**18. D**	**19. A**	**20. A**	**21. C**	**22. E**	**23. B**	**24. D**	**25. D**	**26. A**	**27. D**	**28. B**
29. B	**30. E**	**31. E**											

ANSWER EXPLANATIONS

1. **C.** As can be seen in the diagram, B is directly above the point where the x and y axis meet at the origin. Thus, B's x and y values are 0. Since B is 4 cubes up along the z axis its z coordinate is 4, thus its (x,y,z) coordinate is $(0,0,4)$.

2. **C.** We can construct a right triangle whose hypotenuse is the line shown by drawing an altitude from the line up to the x-axis. Since the slope of the line is $-\sqrt{3}$, it moves down $\sqrt{3}$ units for every 1 unit it moves to the right. Labeling the corresponding legs of the right triangle we drew, it becomes apparent that this is a $30° - 60° - 90°$ right triangle since the length of the sides follow the $1-\sqrt{3}-2$ pattern. The angle whose vertex is at the origin is $60°$, so the exterior angle Φ is $300°$.

3. **A.** The area of a triangle is $A=\frac{1}{2}bh$. The base of this triangle, if we define it as the side parallel to the x-axis, is $11-3=8$, and the height is the distance from the base to the top of the triangle, $9-4=5$. It doesn't matter what the x-value of the top of the triangle is. Plugging in the base and height into our formula yields $A=\frac{1}{2}\times 8\times 5=20$.

4. **A.** The line tangent to the edge of a circle has a slope perpendicular to the radius of the circle at that point and passes through the same tangency point. Find the slope of the radius: divide the change in y by the change in x using the tangent point $(0,-8)$ and the center of the circle $(-6,-4)$: $\frac{(-4)-(-8)}{(-6)-(0)}=-\frac{2}{3}$. The slope tangent to this is the opposite reciprocal: $\frac{3}{2}$. We know the y-intercept, -8, and plug this and the slope into the slope-intercept form: $y=mx+b$. We get $y=\frac{3}{2}x-8$.

5. **D.** The question asks for all the points where $x-2<(x-2)^3$. This represents all values of x for which the corresponding y-value of the line in the graph is lower than the corresponding y-value of the cubic function in the graph. In other words, find the span of x values when the curvy line is vertically above the straight line. Looking at the graph, we can tell that this is between 1 and 2 as well as every value greater than 3. This becomes $1<x<2$ and $x>3$.

6. **B.** The shape in the graph is formed by a rectangle and two semicircles of *different* radii. Draw the base and height of the rectangle to find they are 16 and 14; the radii of the circles are half of their diameters: 7 and 8, respectively. The area of a rectangle is $A=bh$ (or $16\times 14=224$) and the area of a semicircle is half of a full circle, giving us: $A=\frac{1}{2}\pi r^2$. Plugging in 7 and 8, we get each semicircle area: $\frac{1}{2}\pi(7)^2=\frac{49}{2}\pi$ and $\frac{1}{2}\pi(8)^2=\frac{64}{2}\pi=32\pi$ Now add these three pieces together: $224+\frac{113}{2}\pi$. If you struggle to add fractions, use your calculator (see tips in Fractions chapter, Book 2).

7. **B.** The point generated by rotating F $90°$ counterclockwise about G will be at the same height and slightly to the left of G. The same height means we have the same y-value, g. The distance to the left is the difference between the heights of the original F point and G: $f-g$. Since this is the distance the new point is to the left of G, the x-value of the new point will be G's x-value minus this, or, $m-(f-g)$. Simplification yields $m-f+g$. We put our x and y values into a coordinate pair to get $(m-f+g,g)$.

8. **B.** The area of a triangle is $\frac{1}{2}bh$. Multiplying each vertex point of the triangle by $\frac{3}{2}$ will yield a base and height $\frac{3}{2}$ their original measure, because the base and height equal vertex point values that form the triangle. The new area is $\frac{1}{2}\left(\frac{3}{2}b\right)\left(\frac{3}{2}h\right)$. Plugging in the original base and height, we get $\frac{1}{2}\left(\frac{3}{2}\times 4\right)\left(\frac{3}{2}\times 3\right)$. Simplification yields $\frac{27}{2}=13.5$.

9. **C.** The height of the triangle is not changed if the point is moved parallel to the horizontal line $\overline{AB}$. The slope of $\overline{AB}$ is 0, given the identical y-values of A & B, and its slope is equal to the slope of the line along which point C can be moved.

10. C. The conditions given all form the triangle in the picture. Plug in each vertex of any shape that forms the potential range/ domain to find maximums or minimums. The point $(-2,4)$, when plugged in, yields S(x,y)=3(-2)+4=-2. The point (2,2) when plugged in, yields (S(x,y)=3(2)+2=8, and the point (0,0) yields S(x,y)=3(0)+0=0. The minimum of these is $S = -2$.

11. C. In coordinate geometry problems, I always look for vertical and horizontal distances as these are the fastest to calculate. Here, I cut ΔABC into two triangles that have horizontal and vertical bases and heights, and use the triangle area formula $area = \frac{1}{2}(base)(height)$. Draw a vertical line from (0,0) to (0,3) (the y-intercept of y=-1/2x+3). We know the line has length 3 because the y-intercept is 3 (i.e. the "b" in y=mx+b form). Now, we have cut the large triangle into two smaller triangles each with this vertical line as a base. To find the horizontal line "height" of these triangles, we figure out the x-movement from 0 to the vertices to the right and left. We move 2 to reach (2,2) so the right small triangle has a base of 3 and height of 2, or area of (1/2)(3)(2)=3. Now the left triangle's vertex opposite our vertical line base has an x-value of -2, so for a height of 2 also. Again its area is (1/2) (3)(2)=3. Now add these two areas together to get the area of ΔABC 3+3=6.

12. A. $\angle ABD$ is coterminal with the angle formed by $\overline{AB}$ and the x-axis. We can construct a right-triangle that contains this co-terminal reference angle by drawing an altitude from point A directly down to the x-axis. The right triangle formed by $\overline{AB}$, the altitude (length 4) we just drew, and a portion of the x-axis (length 2) can be used to solve for the cosine of the angle. Remember cosine (CAH of SOHCAHTOA) equals the adjacent leg divided by the hypotenuse. We'll need the hypotenuse, which we find by plugging the legs, 2 and 4, into the Pythagorean theorem: $2^2+4^2=c^2 \rightarrow 4+16=c^2 \rightarrow 20=c^2 \rightarrow \sqrt{20}=c$. Now we take the adjacent leg of the reference angle , -2, and put this over the hypotenuse: $\cos(\angle ABD)=\frac{-2}{\sqrt{20}}=-\frac{1}{\sqrt{5}}=-\frac{\sqrt{5}}{5}$. We know the value should be negative because the cosine of all angles in the 2nd quadrant of the unit circle are negative, as ***cos*** corresponds to x values. (See Trigonometry/Book 2 for more.)

13. B. Simply plug in $x=5$ into the function to get $f(5)=\frac{5^2-16}{5-3}=\frac{25-16}{5-3}=\frac{9}{2}=4.5$.

14. E. The range of the function is all y-values of the function. Looking at the graph, we see the function extends infinitely in the positive and negative directions along the y-axis with no holes or horizontal asymptotes. Thus, its range is all real values.

15. B. From examining the graph, we can see that there is only one vertical asymptote. The slant asymptote is not included. The vertical asymptote is created when the bottom of the fraction=0, so long as no factors are common to the numerator and denominator; set x-3=0 to find x=3. See Graph Behavior for more in depth coverage of how to find asymptotes.

16. A. Our graph must be a discrete function (x values are only allowed at certain values, typically integers or whole numbers) since it would not make sense to have half a passenger. The only choice that constrains our domain to whole numbers is A.

17. E. Think it through: the amount of money that Tina must recoup decreases by $5 with each shirt she sells. If she sells zero shirts, she needs $50 to recoup her costs. If she sells 2 shirts, she needs $40. Her starting amount is her $50 set up cost. This will create a linear relationship with a negative slope and a y-intercept of 50. The only linear graph is E.

18. D. We can determine which angles have the same terminal side by adding $360°$ and checking whether it corresponds to the given list of angles and repeating the process until we've gone to or past the furthest angle. In this case, with $-2493°$ we get $-2133°, -1733°, -1413°, -1053°, -693°, -333°, 27°$.

19. A If you have a graphing calculator, you can solve for y and graph the result. However, it is faster to recognize that from the general form for circles $(x-h)+(y-k)=r^2$ where r is the radius and (h,k) is the center. Since (h,k)=(0,0), and 169 is 13 squared, this equation represents a circle with an origin at $(0,0)$ and a radius of 13. Thus, it crosses the x-axis 13 units right and left of the origin, at $(-13,0)$ and $(13,0)$. See Conics, Chapter 21, in this book for much more like this.

20. A. The area of a circle is πr^2. We can solve for r by calculating the distance between the center and the point the circle passes through using the Pythagorean theorem. Our vertical rise is found by taking the difference in y-values: $|8--2|=10$ and our horizontal length by taking the difference in x-values: $|5-2|=3$, so $10^2+3^2=c^2 \rightarrow 109=c^2 \rightarrow \sqrt{109}=c$ This is our radius, so r equals $\sqrt{109}$. Plugging this into the circle area equation yields $A=\pi\left(\sqrt{109}\right)^2=109\pi$.

21. **C.** The leftmost point of the circle is tangent to (barely touching) the y-axis. The distance between the tangent point and the center is the radius, which equals the x-value of the center, 4. We know the lowest point on the circle is directly below the center, one radius distance down, with a y-value of $7-4=3$. So the circle touches its lowest point at $(4, 3)$. The point $(3, 1)$ has a y-value (1) below what we have calculated as the lowest possible y-value (3), so it must be outside the circle. Alternatively use the distance formula; find the distance between the center, $(4, 7)$ and the point $(3, 1)$. We find the vertical change $(7-1=6)$ and horizontal $(4-3=1)$. Now run the Pythagorean theorem: $6^2+1^2=c^2 \rightarrow 36+1=c^2 \rightarrow 37=c^2 \rightarrow \sqrt{37}=c$ This point is greater than 6 away from the center, and my radius is only 4, so the point is outside the circle.

22. **E.** Sketching the triangle shows that the two legs of the triangle have lengths 10 and 7, respectively. Plug these values into the Pythagorean theorem, $10^2+7^2=c^2 \rightarrow 149=c^2$, and simplify to get $\sqrt{149}$.

23. **B.** The altitude of an isosceles triangle bisects the base. Since the base is horizontal, the triangle's peak, Q, lies on the vertical line through the midpoint of $\overline{PR}$. The x-value of the point is thus the average of the x-values of P and R: $\frac{-5+1}{2}=-2$.

24. **D.** Answer choices A, B, and C can be disproved by moving 5 units right, up, or left from $(3,4)$. Answer choice E can be disproved since its distance from $(3,4)$ is 5, as can be shown in the Pythagorean theorem. This leaves answer choice D.

25. **D.** The diagonals of a rectangle are congruent. Thus, we can solve for ZY, which has the same value as XW. Plugging in the difference in x-values and difference in y-values into the Pythagorean theorem yields $\sqrt{34}$.

26. **A.** This problem is simply the rote work of adding together the distance between each set of adjacent points of the quadrilateral, using the formula distance $=\sqrt{(x_2-x_1)^2+(y_2-y_1)^2}$, the Pythagorean theorem. Plugging in each pair of adjacent points, we find that the distance of each side of the quadrilateral is $\sqrt{10}$. Adding all of them together, we get $4\sqrt{10}$.

27. **D.** The easiest way to solve this problem is to first express line M in point-slope form. We isolate the y term from the rest of the equation by moving the x term to the other side and get $-8y=-6x+11$. We then isolate y completely by dividing by its coefficient and get $y=\frac{-6}{-8}x+\frac{11}{-8}$. This simplifies to $y=\frac{3}{4}x-\frac{11}{8}$. Now we find the equation for line N by expressing a line with half the slope and a y-intercept for higher than in line M. The slope of line M is $\frac{3}{4}$, so the slope of line N is half this, $\frac{3}{8}$. The y-intercept of line N is 4 greater than line M's: $-\frac{11}{8}+4=-\frac{11}{8}+\frac{32}{8}=\frac{21}{8}$. Thus, the equation for line N is $y=\frac{3}{8}x+\frac{21}{8}$, or $y=\frac{3x+21}{8}$.

28. **B.** Imagine a tic-tac-toe board. It is made of 2 horizontal and 2 vertical lines. There is only one box that is completely bounded on all sides, and thus has finite, nonzero area. If you drew a larger tic-tac-toe board by using 3 horizontal and 3 vertical lines, you can see quite easily by sketching it that it contains 4 completely bounded boxes (2^2). Now draw 4 horizontal and 4 vertical lines, and you'll see 9 boxes (3^2). The general rule we can deduce is that the number of bounded regions is the square of 1 less than the number of lines that are horizontal and vertical. In this case, there are 14 horizontal and vertical lines, so the solution is $(14-1)^2=169$.

29. **B.** A plane in geometry is constrained by at least three non-collinear points. This is all you need to know to answer this question since all of the correct answers are some extension of this. For example, 2 lines that intersect at only one point have infinitely many non-collinear points, since there are an infinite number of points on one line that are not on the other line. The same is true of 2 parallel lines that do not touch. The 3 vertices of a scalene triangle, of any triangle, are non-collinear points. The only answer choice that does not contain 3 non-collinear points is B. 3 collinear points.

30. **E.** There are an infinite number of points on a line segment, and by extension, there are an infinite number of points for a perpendicular line to cross through. Crossing through a different point makes these lines unique, so there are an infinite number of perpendicular lines.

31. **E.** We know that the ball, when dribbled, will return to the basketball player's hand, so the correct graphical representation will have the height of the ball return to the same level as in the beginning. This rules out choices D and B as the height of the ball is lower than in the beginning. A is incorrect because the ball can not travel back in time and C is incorrect as the height of the ball in the graph given its time is illogical. Therefore, E is the best answer.

PART TWO: ADVANCED ALGEBRA

CHAPTER

14 ABSOLUTE VALUE

SKILLS TO KNOW

- Basic definition of absolute value and how to apply it
- How to solve basic absolute value equations and inequalities
- How to solve "properties of numbers" problems involving absolute value (Could be true, must be true, etc.)
- How to set up word problems using the idea of "open sentence" equations and inequalities

NOTE: For more absolute value problems, see chapters 22 and 23 in this book on **Graph Behavior** and **Translations and Reflections**. Many of the problems in this practice set are also **Properties of Numbers** Questions. See Chapter 2 in Book 2 for more tips, strategies and practice on the inequalities, could be true/must be true style questions.

THE BASICS

We denote absolute value using **a pair of vertical bars**.

TIP: When you see absolute value bars, make the value between them positive. If the value between the bars is already positive, just take away the bars and you're done. $(|7|=7)$.
If the value is negative, you make it positive. $(|-2|=2)$.

By definition, absolute value means the distance between a number and zero. So if you are asked to solve $|-4|$ you could ask yourself, "how far from zero is -4?" As you can see in the number line below, it's four away from zero.

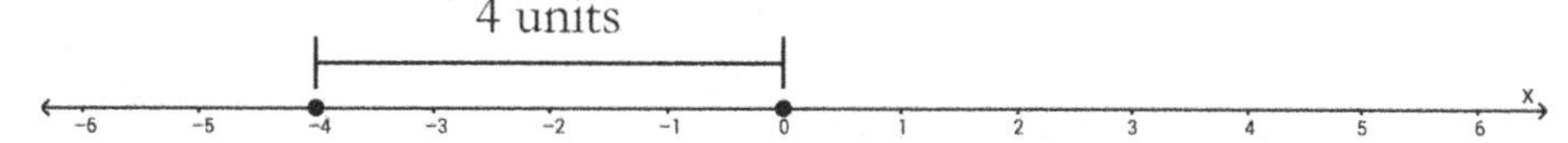

In any case, any absolute value **is always zero or positive—it cannot be negative.**

Absolute value signs are like a form of parentheses—in PEMDAS you treat them on the same level as you would treat parentheses, i.e., complete all work within them first, then simplify once you have a single numeric value between the bars (if possible). For example:

$$|-2-3|-4=?$$

$$|-5|-4=$$

$$5-4=1$$

Answer: **1**.

SOLVING ABSOLUTE VALUE EQUATIONS

To solve absolute value equations, follow these four steps:

STEP 1: ISOLATE

Isolate the part of the equation that has absolute value in it (often it will already be isolated).

How many real solutions exist for the equation: $|x^2-5|-9=0$

$|x^2-5|-9=0$ The absolute value part is not isolated yet!

$|x^2-5|=9$ Add 9 to both sides to isolate the absolute value.

STEP 2: CHECK FOR THE IMPOSSIBLE

Remember absolute values cannot be negative! If your answer is negative at this stage, you're done. The answer would be no solution. For example, $|x^2-4|=-12$ is impossible! No absolute value can ever be negative!

Our problem $|x^2-5|=9$, however, sets an absolute value equal to a positive—so we move on.

STEP 3: SPLIT THE PROBLEM

Split the problem into two cases. **Case 1:** When the element between the bars is positive. Think about it—what you're doing is changing the sign on what the answer would be if there were no bars—as such you need to simply REMOVE the bars, and **Case 2:** Multiply ONE side of the equation by -1. I typically multiply whichever side is simplest (i.e. if a plain integer is on one side, that's the side I'll choose so I don't have to distribute the negative).

Again, our equation is $|x^2-5|=9$.

Case 1: Remove the bars	**Case 2: Multiply one side of the equation by -1**
$x^2-5=9$	Remember to distribute the negative if necessary
$x^2=14$ Add 5 to both sides	$x^2-5=(9)(-1)$ Multiply by negative one
$x=\pm\sqrt{14}$ Take the square root of both sides	$x^2-5=-9$ Add 5 to both sides
	$x^2=-4$ No real numbers squared equal a negative.
We have two real, irrational solutions: $\sqrt{14}$ and $-\sqrt{14}$	Given that the question does not want non-real solutions, there are no solutions from this case.

STEP 4. DOUBLE CHECK!

It's always a good idea to double check absolute value questions if you have time—though if you don't forget step 2, you'll probably be fine. In any case, I find that plugging in at the end of these problems helps eliminate extraneous solutions and careless errors, particularly on more challenging problems (i.e. in the last 15 or so questions on the test.) For this example, when I plug in $\sqrt{14}$ and $-\sqrt{14}$, I get $14-5=9$—that's correct so we're good.

The answer to the question is **2**—there are **two real solutions. Remember to always reread the question** before you put an answer. Often you'll be asked not for the numeric solution, but rather the *number* of possible solutions, the sum of the solutions, or some other value! This **"safety net"** step is particularly important on absolute value problems!

ABSOLUTE VALUE INEQUALITIES

Here's the deal—when approaching absolute value inequalities you can usually turn them into equations and then test points after you've solved the thing down. That method works and there's nothing really wrong with it—though testing points is sometimes time consuming. The great thing about this method is that it's easy to learn—and your prep time is limited. These come up less often than other problems, so this is the method I recommend for its simplicity (it's just less likely you'll mess it up!) for anyone aiming for under a 31-32:

Step 1: Make the inequality sign into an equals sign
Step 2: Solve as an equality
Step 3: Take your "hinge points" create regions, then test regions

For simplicity's sake, let's take our previous problem and make it an inequality: $|x^2-5|-9\geq 0$.

We make the greater than sign an equals sign, solve it down in the same way—and get our "hinge" points as $\sqrt{14}$ *and* $-\sqrt{14}$ (see above example if you want to see how I got these numbers).

SPEED TIP: You can often skip steps 1 & 2 above if all the answer choices show similar "hinge" points. I.e. look at your answer options, use some logic, and head straight to testing regions as I do below. This is similar to the **"Pluck Points"** strategy presented in **Chapter 8, Inequalities: Advanced** in this book.

At this point we need to test three regions, find a test point in each, and plug it into $|x^2-5|\geq 9$.

Region		Test Point		Plug into	
Less than $-\sqrt{14}$	→	test point of $-\sqrt{16}=-4$	→	$\left|(-4)^2-5\right|=|16-5|=11$	→ Is $11>9$? YES
Between $\sqrt{14}$ and $-\sqrt{14}$	→	test point of 0	→	$|0^2-5|=|-5|=5$	→ Is $5>9$? NO
More than $\sqrt{14}$	→	test point of $\sqrt{16}=4$	→	$|4^2-5|=|16-5|=11$	→ Is $11>9$? YES

As you can see, the first and third test regions are greater than 9, so these adhere to the inequality and make it true. As such the answer is $x\leq-\sqrt{14}$ or $x\geq\sqrt{14}$. We can also write this as $x\leq-\sqrt{14}\cup x\geq\sqrt{14}$ or can represent graphically:

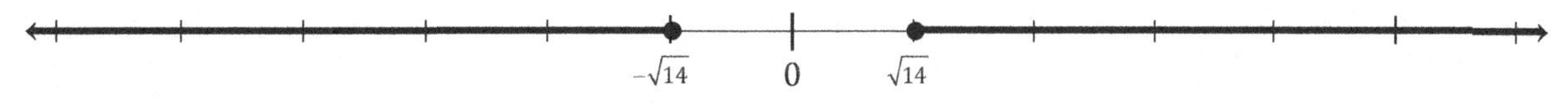

Sometimes, that little method won't work. Occasionally, on very tough problems, you're stuck using your brain, using logic (or memorization) to keep the inequality sign in—but that takes more study time to learn. Aim to do these if your goal is a 32+ on the math section.

The solution set for the inequality $|3x+a|<5$ is $\left\{x|-\frac{1}{3}<x<3\right\}$. What is the value of a?

Here, I can set this equal to 5 and then solve down my two cases:

$3x+a=5$ or $3x+a=-5$ but now with that last method I get stuck—I could try to plug in the two x "hinge points" but I'm not sure which equation each goes with and the whole process could get time consuming if I have to test each of the two values in each of the two equations—(that's four tests!) not to mention we don't know the sign direction. Instead I'll keep the inequality sign in and think about what absolute value means.

$|3x+a|<5$ *means that* $|SOMETHING|<5$—in other words, that SOMETHING is "less than 5 away from zero"—the distance between that something and zero is less than five. With that meaning in mind, I know that SOMETHING is either five away to the left (at negative five) or five away to the right (at five) or somewhere in that zone even closer to zero. I get to all this through logical reasoning, and now I have something to solve down—with the goal of isolating the x in the middle.

$$-5<3x+a<5$$

$$-5-a<3x<5-a$$

$$\frac{-5-a}{3}<x<\frac{5-a}{3}$$ This looks familiar—let's match it up with our condition.

$$\downarrow \quad \downarrow$$

$$-\frac{1}{3}<x<3$$ Now we'll turn this coincidence into a couple of equations.

$$\frac{-5-a}{3}=-\frac{1}{3} \qquad \frac{5-a}{3}=3$$

$$-5-a=-1 \qquad 5-a=9$$

$$-a=4 \qquad -a=4$$

$$a=-4 \qquad a=-4$$

NOTE: It's very important that you solve out BOTH equations—the two a values might not be equal and in that case there would be no solution (or you made a mistake)!

In any case, the answer here is $a=-4$.

PROPERTIES OF NUMBERS ABSOLUTE VALUE QUESTIONS

These problems are the kind that you don't likely see much in math at school, and that is what makes them a challenge. (NOTE: for similar problems see **Chapter 2: Properties of Numbers** in Book 2.)

In these problems you often have a parameter—such as "*For all* $x>0$" that limits the input of what values we're talking about. When you have that limitation, you can't simply "solve" the problem algebraically—you can either make up a number or use a combination of algebra and logic.

TECHNIQUE #1: Make up a number

Make up a number when you have variables in the answer choices and have no algebraic way to simplify.

Though in other parts of the book, I may warn you that making up numbers is a slow way to go; with absolute value problems, there is often no straight algebraic way out—only a logical one—and logic can strain your brain. If you're aiming for a 33+, I recommend you also learn to solve these problems algebraically and logically. But otherwise, you can likely get away with making up numbers.

If $x \geq 7$, then $|7-x| = ?$

A. $7-x$ **B.** $x-7$ **C.** $x+7$ **D.** 0 **E.** $-7-x$

Here, the easiest way to the answer is likely to plug in **8**. First, we plug into the expression $|7-x|$ in the question stem: $|7-x|=|7-8|=|-1|=1$.

Now I plug in 8 to all the answers and look for a result of "1"—I have to try EVERY answer though in case I randomly chose a number that works twice (it happens—and is a good reason to avoid choosing any number already in the problem).

A. $7-x=7-8=-1$ *NO*
B. $x-7=8-7=1$ *YES*
C. $x+7=8+7=15$ *NO*
D. 0 *NO*
E. $-7-x=-7-8=-15$ *NO*

B is the correct answer, as it equals 1.

TECHNIQUE #2:

Solve with a combination of algebra & logic—it's sometimes faster, but only if your brain is practiced enough to see the logical connections quickly.

Let's try splitting the problem into two cases—

$|7-x|$ is either going to be equal to $7-x$ or $-(7-x)$.

When I split this into two cases, I don't have the luxury of an equation—I only have an expression, and that means I must distribute the negative in the second case: $-(7-x)=7+x=x-7$

Now I have my two cases:

$7-x$ and $x-7$

Instantly we know the answer is either (A) $7-x$ or (B) $x-7$.

But here I'm not done and I need to use logic, considering the restriction on x. First, I consider the idea that in BOTH of these two cases, if these are what this expression equals, each MUST BE non-negative. That means for the first case, x has to be smaller than 7 to create a non-negative answer. Once x grows to a larger number, the expression would be negative, and an absolute value can't be negative.

In the second case, I can imagine how x MUST BE positive and at least 7—else subtracting 7 is going to send the number into the region of negatives. Since I know the question states that $x \geq 7$, that means the second case is the correct case.

As you can see, all this thinking is a bit confusing—true, I get an algebraic answer—true, if you're aiming for a 36 it's good to understand these problems this well, but for many this method is overkill. The bottom line with doing these the logical way is that you're looking to see the general behavior of numbers and expressions.

ABSOLUTE VALUE INEQUALITIES: WORD PROBLEMS

NOTE: These don't show up all that often on the test—I would recommend practicing them if you're aiming for above a 30 and prepping for at least a few weeks.

TIME SAVER TIP: The other thing to remember about absolute value is that it can also be used to represent the phrase: **"the difference between x and a number is"** or in coordinate geometry, **"the distance between a and b is."** For example:

$$|x-3| \leq 7$$

Can be translated as:

"The distance between x and 3 is no more than 7."

OR

"x and 3 are no more than 7 apart."

Think about it—when you subtract 3 from x, you do what you do when you calculated slope. You find the difference in two numbers—one of the y's and the other y, which is the same as the distance. Remember rise over run? You can use the slope formula, or you can measure it out with little boxes on the page as a distance. **When you take the absolute value of a difference, you find the physical distance between two points.** Slope formulas still use the sign (you're figuring out a rate, really) but the concept of distance is the same.

You can also think of these as a way to represent margin of error. Let's say you're building a bookshelf and it's supposed to be $36''$ high, but can have a margin of error of up to a quarter inch. **The difference between** the <u>ideal</u>, 36, and the <u>actual</u> height, d, can be **no more than** a quarter inch.

$$|36-d| \leq .25$$

Because absolute value subtracts one thing from another thing, but then gets rid of the "negative" sign, **it's the same as finding the DISTANCE between two values or numbers**. This idea can come in handy on the occasional word problem. I recommend you commit to memory that phrase—**"the distance/difference between"**—to think of absolute value subtraction problems!

The diameter, d, of the plastic pipes that a hardware stores sells must satisfy the inequality $|4-d| \leq .005$. What is the maximum diameter, in centimeters, that a plastic pipe may have?

Let's talk about what all this represents—

The part in the absolute value sign means that the **distance between 4 and d is no greater than $.005$. 4 and d are no more than $.005$ apart.**

That means d can be $.005$ more than 4, $.005$ less than 4 or in between, even closer to 4.

If we need the greatest diameter, that would mean we'd add that miniscule margin of error to our "ideal" of 4—

$$4+.005=4.005$$

Answer: 4.005

Now you could back-solve these type of problems, meaning you could plug in all the answer choices and see what works—but that method is slow. These are called "open sentence" inequalities. They're found in most Algebra I books (I know!) so in all likelihood you learned these things back in 8th grade. Seriously, if you want more practice on these, find your younger sister's math book… My best advice is just to memorize the phrase—

THE DISTANCE BETWEEN ( ) and ( ) is (no more than/less than/greater than/at least).

If you can translate these equations to that idea, you'll be able to solve these problems extremely quickly.

1. $|8-3|-|3-8|=?$

A. -10
B. 10
C. 0
D. -6
E. -5

2. $-4|-13+2|=?$

A. 44
B. -15
C. -60
D. 60
E. -44

3. What is the value of $|x-y|+(x-2y)^2$ when $x=2$ and $y=4$?

A. 34
B. 38
C. 2
D. -4
E. 4

4. If $x \geq 9$, then $|9-x|=?$

A. $9-x$
B. $x-9$
C. $x+9$
D. 0
E. $-9-x$

5. If $|x+7|=32$ what are all the possible values for x?

A. 25 and -25
B. 39 and 25
C. -7 and 7
D. 25 and -7
E. 25 and -39

6. If $|x-3|=14$ what are all the possible values for x?

A. 17 and -11
B. -17 and -11
C. 17 and 11
D. 17 and -17
E. 3 and -3

7. If $|x-7|=|-2|$ how many different values are possible for x?

A. 0
B. 1
C. 2
D. 4
E. Infinitely many

8. How many real solutions are possible for m in the equation $|3m+5|=-4$?

A. 0
B. 1
C. 2
D. 4
E. Infinitely many

9. Considering that d and g are two real numbers, and $d<0$ while $g<0$, which of the following is equivalent to $|d-g|$?

A. $d+g$
B. $|d+|g||$
C. $|d|+|g|$
D. $|d|-g$
E. $d-g$

10. For all $n<0$, $\left|-n^5\right|-\left(-|2|^3\right)=?$

A. n^5+6
B. $-n^5+8$
C. n^5-8
D. n^5+8
E. $-n^5+6$

11. What are all the real solutions to the equation $|x^2|+3|x|-10=0$?

A. ± 2
B. ± 5
C. -5 and 2
D. -5 and -2
E. ± 2 and ± 5

12. If $a>b$ then $|a-b|=?$

A. $\sqrt{a-b}$
B. $-(a-b)$
C. $-(a+b)$
D. $a-b$
E. 0

13. For all non-zero real numbers a and b such that $\left|\frac{a}{b}\right| = \left|\frac{b}{a}\right|$, which of the following COULD be TRUE?

I. $a = -\sqrt{b^2}$
II. $a - b = 2$
III. $ab = -ab$

A. I only
B. II only
C. I & II only
D. I & III only
E. I, II, and III

14. At a newsstand, it costs n dollars for a newspaper, and m dollars for a magazine. The difference between the cost of 15 newspapers and 18 magazines is \$48. Which of the following equations represents the relationship between n and m?

A. $\frac{15n}{18m} = 48$
B. $270nm = 48$
C. $|15n - 18m| = 48$
D. $|15n + 18m| = 48$
E. $18n - 15m = 48$

15. For all real values of x, y, and all values of a such that $a \geq 0$, $|x| = |y| = -a$ for how many (x, y) solutions?

A. 0
B. 1
C. 2
D. 3
E. 4

16. For how many pairs (a, b) is the following equation true?

$$\left|\frac{a}{b} - \frac{b}{a}\right| = \left|\frac{b}{a} - \frac{a}{b}\right|$$

A. 0
B. 1
C. 2
D. 4
E. Infinitely many

17. If $x \geq 5$, then $|x - 5| = ?$

A. 0
B. $x - 5$
C. $x + 5$
D. $-x + 5$
E. $-x - 5$

18. If $x - |x| = 0$ then x is:

A. always negative
B. sometimes positive
C. always positive
D. always zero
E. sometimes negative

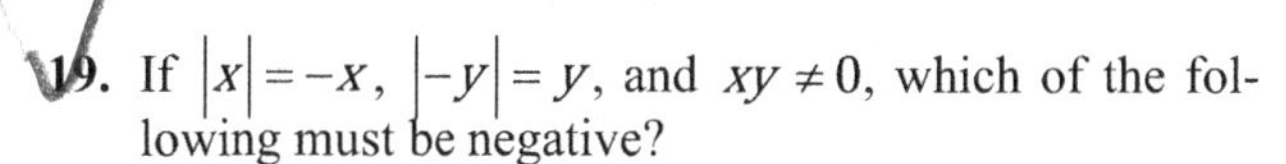

19. If $|x| = -x$, $|-y| = y$, and $xy \neq 0$, which of the following must be negative?

A. x^y
B. y^x
C. $x - y$
D. $x + y$
E. $y - x$

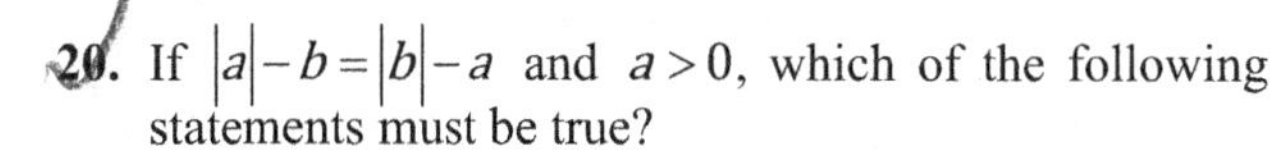

20. If $|a| - b = |b| - a$ and $a > 0$, which of the following statements must be true?

A. $a + b = 0$
B. $ab < 0$
C. $a = b$
D. $a = 0$ or $b = 0$, but not both
E. $a = -b$

21. If $|x| > x$, then which of the following must be true?

A. $-x \leq x$
B. $x = 0$
C. $x^3 < 0$
D. $x \geq 0$
E. $2x > x$

22. Which of the following equations could be used to represent "the distance between x and n" if there are two solutions for x, if n is the mean of the two solutions, and the two solutions are 10 units apart on the number line?

A. $|x + n| = 10$
B. $|x - n| = 10$
C. $|x - n| = 5$
D. $|x + n| = 5$
E. $|x - 5| = n$

23. On an amusement park ride, riders must be between 42 and 72 inches in order to ride. Which of the following equations could be used to represent the possible heights, h, of a potential rider, in inches?

A. $|h-57| \geq 15$

B. $|h-57| \leq 15$

C. $|h-72| \leq 30$

D. $|h-42| \leq 30$

E. $|h+42| \leq 30$

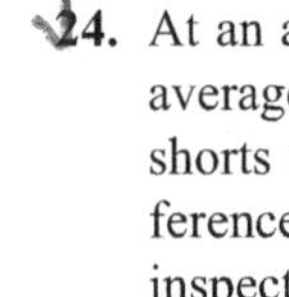

24. At an apparel factory, a pair of medium shorts has an average waist circumference of 29 inches. These shorts must not vary from the average waist circumference by more than .25 inch to pass quality control inspections. Which of the following equations could be used to represent the range of possible waist circumferences, c, that would pass quality control inspections for a pair of medium shorts?

A. $|29-c| \leq 0.25$

B. $|29-c| \geq 0.25$

C. $|c-0.25| \geq 29$

D. $\frac{c}{29} \leq 0.25$

E. $|29+c| \geq 0.25$

25. The solution to which of the following equations is the set of real numbers that are 3 units away from 7?

A. $|x-7| = 3$

B. $|x+3| = 7$

C. $|x-3| = 7$

D. $|x+7| = 3$

E. $|x-7| = -3$

26. Which irrational number is the solution to $|x^2-18|-7=0$?

A. $\sqrt{11}$

B. $\sqrt{5}$

C. 2.5

D. $3\sqrt{2}$

E. $4\sqrt{2}$

27. Which of the following expressions, if any, are equal to each other for all real numbers x?

I. $-\sqrt{(-x)^2}$

II. $|-x|^3$

III. $-|-x|$

A. I and II only

B. II and III only

C. I and III only

D. I, II, and III

E. None of the above

28. Which of the following graphs represents the solution set of the inequality $|x| \leq 1$?

A. 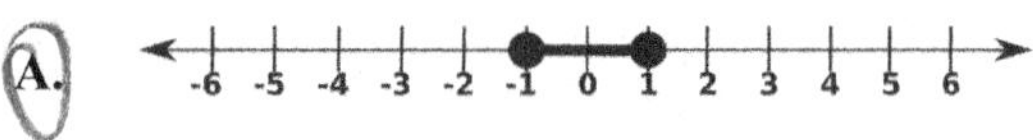

B. -6 -5 -4 -3 -2 -1 0 1 2 3 4 5 6

C. -6 -5 -4 -3 -2 -1 0 1 2 3 4 5 6

D. -6 -5 -4 -3 -2 -1 0 1 2 3 4 5 6

E. -6 -5 -4 -3 -2 -1 0 1 2 3 4 5 6

29. Whenever $a > 0$, which of the following real number line graphs represents the solution for x to the inequality $|x-a| > 3$?

A. -a-3 a-3 a a+3

B. -a-3 a-3 a a+3

C. -a-3 a-3 a a+3

D. -a-3 a-3 a a+3

E. -a-3 a-3 a a+3

30. Which of the following is the solution set for $-2|x+4| \geq -12$?

A. 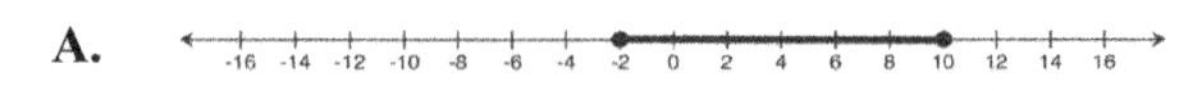

B. -16 -14 -12 -10 -8 -6 -4 -2 0 2 4 6 8 10 12 14 16

C. -16 -14 -12 -10 -8 -6 -4 -2 0 2 4 6 8 10 12 14 16

D. -16 -14 -12 -10 -8 -6 -4 -2 0 2 4 6 8 10 12 14 16

E. -16 -14 -12 -10 -8 -6 -4 -2 0 2 4 6 8 10 12 14 16

ANSWER KEY

1. C 2. E 3. B 4. B 5. E 6. A 7. C 8. A 9. B 10. B 11. A 12. D 13. C 14. C
15. B 16. E 17. B 18. B 19. C 20. C 21. C 22. C 23. B 24. A 25. A 26. A 27. C 28. A
29. C 30. C

ANSWER EXPLANATIONS

1. C. $|8-3|-|3-8|=|5|-|-5|\rightarrow 5-5=0$

2. E. $-4|-13+2|=-4|-11|\rightarrow -4(11)\rightarrow -44$

3. B. Plugging in $x=2$ and $y=4$, we get $|2-4|+(2-2(4))^2\rightarrow|-2|+(2-8)^2\rightarrow 2+(-6)^2=2+36\rightarrow 38$.

4. B. $|9-x|$ is the positive difference between x and 9. $x-9$ is also equal to the positive difference between x and 9 because $x\geq 9$. Notice that $9-x$ will yield a negative answer and will not be equivalent to $|9-x|$. You can also make up numbers and test them to solve this problem. For example, let x=10 and plug into all solutions to find only B equals 1.

5. E. $|x+7|=32$ means that $x+7=32$ or $x+7=-32$. Solving these two inequalities separately, we get $x=32-7\rightarrow 25$ or $x=-32-7\rightarrow -39$. So, the possible values for x are 25 and -39.

6. A. $|x-3|=14$ means that $x-3=14$ or $x-3=-14$. Solving these two inequalities separately, we get $x=14+3\rightarrow 17$ or $x=-14+3\rightarrow -11$. So, the possible values for x are 17 and -11.

7. C. $|x-7|=|-2|$ means $|x-7|=2$. So, $x-7=2$. or $x-7=-2$. Solving these two inequalities separately, we get $x=7+29$ or $x=7-2\rightarrow 5$. So, there are two possible solutions.

8. A. The absolute value of something is always positive, so there is no solution for $|3m+5|=-4$ because -4 is negative.

9. B. To solve, make up numbers. Let $d=-5$ and $g=-3$. $|-5--3|=|-5+3|=|-2|=2$ Now plug $d\,\&\,g$ into the answers and look for 2. A) $-5-3=-8$ (no) B) $|-5+|-3||=|-5+3|=2$ (yes) C) $5+2=7$ (no) D) $5--3=8$ (no) E) $-5--3=-2$ (no). Or solve with logic: We know g is negative. If I subtract a negative number, I am adding its absolute value, because two negatives make a positive. Thus "$-g$" is equivalent to adding $|g|$. This corresponds with choice B.

10. B. We know n is negative, so n^5 is negative and thus the absolute value of the opposite of n^5 is equal to negative one times the already negative n^5 . We can thus simplify the first part of the equation to $-n^5$ The second part $-\left(-|2|^3\right)=-(-(8))\rightarrow +8$. Working from the inside out, our negative signs cancel to make positive eight. The negative sign next to the absolute value is applied AFTER the exponent. Always apply negative signs NOT in a grouping symbol (such as absolute value) after applying the exponent to everything in the grouping symbol. Putting these together: $-n^5+8$

11. A. We can back-solve as absolute value questions tend to generate extraneous solutions. Plug in each answer into the equation to see what works. We'll see that only A has answer choices that work. Alternatively, let $|x|=n$ and solve $n^2+3n-10=0$. This factors to be $(n+5)(n-2)$. So, the solutions that make this equal to zero are $n=-5$ and $n=2$. Since the solution for n (which equals $|x|$) must be positive, $|x|$ must equal 2 , so $x=\pm 2$.

12. D. If $a>b$ then $a-b$ is positive. So, $|a-b|=a-b$. If necessary, make up numbers (i.e. a=5, b=4) and plug into all choices to verify this is true. 5-4=1, a positive number, so that is the same as its absolute value.

13. C. If $\left|\frac{a}{b}\right|=\left|\frac{b}{a}\right|$ then multiplying both sides by $|a||b|$, we get $|a^2|=|b^2|$. Since all squares are positive, $a^2=b^2$. Taking the square root on both sides, we get $a=\pm\sqrt{b^2}$. So, I could be true. For II, we have $a-b=2\rightarrow a=b+2$ and plugging in $a=b+2$ to $\left|\frac{a}{b}\right|=\left|\frac{b}{a}\right|$ we get $\left|\frac{b+2}{b}\right|=\left|\frac{b}{b+2}\right|$. This statement could be true if $b=-1\rightarrow\left|\frac{-1+2}{-1}\right|=\left|\frac{-1}{-1+2}\right|\rightarrow\left|\frac{1}{-1}\right|=\left|-\frac{1}{1}\right|\rightarrow 1=1$. For III, we have $ab=-ab\rightarrow a=-a$ or $b=-b$. This is only true if a or b equals zero. However, it is given that a and b are non-zero, so III cannot be true.

14. C. The cost of 15 newspapers can be represented by $15n$ and the cost of 18 magazines can be represented by $18m$. The difference between these two prices is 48, but we don't know if $15n$ is greater or if $18m$ is greater. So, $|15n-18m|=48$.

15. B. $|x|$ and $|y|$ are always positive or zero, so $|x|=|y|=-a$ is only true when x, y and a are zero. $|0|=|0|=-0=0$.

16. E. Because $|x|=|-x|$, we can manipulate one side of the expression by multiplying it by -1. This gives us $\left|\frac{a}{b}-\frac{b}{a}\right|=\left|(-1)\left(\frac{b}{a}-\frac{a}{b}\right)\right|$ and therefore that $\left|\frac{a}{b}-\frac{b}{a}\right|=\left|\frac{a}{b}-\frac{b}{a}\right|$. This is true for all values of a and b given that neither of them are 0, so there are infinitely many solutions.

17. B. If $x \geq 5$ then $x-5$ is positive. So, $|x-5|=x-5$. Plug in values to check if you're not sure.

18. B. If $x-|x|=0$ then $x=|x|$. So, $x \geq 0$. Since 0 is not positive, x is only sometimes positive.

19. C. If $|x|=-x$ then x must be negative because any absolute value is positive, and the negative of a negative number is positive. Likewise, $|-y|=y$ implies that y is positive. So, the only answer choice that gives us a negative value is $x-y$ because it is a negative number minus a positive number. Note that other answer choices that can yield negative values also yield positive values. $x-y$ is the only choice that guarantees a negative answer.

20. C. If a is positive, then $|a|=a$. So, we can rewrite the equation as $a-b=|b|-a$. Adding a on both sides and adding b on both sides, we get $2a=|b|+b$. Since we know $2a$ is positive and $|b|$ has to be positive because it is an absolute value, we know that b also has to be positive in order for the statement to be true. This means that $2a=2b \rightarrow a=b$.

21. C. If $|x|>x$, then that means x is negative because making x positive makes it bigger. Hence, x^3 is also negative.

22. C. The distance between x and n is $|x-n|=d$. If the average of the two solutions is n, then their sum is $2n$. The solutions are, $x-n=d$ or $x-n=-d \rightarrow x=n+d$ or $x=n-d$. Their sum is verified to equal $2n$ because $n+d+n-d=2n$. Now, we are given that the two solutions are 10 units apart, so $n+d-(n-d)=10 \rightarrow 2d=10 \rightarrow d=5$. So, the distance between x and n can be expressed as $|x-n|=5$.

23. B. The heights must not be over 72 and not be below 42, so the heights can be represented as within the range of the mean of the heights $\pm$ half the range of the heights. This can be expressed as $h \leq \frac{72+42}{2} \pm \frac{72-42}{2} \leq \frac{114}{2} \pm \frac{30}{2} \leq 57 \pm 15$. This is re-written as $h-57 \leq \pm 15$ or $|h-57| \leq 15$. We can also average $72+42$ to get 114, divide by 2 to get 57, our mean or "ideal" and remember that this mean or ideal is the number subtracted from the h. The difference between 72 and 57 is 15, which then goes on the right of the inequality. We ideally want 57, but can be up to 15 away from it.

24. A. The range of waist circumferences that will pass the inspections is within the range of the average 29 inches $\pm\frac{1}{4}$ inches. So, $c-29 \leq \pm\frac{1}{4}$. This means $|c-29| \leq \frac{1}{4}$. This is equivalent to $|29-c| \leq \frac{1}{4}$.

25. A. "The distance between a number and 7 is 3" translates to $|x-7|=3$. See the end of the lesson for this concept.

26. A. Adding 7 on both sides of the equation, we get $|x^2-18|=7$ so $x^2-18=7$ or $x^2-18=-7$. Adding 18 on both sides of both equations, we get $x^2=23$ or $x^2=11$. So, $x=\pm\sqrt{23}$ or $\pm\sqrt{11}$. $\pm\sqrt{11}$ is the only solution for x that is in the answer choices. Plug in to double check it's not extraneous.

27. C. I. simplifies to $-\sqrt{(-x)^2}=-\sqrt{x^2} \rightarrow -x$. II. Simplifies to $|-x|^3=x^3$, and III. Simplifies to $-|-x|=-x$. So, only expressions I and III are equal for all x.

28. A. $|x| \leq 1$ means $-1 \leq x \leq 1$, which is represented by the graph in answer choice A.

29. C. $|x-a|>3$ means that $-3>x-a>3$, which when we add a to all three parts becomes: $-3+a>x>3+a$ (this can also be written as $x<a+3$ and $x>a+3$). The graph in answer choice C correctly reflects this.

30. C. First isolate the absolute value sign by dividing both side by -2 and flipping the sign to get $|x+4| \leq 6$. Now we can treat this like a regular inequality and say that $-6 \leq x+4 \leq 6$. When we subtract 4 from all three parts, we get $-10 \leq x \leq 2$, which is the same as $x \leq 2$ and $x \geq -10$, which is shown in answer C.

CHAPTER

15 EXPONENTS AND RADICALS

SKILLS TO KNOW

- How to use exponents and how radicals relate to exponents
- Applying exponent rules
- Applying radical rules
- How to add, subtract, multiply, and divide with exponents

RADICAL AND EXPONENT RULES

EXPONENT RULES:

Zero power rule: $a^0 = 1$

Negative exponent rule: $a^{-b} = \frac{1}{a^b}$

Power of a power: $\left(a^b\right)^c = a^{bc}$

Product of powers: $a^b a^c = a^{b+c}$

Power of a product: $(ab)^c = a^c b^c$

Power of a quotient: $\frac{a^b}{c^b} = \left(\frac{a}{c}\right)^b$

Quotient of powers: $\frac{a^b}{a^c} = a^{b-c}$

RADICAL RULES:

Fractional exponent conversion: $\sqrt[c]{a^b} = a^{\frac{b}{c}}$

Product of radicals: $\sqrt[c]{a}\sqrt[c]{b} = \sqrt[c]{ab}$

Quotient of radicals: $\frac{\sqrt[c]{a}}{\sqrt[c]{b}} = \sqrt[c]{\frac{a}{b}}$

On the ACT®, you'll need to know how to apply all the basic rules listed above. W**ork to understand the principle behind each formula ,** so even if you're stuck, you can make up some actual numbers and step through the function to remember the rule you've forgotten. I call this making up "mini proofs." Ideally you won't need to do this on the ACT® itself, but you may need to as you work through this chapter to refresh your memory. Below I offer "mini proofs" of each of these rules—create your own when you get stuck! You can also make up little mnemonics (memory tricks) to remember rules.

POWER OF A POWER

$$\left(a^b\right)^c = a^{bc}$$

To remember this rule, think of how when the exponents are separated only by a parenthesis, they are very close together so think multiply—in multiplication you can omit the multiplication sign and separate items with parentheses. Below is a mini-proof that shows this rule to be true:

$$\left(5^2\right)^3 = \left(5\times5\right)^3 = \underbrace{\overbrace{(5\times5)}^{2\ 5's}\overbrace{(5\times5)}^{2\ 5's}\overbrace{(5\times5)}^{2\ 5's}}_{3\text{ times}} = 5^6 = 5^{(2\times3)}$$

Simplify $\left(x^4\right)^3$

We simply apply the rule: $\left(x^4\right)^3 = x^{4\times3} = x^{12}$

If the rule is unclear, or you forget it, write everything out:

$$\left(x^4\right)^3 = \left(x^4\right)\left(x^4\right)\left(x^4\right) = \underbrace{\overbrace{(x\times x\times x\times x)}^{4x's}\overbrace{(x\times x\times x\times x)}^{4x's}\overbrace{(x\times x\times x\times x)}^{4x's}}_{3\text{ sets}} = x^{3\times4} = x^{12}$$

PRODUCT OF POWERS

$$a^b a^c = a^{b+c}$$

For this one, to help you remember, notice that the b and c seem farther apart. Imagine that there's room for an addition sign between them. Here's a mini proof/example:

$$7^3 7^5 = \overbrace{(7\times7\times7)}^{3\ 7's}\overbrace{(7\times7\times7\times7\times7)}^{5\ 7's} = 7^{3+5} = 7^8$$

TIP: **CREATE A MINI PROOF IF MEMORY FAILS!**

If a problem is too challenging to write out by hand, say $x^{99}x^4$, and you aren't 100% sure of the formula, write a "mini proof" as I do above using smaller numbers such as 7, 3 and 5—they're more manageable numbers than 99! You'll soon remember the formula and can apply it (the answer is x^{103}).

Which of the following expressions is equivalent to $x^9 x^{(-2)^2}$?

A. $x^{\frac{9}{4}}$ **B.** x^5 **C.** x^{13} **D.** x^{36} **E.** $x^9 + x^4$

A couple of rules are at play—first start with the exponent $x^{(-2)^2}$ and simplify i.e. $(-2)^2 = 4$, so this simplifies to x^4. (Remember that when negative numbers are in parentheses and squared (or taken to any even power), the answer is always positive—so the answer is NOT negative four!)

$$x^9 x^{(-2)^2} = x^9 x^4$$

Now we apply our **Product of Powers** rule:

$$x^9 x^4 = x^{9+4} = x^{13}$$

Answer: **C.**

POWER OF A PRODUCT

$$(ab)^c = a^c b^c$$

Think the distributive property—but for exponents.

Proof: $$(ab)^c = \underbrace{(ab)(ab)...(ab)}_{c\text{ times}} = \underbrace{(a)(a)...(a)}_{c\text{ times}} \times \underbrace{(b)(b)...(b)}_{c\text{ times}} = a^c \times b^c = a^c b^c$$

Mini-proof: $$x^3x^5=(x\times x\times x)(x\times x\times x\times x\times x)=x^8$$

$(2xy)^4$ is equivalent to:

A. $2x^4y^4$ **B.** $8x^5y$ **C.** $16x^4y^4$ **D.** $2xy^4$ **E.** $\frac{x^4}{2}$

Here we distribute the 4 to the 2, x, and y then simplify: $(2xy)^4=2^4\times x^4\times y^4=16x^4y^4$

Answer: **C.**

MISTAKE ALERT: Don't forget to distribute to EACH piece, and to write parentheses down when you are recopying problems. Choice (A) doesn't distribute to the 2! (D) doesn't distribute at all!

$\left[(x^3)(x^2)^2\right]^3$ is equivalent to:

A. x^{10} **B.** x^{21} **C.** x^{27} **D.** x^{30} **E.** x^{36}

To solve this, we use the three rules above. Don't get overwhelmed! Just apply one rule at a time. We'll start with the **power of a product,** $(ab)^c=a^cb^c$, and eliminate the outer most brackets. We'll distribute that three exponent to each multiplied piece:

$$\left[(x^3)(x^2)^2\right]^3 \text{ becomes } (x^3)^3(x^2)^{2\times3} \text{ which equals } (x^3)^3(x^2)^6$$

Now we apply the **Power of a Power** rule, $(a^b)^c=a^{bc}$ and simplify:

$$(x^3)^3(x^2)^6=(x^{3\times3})(x^{6\times2})=(x^9)(x^{12})$$

Finally, we apply the **Product of Powers** rule, $a^ba^c=a^{b+c}$.

$$(x^9)(x^{12})=x^{9+12}=x^{21}$$

Answer: **B.**

POWER OF A QUOTIENT

$$\frac{a^b}{c^b}=\left(\frac{a}{c}\right)^b$$

Proofs: $\frac{a^c}{b^c}=a^cb^{-c}=(a^1b^{-1})^c=\left(\frac{a}{b}\right)^c$ OR $\frac{8^2}{4^2}=\left(\frac{8}{4}\right)^2=(2)^2=4$, which we verify: $\frac{8^2}{4^2}=\frac{64}{16}=4$

As with **power of a product**, think of the distributive property. Distribute to the top and bottom.

QUOTIENT OF POWERS

$$\frac{a^b}{a^c} = a^{b-c}$$

Mini proof: Simplify $\frac{3^5}{3^3}$

$$\left(\frac{3\times3\times3\times3\times3}{3\times3\times3}\right) = \left(\frac{3\times3\times\cancel{3}\times\cancel{3}\times\cancel{3}}{\cancel{3}\times\cancel{3}\times\cancel{3}}\right) = \frac{3\times3}{1} = 9$$

$$\frac{3^5}{3^3} = 3^{5-3} = 3^2 = 9$$

For all nonzero a and b, $\frac{\left(6a^5b^4\right)\left(8a^3b^7\right)}{4a^7b^9} = ?$

A. $12ab$ **B.** $12ab^2$ **C.** $12a^{\frac{8}{7}}b^{\frac{11}{9}}$ **D.** $12a^{\frac{15}{7}}b^{\frac{28}{9}}$ **E.** $12a^8b^{19}$

We'll have to apply a few rules—let's start by simplifying the top of the fraction:

$$\left(6a^5b^4\right)\left(8a^3b^7\right)$$

We can group like terms together and simplify using the commutative property:

$$\left(6\times8\right)\left(a^5a^3\right)\left(b^4b^7\right)$$
$$= 48\left(a^5a^3\right)\left(b^4b^7\right)$$

Now we can use the **product of powers** property $\left(a^ba^c = a^{b+c}\right)$ to add the appropriate exponents.

$$48\left(a^5a^3\right)\left(b^4b^7\right)$$
$$= 48a^{5+3}b^{4+7}$$
$$= 48a^8b^{11}$$

Now insert this simplified expression back into the numerator: $\frac{48a^8b^{11}}{4a^7b^9}$

Because the answers have no fractions, we use the **quotient of powers** on all like terms, $\frac{a^b}{a^c} = a^{b-c}$:

$$\frac{48}{4}\times\frac{a^8}{a^7}\times\frac{b^{11}}{b^9}$$
$$= 12\times a^{8-7}\times b^{11-9}$$
$$= 12a\times b^2$$
$$= 12ab^2$$

Answer: **B.**

TIP: **GET YOUR BASES TO MATCH!** To solve many exponent problems, your best strategy is to manipulate the situation to get your bases to match. **Once bases match, you can set the exponents equal to each other to solve.** For example. If $x^{2n} = x^6$, solve by setting the exponents equal to each other and "dropping" the bases to get $2n = 6$. If you have $27^5 = 9^{n+2}$, factor the bases down and make them match. 27 and 9 are both 3 to a power, so we can express this all in terms of three: $(3^3)^5 = (3^2)^{n+2}$. Now we can apply the power of a power rule to simplify $3^{3\times5} = 3^{2(n+2)}$. Finally we can "drop" the bases and set the exponents equal to each other: $3\times5 = 2(n+2)$. From here it's all basic algebra. $15 = 2n+4$, $11 = 2n$, and thus $n = 5.5$.

RADICAL RULES

We review basic simplification of radicals in **Chapter 6: Distance & Midpoint**, page 63.

A$_1$ fractional exponent is the same as the reciprocal root, i.e. $a^{\frac{1}{b}} = \sqrt[b]{a}$, or $4^{\frac{1}{2}} = \sqrt{4}$, $4^{\frac{1}{3}} = \sqrt[3]{4}$ etc. **To solve complex questions involving roots, convert to fractional exponents.**

FRACTIONAL EXPONENT CONVERSION

$$\sqrt[c]{a^b} = a^{\frac{b}{c}}$$

Proof: $\sqrt[c]{a^b} = \left(a^b\right)^{\frac{1}{c}} = a^{\frac{b}{c}}$

Once you convert to fractional exponents, apply the exponent rules to solve any questions with radicals!

PRODUCT OF RADICALS / QUOTIENT OF RADICALS

$$\sqrt[c]{a}\sqrt[c]{b} = \sqrt[c]{ab} \qquad \sqrt[c]{\frac{a}{b}} = \frac{\sqrt[c]{a}}{\sqrt[c]{b}}$$

Proof: $\sqrt[c]{a}\sqrt[c]{b} = a^{\frac{1}{c}} \times b^{\frac{1}{c}} = \left(ab\right)^{\frac{1}{c}} = \sqrt[c]{ab}$

Which is a solution of $\sqrt[5]{32x^{10}} = 800$?

A. 4 **B.** 20 **C.** 40 **D.** 400 **E.** $\sqrt[10]{400}$

Using the rule $\sqrt[c]{a^b} = a^{\frac{b}{c}}$, and the rule $\left(a^b\right)^c = a^{bc}$, we know that $\sqrt[5]{32x^{10}}$ can be rewritten as $32^{\frac{1}{5}} x^{\frac{10}{5}}$. (Alternatively, use the product of radicals rule above to get $\sqrt[5]{32}\sqrt[5]{x^{10}}$).This simplifies to $32^{\frac{1}{5}} x^2$. We can use our calculator to find out that $32^{\frac{1}{5}} = 2$. Now we subtitute this in and solve:

$$2x^2 = 800$$
$$x^2 = 400$$
$$x = 20$$

Answer: **B.**

Which expression is equivalent to $\frac{1}{\sqrt[4]{81x^7}}$?

A. $3x^3$ **B.** $\frac{1}{3}(x^{-7/4})$ **C.** $3x^7$ **D.** $81x^3$ **E.** $81x^{\frac{7}{4}}$

Take this one step at a time. First apply the rule $\sqrt[c]{a}\sqrt[c]{b} = \sqrt[c]{ab}$ to the denominator. We can rewrite $\sqrt[4]{81x^7}$ to $\sqrt[4]{81}\sqrt[4]{x^7}$ and then we can use the calculator to figure out that $\sqrt[4]{81} = 3$; using the rule $\sqrt[c]{a^b} = a^{\frac{b}{c}}$, we can rewrite $\sqrt[4]{x^7}$ as $x^{\frac{7}{4}}$. Thus, our denominator is $3x^{\frac{7}{4}}$. Because we take the reciprocal of this, if we make the exponent negative we can "move" the x element to the top of the fraction.
This is essentially the **negative exponent rule:** $a^{-b} = \frac{1}{a^b}$. If you change the sign on an exponent (negative to positive or positive to negative) you can then move that piece up or down to the numerator or denominator. So $\frac{1}{3x^{7/4}} = \frac{1(x^{-7/4})}{3}$ or $\frac{1}{3}(x^{-7/4})$

Answer: **B.**

TIP: **ISOLATE roots to clear them.** Sometimes an algebraic equation may involve roots, such as $\sqrt{x+6} - x = 0$. Isolate the radical and then square both sides (or take each side to the appropriate power). Then solve, but check for extraneous solutions (i.e. you must plug in your answers at the end to make sure they work)! Remember whatever is under an even radical cannot be less than zero!

WARNING: **Beware of extraneous solutions.** Whenever you apply an EVEN power to both sides of an equation (for example, to clear a root or fractional exponent) be vigilant. Unless your variable is restricted to positive values, you may introduce faulty solutions. Plug in to double check your answers in any situation when you must apply an even power or root to both sides of an equation.

TIP: **Clear exponents by taking them to the reciprocal power.** You can "clear" isolated exponents by taking them to a reciprocal power. For example, $x^{2/3} = 8$ can be solved by taking both sides to the 3/2 power so the power on the left cancels. $(x^{2/3})^{3/2} = 8^{3/2}$ so $x = 8^{3/2}$. Remember your calculator can simplify this if you're stuck.

WARNING: **(overachievers only): Beware of cancelling radicals against even exponents!** Even exponents make negative signs disappear, so if these are under a radical sign, and you cancel powers to simplify, **be sure your result either has an absolute value bracket or an even exponent still applied**, or you could introduce faulty solutions. $\sqrt[n]{x^n} = |x|$ when n is an even number and x could be negative. i.e., $\sqrt{x^2} = |x|$. Needing to know this is RARE on the ACT as most questions will limit the domain of the function (i.e. for all "positive" values of x); pay attention to these domain cues.

1. If $3^4\left(81^2\right)=3^n$, what is n?

A. 6

B. $\frac{1}{6}$

C. $\frac{1}{12}$

D. 12

E. 2

2. For what value(s) of x does $\sqrt[3]{x^2-12x}=4$?

A. 16 and −4
B. 16 and 4
C. −16 and −4
D. −16 and 4
E. 16

3. What real value of k satisfies the equation $25^{2k}=\frac{1}{625^{k-1}}$?

A. 1
B. 2
C. $\frac{1}{4}$
D. $\frac{1}{2}$
E. 4

4. For all nonzero x and y, $\frac{\left(32x^4y^{-5}\right)\left(-3x^2y^3\right)}{8x^5y^{-6}}=$?

A. $12xy^4$

B. $12xy^2$

C. $-12xy^4$

D. $12xy^{-4}$

E. $-12xy^{-4}$

5. $-\left(-ab\right)^4\left(a^3b^4\right)^5$ is equal to?

A. $a^{19}b^{24}$

B. $-a^{19}b^{24}$

C. $-a^7b^8$

D. $-a^{60}b^{80}$

E. $-a^7b^9$

6. When c and d are nonzero values, $\left(\frac{c^4d}{d^{-3}}\right)\frac{c^{-2}d^{-5}}{cd}$ is equivalent to?

A. $\frac{c}{d^2}$

B. c^5d^2

C. $\frac{c}{d^8}$

D. $\frac{c^5}{d^8}$

E. cd

7. Which of the following is equal to $\frac{1}{100^{100}}-\frac{1}{100^{99}}$?

A. $\frac{99}{100^{100}}$

B. $\frac{99}{100^{99}}$

C. $\frac{-99}{100^{100}}$

D. $\frac{-99}{100^{99}}$

E. $\frac{1}{100^{200}}$

8. Given that $x=3-y$, what is $\left(x+y\right)^5$?

A. 243
B. −243
C. 3
D. 81
E. −81

9. If $-x^t>\left(-x\right)^s$ and x,t,s are all positive integers, which of the following must be true?

A. s is even
B. $t>s$ and s is even
C. $t<s$ and s is even
D. $t<s$ and s is odd
E. $t>s$ and s is odd

10. If $a=\sqrt[3]{b}$ and $c=b^5$, what is c in terms of a?

A. $a^{\frac{5}{3}}$
B. $a^{\frac{3}{5}}$
C. a^3
D. a^{15}
E. a^5

11. If a and b are nonzero numbers, $\sqrt{\frac{b}{a}}-\sqrt{\frac{a}{b}}$ is equal to which of the following?

A. $\sqrt{\frac{b-a}{a-b}}$
B. $\frac{b-a}{\sqrt{ab}}$
C. x and $1s$
D. $\frac{b-a}{ab}$
E. $\sqrt{\frac{b^2-a^2}{ab}}$

12. For all real values of b, which of the following is equivalent to $\sqrt[4]{16b^{12}}$?

A. $2b^3$
B. $2b^2$
C. $2|b^3|$
D. $4b^2$
E. $16b^3$

13. If $2\sqrt{r}+\sqrt{s}=5\sqrt{s}$, what is r in terms of s?

A. $2s$
B. $2\sqrt{s}$
C. $4\sqrt{s}$
D. $4s$
E. $2s^2$

14. If for all x, $(x^{3a+2})^2=x^3$, then $a=$?

A. -1
B. $\pm\frac{\sqrt{3}-2}{3}$
C. $-\frac{1}{3}$
D. 1
E. $-\frac{1}{6}$

15. For what real number value of n is the equation $\left(x^{1/2}\right)^4\left(x^{-1}\right)^6=x^{n/2}$ true?

A. -24
B. -8
C. -4
D. 16
E. 19

16. If $\frac{x^{a/2}}{x^{b/2}}=x^6$ for all $x\neq 0$, which of the following must be true?

A. $a-b=12$
B. $\frac{a}{b}=6$
C. $a-b=3$
D. $\frac{ab}{4}=6$
E. $a-b=6$

17. $\frac{1}{2}x^2y^2(4x)(6xy^3)$ is equivalent to:

A. $12x^4y^5$
B. $12x^2y^6$
C. $12x^2y^3$
D. $\frac{21}{2}x^2y^6$
E. $\frac{21}{2}x^6y^5$

18. Which of the following is equivalent to $\left(5x^4\right)^3$?

A. $5x^{12}$
B. $15x^7$
C. $15x^{12}$
D. $125x^7$
E. $125x^{12}$

19. For all real values of x, which of the following expressions is equal to $x^n\times x^n\times x^n\times x^n\times x^n$?

A. x^{n^5}
B. $5x^n$
C. $n(x^5)$
D. x^{5n}
E. $(5x)^n$

20. What is the value of $f(x)=16^{1/x}-6$ when $x=4$?

A. 65530
B. -2
C. 2
D. 4
E. -4

21. For any positive value of y, $\left(y^{-2}\right)^{1/5}=?$

A. $\frac{1}{y^{9/5}}$
B. $y^{11/5}$
C. $\frac{1}{y^{2/5}}$
D. $y^{2/5}$
E. $\frac{1}{y^{10}}$

22. For all values of x, $\left[\frac{27(x+1)^2}{3}\right]^{1/2}=?$

A. $81(x+1)$
B. $9x^2+18x+9$
C. $3x+1$
D. $3|x+1|$
E. $9x^2-18x+9$

23. For all $a>0$, $\frac{\left(3a^{5/2}\right)^2+5\left(a^4\right)^2+4a^8}{3a^2}=?$

A. $3\left(a^3+a^6\right)$
B. a^3+a^6
C. a^3+3a^6
D. $3a^{5/2}+5a^4+4a^6$
E. $3a^{5/2}+\frac{5}{3}a^4+\frac{4}{3}a^6$

24. Which of the following is equivalent to $\left(-2x^3y^7\right)^3$?

A. $8x^9y^{21}$
B. $-8x^9y^{21}$
C. $8x^6y^{10}$
D. $6x^9y^{21}$
E. $-6x^6y^{10}$

25. If x and y are real and $\sqrt{4\frac{x^3}{y}}=2$, then what must be true?

A. x and y must both be positive
B. x and y must both be negative
C. $x^3=y$
D. $x=y^3$
E. x and y must be the same number

26. A certain perfect square has exactly 5 digits (that is, it is an integer between 10,000 and 99,999). What is the maximum number of digits the positive square root of that perfect square could have?

A. 2
B. 3
C. 4
D. 316
E. 317

27. Given $13^{\frac{(x+1)^2}{(x^2+1)}}=1$, $x=?$

A. 1
B. -1
C. 1 and -1
D. 0
E. $\frac{1}{2}$

28. For all positive values of a and b, which of the following expressions is equivalent to $\left(a^2\sqrt[3]{b^7}\right)\left(b^4\sqrt[6]{a^8}\right)$?

A. $a^3b^5\sqrt[6]{ab}$
B. $a^3b^6\sqrt[3]{ab}$
C. $a^6b^3\sqrt[6]{ab^2}$
D. $a^2b^4\sqrt[6]{ab}$
E. $a^2b^4\sqrt[3]{ab}$

29. For how many integers x is the equation $8^{x-1}=2^{x^2-1}$ true?

A. 0
B. 1
C. 2
D. 4
E. 5

30. What is the value of the expression $\sqrt[3]{\frac{3m}{2n+2}}$ if $m=6$ and $n=-4$?

A. $i\sqrt[3]{3}$
B. $-\sqrt{3}$
C. $i\sqrt{3}$
D. $\sqrt[3]{3}$
E. $-\sqrt[3]{3}$

31. The square root of a certain number is approximately 6.324555. The certain number is between what two integers?

A. 2 and 3
B. 6 and 7
C. 25 and 37
D. 35 and 50
E. 63 and 82

32. For nonzero values of x and y, which of the following expressions is equivalent to $-\frac{35x^5y}{7x^3y^4}$?

A. $-28x^2y^{-3}$
B. $-42x^2y^{-3}$
C. $-5x^8y^4$
D. $-5x^2y^{-3}$
E. $-5x^2y^3$

33. What is the positive solution to the equation $14c^2=25$?

A. $\frac{25}{14}$
B. $\sqrt{\frac{14}{25}}$
C. $\sqrt{\frac{25}{14}}$
D. $\frac{30\sqrt{14}}{14}$
E. $\left(\frac{25}{14}\right)^2$

34. If $16\times2^{x-4}=4^{y+3}$ and $y=4$, what is the value of x ?

A. $\frac{1}{2}$
B. $\frac{15}{2}$
C. 7
D. 14
E. $\frac{34}{5}$

35. In the realm of real numbers, what is the solution of the equation $9^{2x-1}=3^{1+x}$?

A. 0
B. $\frac{2}{3}$
C. -1
D. 2
E. 1

ANSWER KEY

1. D	2. A	3. D	4. C	5. B	6. A	7. C	8. A	9. D	10. D	11. B	12. C	13. D	14. E
15. B	16. A	17. A	18. E	19. D	20. E	21. C	22. D	23. A	24. B	25. C	26. B	27. B	28. B
29. C	30. E	31. D	32. D	33. C	34. D	35. E							

ANSWER EXPLANATIONS

1. **D.** Our goal is to get the bases to match. Once they do, we can "drop the bases" and set the two exponents equal to each other to solve for ***n***. $81=3^4$, so let's plug this into the expression on the left side of the equation. We'll then simplify using the **power of a power** $\left(a^b\right)^c=a^{bc}$ and **product of powers** $a^b a^c=a^{b+c}$ rules: $3^4\left(81\right)^2\rightarrow 3^4\left(3^4\right)^2\rightarrow 3^4\left(3^8\right)\rightarrow 3^{12}$. Now we plug this back in for the left side of the equation to get $3^{12}=3^n$, so dropping the bases, $n=12$.

2. **A.** To solve equations with radicals, first isolate the radical on one side of the equation. Then take the radical to a power that makes that radical cancel (beware extraneous solutions when the radicals are even). Here, cube both sides to get $x^2-12x=64$, and subtract 64 on both sides: $x^2-12x-64=0$. To factor we need two numbers whose product is -64 and whose sum is -12. -16 and 4 work: $(x-16)(x+4)=0$. Set each factor equal to zero, to find $x=-4$ or $x=16$.

3. **D.** Multiplying 625^{k-1} on both sides, we get $25^{2k}\left(625\right)^{k-1}=1$. Remember to get your bases to match when possible. Since $625=25^2$, we can substitute this value in and then apply the **product of powers** $a^b a^c=a^{b+c}$ rule: $25^{2k}\left(25^2\right)^{k-1}=1\rightarrow 25^{2k}\left(25^{2k-2}\right)=1\rightarrow 25^{2k+2k-2}=1\rightarrow 25^{4k-2}=1$. For a number other than 1 raised to a power to equal 1, its exponent must equal 0, i.e. $25^0=1$. Thus, $4k-2=0$, which solved tells us that $k=\frac{2}{4}\rightarrow\frac{1}{2}$.

4. **C.** Combining the like terms in the numerator, and applying the **product of powers** $a^b a^c=a^{b+c}$, we get $\frac{32(-3)x^4x^2y^{-5}y^3}{8x^5y^{-6}}=\frac{-96x^6y^{-2}}{8x^5y^{-6}}$. Now, simplifying $\frac{-96}{8}$ and applying the **quotient of powers** rule $\frac{a^b}{a^c}=a^{b-c}$ we get $-12x^{6-5}y^{-2-(-6)}$, which simplifies to $-12xy^4$.

5. **B.** Be careful! You can't distribute the negative first because of the exponent! First apply **power of a product** $\left(ab\right)^c=a^c b^c$ to "distribute" the exponents: $-\left((-1)^4a^4b^4\right)\left(a^{15}b^{20}\right)$. Now apply the **product of powers** $a^b a^c=a^{b+c}$ to get $-\left(a^4a^{15}b^4b^{20}\right)=-a^{19}b^{24}$.

6. **A.** Rearrange the terms using the commutative property and apply the **product of powers** $a^b a^c=a^{b+c}$ across the numerator and denominator to get $\frac{c^4\times c^{-2}\times d\times d^{-5}}{d^{-3}\times d\times c}\rightarrow\frac{c^2d^{-4}}{d^{-2}c}$. Applying the **quotient of powers** rule $\frac{a^b}{a^c}=a^{b-c}$ we get $c^{2-1}d^{-4-(-2)}=cd^{-2}$ and then we apply the **negative exponent** rule $a^{-b}=\frac{1}{a^b}$ to get $\frac{c}{d^2}$.

7. **C.** Think: all you need to do is subtract two fractions. Find the common denominator. Multiply by 100 over 100 to turn $\frac{1}{100^{99}}$ into $\frac{100^1}{100^{100}}$, which allows us to rewrite the expression as $\frac{1}{100^{100}}-\frac{100}{100^{100}}$. This simplifies to $-\frac{99}{100^{100}}$.

8. **A.** This is the **equation-expression** pattern we discussed in Chapter 1. Simply plug in for x in the expression and trust it will work. X is already isolated: $x=3-y$, so plugging into $\left(x+y\right)^5$, we get $\left(3-y+y\right)^5$, which simplifies to $3^5=243$.

9. **D.** First, $-x^t$ will have a negative value no matter what because the negative sign is not raised to the power of t, so it cannot be canceled out even when t is even. Thus, in order for $-x^t>\left(-x\right)^s$, the expression $\left(-x\right)^s$ must also be negative. Since the negative sign is raised to the power of s in $\left(-x\right)^s$, s must be odd in order for $\left(-x\right)^s$ to be negative. This eliminates choices A, B, and C. Lastly, if $t>s$ then $-x^t$ will have a greater magnitude and therefore be more negative than $\left(-x\right)^s$, yielding $-x^t<\left(-x\right)^s$. So, $t<s$ and s must be odd. See Book 2 Chapter 2 for similar problems.

10. **D. Remember ISOLATE TO ELIMINATE! We discuss this in Chapter 2.** We want c in terms of a so we must

"eliminate" b. To do so, we must isolate it! Cube both sides of the equation $a=\sqrt[3]{b}$, to get $a^3=b$. Substituting this value of b into $c=b^5$, and applying the **power of a power** $\left(a^b\right)^c=a^{bc}$, we get $c=\left(a^3\right)^5=a^{15}$.

11. B. $\frac{\sqrt[c]{a}}{\sqrt[c]{b}}=\sqrt[c]{\frac{a}{b}}$ (**quotient of radicals rule**), so we can split up these "fractions" and find a common denominator to combine them. $\sqrt{\frac{b}{a}}-\sqrt{\frac{a}{b}}\rightarrow\frac{\sqrt{b}}{\sqrt{a}}-\frac{\sqrt{a}}{\sqrt{b}}\rightarrow\frac{\sqrt{b}}{\sqrt{a}}\frac{\sqrt{b}}{\sqrt{b}}-\frac{\sqrt{a}}{\sqrt{b}}\frac{\sqrt{a}}{\sqrt{a}}\rightarrow\frac{b}{\sqrt{ab}}-\frac{a}{\sqrt{ab}}\rightarrow\frac{b-a}{\sqrt{ab}}$.

12. C. $16=2^4$. So, substituting and using **power of a product** $(ab)^c=a^cb^c$, we can simplify the expression to be $\sqrt[4]{2^4\left(b^3\right)^4}=\sqrt[4]{\left(2b^3\right)^4}$. Now we cancel the fourth power with the fourth root, BUT because we have even root with a variable to an even power, we must insert the variable portion to any odd exponent power in absolute value brackets. Because even exponents make negative signs disappear, the original expression will only output a positive number; here if variable b is negative, $2b^3$ would be negative (and thus wrong). Thus $\sqrt[4]{\left(2b^3\right)^4}\rightarrow 2\left|b^3\right|$.

13. D. Subtracting $\sqrt{s}$ on both sides, $2\sqrt{r}=4\sqrt{s}$. Squaring both sides, $4r=16s$. Dividing both sides by 4, we get $r=4s$.

14. E. Given $(x^{3a+2})^2=x^3$ use **power of a power** $\left(a^b\right)^c=a^{bc}$ and "distribute" the power of 2 to get $x^{6a+4}=x^3$. Because the base is the same on both sides, the exponents are equal, which means $6a+4=3$. Solve to find that $a=-\frac{1}{6}$.

15. B. Use **power of a power** $\left(a^b\right)^c=a^{bc}$ to simplify $\left(x^{1/2}\right)^4\left(x^{-1}\right)^6$ into $(x^{4/2})\left(x^{-6}\right)$, and then add the exponents of the terms with like bases using **product of powers** $a^ba^c=a^{b+c}$: $(x^{4/2})\left(x^{-6}\right)\rightarrow x^{2-6}\rightarrow x^{-4}$. Stopping there would lead to answer (C), which is incorrect. Going back to the original equation, $x^{-4}=x^{\frac{n}{2}}$, and since the two expressions have the same base their exponents are equal: $-4=\frac{n}{2}$. Solving this tells us that $n=-8$, answer (B).

16. A. Knowing $\frac{a^b}{a^c}=a^{b-c}$, we simplify the fraction by subtracting the exponents. $\frac{x^{a/2}}{x^{b/2}}=x^6$ becomes $x^{\frac{a}{2}-\frac{b}{2}}=x^6$; because the bases are the same, the exponents are equal: $\frac{a}{2}-\frac{b}{2}=6$. Combine the fractions: $\frac{a-b}{2}=6$, and simplify to $a-b=12$.

17. A. We rearrange to combine like terms, then add the exponents per the **product of powers** $a^ba^c=a^{b+c}$. $\left(\frac{1}{2}\times4\times6\right)\left(x^2\times x^1\times x^1\right)\left(y^2\times y^3\right)\rightarrow12\left(x^{2+1+1}\right)\left(y^{2+3}\right)\rightarrow12x^6y^5$, answer (A).

18. E. Use **power of a power:** $\left(a^b\right)^c=a^{bc}$ so $\left(5x^4\right)^3=5^3x^{4\times3}=125x^{12}$.

19. D. Using the **product of powers** $a^ba^c=a^{b+c}$, combine same base terms by adding the exponents. $x^n\times x^n\times x^n\times x^n\times x^n=x^{n+n+n+n+n}\rightarrow x^{5n}$

20. E. Plug in the given value for the variable. $16^{1/4}-6=\sqrt[4]{16}-6\rightarrow2-6\rightarrow-4$. We get -4, which is answer (E).

21. C. Use **power of a power** $\left(a^b\right)^c=a^{bc}$ the **negative exponent** rule $a^{-b}=\frac{1}{a^b}$ to simplify: $\left(y^{-2}\right)^{1/5}\rightarrow y^{-2(1/5)}\rightarrow y^{-2/5}\rightarrow\frac{1}{y^{2/5}}$, answer (C).

22. D. Simplify inside the parentheses, then distribute the root. Because we are cancelling out an even exponent with a root (1/2 exponent), we place the answer in absolute value brackets: $\left[\frac{27\left(x+1\right)^2}{3}\right]^{1/2}\rightarrow\left[9\left(x+1\right)^2\right]^{1/2}\rightarrow9^{\frac{1}{2}}\left(x+1\right)^{(2)(\frac{1}{2})}\rightarrow3\left|x+1\right|^1$

23. A. Because a>0 we don't need to worry about faulty/extraneous solutions or absolute value signs. Simplify the numerator first: $\frac{\left(3a^{5/2}\right)^2+5\left(a^4\right)^2+4a^8}{3a^2} \to \frac{3^2\left(a^{(\frac{5}{2})(2)}\right)+5a^8+4a^8}{3a^2} \to \frac{9a^5+9a^8}{3a^2}$. Because the denominator is a monomial, we can split up the expression into fractions by distribution, and then simplify further: $\frac{9a^5+9a^8}{3a^2} \to \frac{9a^5}{3a^2}+\frac{9a^8}{3a^2} \to 3a^3+3a^6 \to 4\left(a^3+a^6\right)$.

24. B. Using **power of a product** $(ab)^c=a^cb^c$, we apply the power of 3 to all terms within the parentheses: $\left(-2x^3y^7\right)^3=(-1)^3 2^3(x^3)^3\left(y^7\right)^3$. Simplify and apply **power of a power** $\left(a^b\right)^c=a^{bc}$: $-1(8)(x^{3\times3})(y^{7\times3}) \to -8x^9y^{21}$

25. C. Square both sides to simplify, then divide by 4: $\sqrt{4\frac{x^3}{y}}=2 \to 4\frac{x^3}{y}=4 \to \frac{x^3}{y}=1$. Now go through the answers and use process of elimination. Answers (A) and (B) are wrong because they say 'must'. True, in order to equal 1, which is positive, x and y must be the same sign, but they could be either both negative or both positive. (C) is correct as it simply rearranges the expression we found, multiplying y to either side. Answer (D) confuses the powers, and answer (E) is wrong because it says 'must'. x and y *could* be the same number (if they both equaled 1), but they don't have to be.

26. B. If we take the square root of 99,999, the largest five-digit number, we get 316.22618. This means that 316^2 is less than 99,999 but 317^2 is greater than 99,999. If 316 is the maximum number that can yield a square with 5 digits, and 316 has 3 digits, then 3 is the maximum number of digits that a number can have and still have a square of five digits or less, which is answer (B). Answers (D) and (E) are akin to actual maximum possible number(s) not digits.

27. B. The **zero power rule** states $a^0=1$ (anything to the zero power equals 1). Thus set the exponent equal to zero: $\frac{(x+1)^2}{\left(x^2+1\right)}=0$. For this to be true, the numerator must equal zero, so $x=-1$. Plug in to check: $\frac{(-1+1)^2}{\left((-1)^2+1\right)} \to \frac{0}{1+1} \to 0$.

28. B. Per the answer choices, we need to "pull out" whole integers and combine the remaining roots. First, factor under the radicals reversing the pattern $a^ba^c=a^{b+c}$ to pull out whole integers. I will do one piece at a time: $\sqrt[3]{b^7} \to \sqrt[3]{b^3 \times b^3 \times b^1} \to \sqrt[3]{b^3}\sqrt[3]{b^3}\sqrt[3]{b} \to (b)(b)\sqrt[3]{b}$. $\sqrt[6]{a^8} \to \sqrt[6]{a^6a^2} \to \sqrt[6]{a^6}\sqrt[6]{a^2} \to a^1\sqrt[6]{a^2}$. Notice that $\sqrt[6]{a^2}$, which is the same as $a^{\frac{2}{6}}$, can be rewritten as $a^{\frac{1}{3}}$, which is $\sqrt[3]{a}$. Plug these pieces into the expression: $a^2b^2\sqrt[3]{b}\cdot b^4a^1\sqrt[3]{a}$. Combine like terms using rules $a^ba^c=a^{b+c}$ and $\sqrt[c]{a}\sqrt[c]{b}=\sqrt[c]{ab}$ to get: $a^3b^6\sqrt[3]{ab}$.

29. C. We want the bases to be the same so that the exponents can be set equal. 8 can be rewritten as 2^3, so our expression becomes $(2^3)^{x-1}=2^{x^2-1} \to 2^{3x-3}=2^{x^2-1}$. Because our bases are the same, the exponents are equal: $3x-3=x^2-1$. Transfer everything to one side so that the sum of the terms equals zero and factor: $x^2-3x+2=0 \to (x-1)(x-2)=0$. There are two solutions to this equation: 1 and 2.

30. E. Plug in the values given and simplify: $\sqrt[3]{\frac{3(6)}{2(-4)+2}} \to \sqrt[3]{\frac{18}{-6}} \to \sqrt[3]{-3} \to \sqrt[3]{-1\cdot3} \to -1\sqrt[3]{3}$.

31. D. Because the square root of this certain number is greater than 6 but less than 7, the number must be greater than 6^2 and less than 7^2: $6.324555^2=x$, the number we want to find, and $6^2<6.324555^2<7^2 \to 36<6.324555^2<49$. If a number is less than 49, it is also less than 50, and if a number is greater than 36 is it also greater than 35, making answer D correct.

32. D. Simplify knowing $\frac{35}{7}=5$ and the **quotient of powers** rule $\frac{a^b}{a^c}=a^{b-c}$ to get $-5x^{5-3}y^{1-4} \to -5x^2y^{-3}$.

33. C. Dividing both sides by 14, we get $c^2=\frac{25}{14}$. Now, taking the square root of both sides gives us $\pm\sqrt{\frac{25}{14}}$. The question asks for the positive solution, so it is $\sqrt{\frac{25}{14}}$.

34. D. Plugging in $y=4$, $16\left(2^{x-4}\right)=4^{4+3}$. Since $16=2^4$, and $4=2^2$ we rewrite this as $2^4\left(2^{x-4}\right)=\left(2^2\right)^7$. On the left side we apply $a^ba^c=a^{b+c}$ and on the right side we apply $\left(a^b\right)^c=a^{bc}$: $2^{4+x-4}=2^{2\times7} \to 2^x=2^{14}$. Matching exponents, x=14

35. E. Since $9=3^2$, we can write $\left(3^2\right)^{2x-1}=3^{1+x}$. Because $\left(a^b\right)^c=a^{bc}$, this is equal to $3^{4x-2}=3^{1+x}$. So, $4x-2=1+x$. Adding 2 and subtracting x to both sides, we get $3x=3 \to x=1$.

CHAPTER

16 FUNCTIONS

SKILLS TO KNOW

- The definition of a function
- Function notation
- Compound (nested) functions—plugging in and evaluating or solving for a variable
- Weird symbol problems (symbol-defined operation problems)
- Inverse function problems

NOTE: We also cover Function Notation in **Chapter 10: Function as a Model** in this book.

FUNCTION BASICS

A function is a mathematical construct that relates each value of a set of input values to a unique output value. More simply, a function relates each value of x to no more than one value of y.

Functions can be written in (x,y) notation, or in function notation ($f(x)$), which we'll cover next. Some common examples of functions are: $y=x^2$, $f(x)=1$, and $x+y=7$.

Some examples of equations that are not functions include: $x^2+y^2=49$, $x=y^2$, and $x=12$.

These are not functions because for some x-values, there is more than one corresponding y-value. For example, in the first equation, the points $(0,7)$ and $(0,-7)$ are both on the curve given by the equation, meaning that the equation does not relate each x value to a unique y value. With a function, every x-value generates a unique y-value.

Sometimes I like to think of functions like a vending machine that masks the inside contents of the machine. When you push a button, you expect to get what the button advertises. For example, if the top button says "Coke," pushing it will always give you a Coke. The position of the buttons is like the x-value: each one is different. But the output is like a y-value. Though you can have more than one button programmed to the same output (say, the top three buttons on a machine say "Coke"), the input (button 1, 2, or 3) will reliably give you the same result every time. You won't randomly get an apple juice if you push button number 1.

A handy trick for deciding whether an equation is a function is called the "vertical line test." To use this test, run a vertical line through the graph at any x point. If the graph is a function, in any position, the vertical line will only intersect the graph at one point at a time. If it intersects at more than one point at a time, the graph is not a function.

Here, you can see that $y=x^2$ is a function.

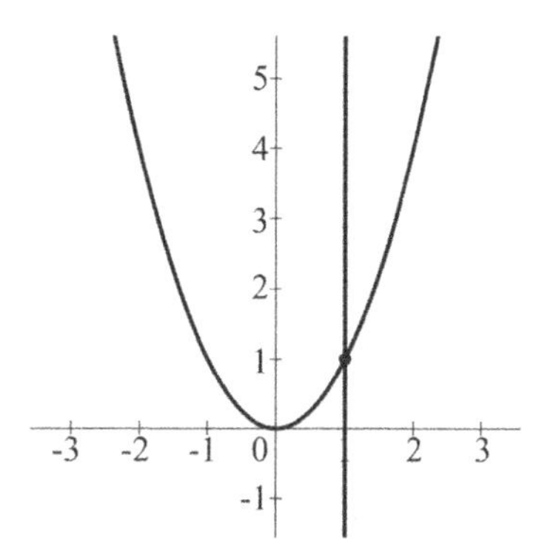

We could move the vertical line anywhere left or right and it would only cross the graph once.

The following graph of $x = y^2$, however, is not a function: a vertical line will cross the graph twice in many places, specifically, for any $x > 0$. For example, $x = 3$ at $y = +\sqrt{3}$, AND $-\sqrt{3}$.

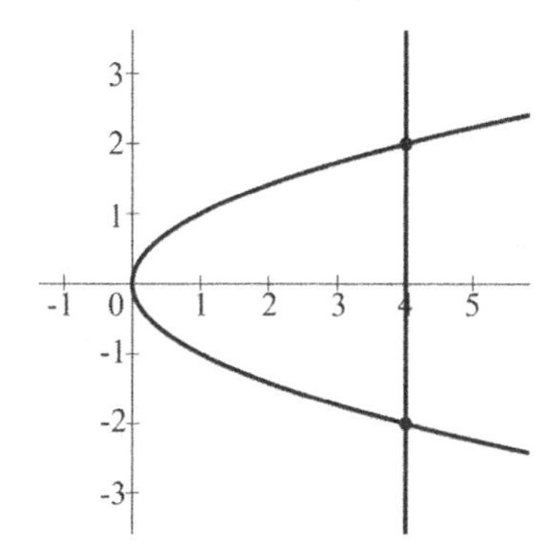

To use the previous analogy, this would be like pushing the same button on our vending machine and having two possible outcomes. Thus, the equation is not a function.

FUNCTION NOTATION

Function notation is a special way of writing the "y" value of a function. The most common function notation is $f(x)$ (pronounced f of x). This does not mean you multiply variable f times x! Writing $f(x)$ is the same as writing a "y." Other variables can be used in place of f or even x. For example, I could say a function P is defined by $p(r) = r + 7$. Here you see I've swapped the f for "p" and the x for "r." You might also see that I can refer to the function as "P," "p of r," or $p(r)$. All of these expressions are identical. Whenever we "define" a function and we have the pattern of a letter followed by a letter in parentheses, we are likely dealing with function notation.

We can think of an ordered pair in function notation:
$(x, f(x))$, where $f(x)$ is equal to y.

We can also turn statements in function notation into ordered pairs. For example, if $f(3) = 7$, this equation basically is saying when $x = 3$, $y = 7$. We can rewrite this fact as the ordered pair $(3,7)$.

More commonly, you'll be asked to plug a value into a given equation. When you see $f(5)$, for example, you'll need to replace the variable in the parentheses with 5.

If $f(x) = 2x + 7$, what is $f(3) - f(2)$?

Here when we see $f(3)$, we want the "y" value of the entire equation when we plug in 3 for x. Rewrite the equation:

$$f(x)=2x+7$$

Plug in 3 for x, because it is in the parenthesis, so should replace the variable originally in that position:

$$f(3)=2(3)+7$$
$$f(3)=6+7$$
$$f(3)=13$$

Then we can find $f(2)$ by the same principle. Plug in 2 for x:

$$f(x)=2x+7$$
$$f(2)=2(2)+7$$
$$f(2)=4+7$$
$$f(2)=11$$

Just as we can substitute in for y values, we can substitute in for $f(3)$ and $f(2)$ once we solve each down by plugging in the given value for x. Though function notation is a bit ugly, it is replaceable in the same way a single variable is. Because $f(3)=13$ and $f(2)=11$, their difference, or $f(3)-f(2)$ is simply $13-11$, or 2.

Answer: 2.

COMPOSITION OF FUNCTIONS

Compound functions (or nested functions) occur when functions are applied to other functions. We say $g(f(x))$ as "g of f of x." Composition of functions can also be written using a hollow circle. For example, $p \circ h = p(h(x))$. To compose a function, you plug the ENTIRE equation from the second named function (or inner nested function) in for the x value of the outer or first named function. That can sound pretty confusing so let's just look at an example.

If $f(x)=x^2$ and $g(x)=\sqrt{x}$, what is $g(f(x))$?

We substitute in for $f(x)$ by plugging in its value, x^2, and we get $g(x^2)$:

$$g(f(x)) \rightarrow g(x^2)$$

Remember, all I do is treat $f(x)$ just as I would a variable, simply replacing it with what I know it equals. Now to solve, we take the item in parentheses and plug it into each instance of the variable in parentheses in our function g. In other words, if $g(x)=\sqrt{x}$, and I need to solve for $g(x^2)$, I can plug in x^2 for x just as if I wanted to find $g(2)$ I would plug "2" in everywhere I see an x. Don't get confused by the fact that x^2 has an x in it. These aren't the same variable even if they look the same. Substitute, treating the term the same as you would an integer.

$$g(x)=\sqrt{x}$$
$$\uparrow \quad \uparrow$$
$$x^2 \quad x^2$$
$$g(x^2)=\sqrt{x^2}$$
$$g(x^2)=|x|$$

I know this is a bit confusing with all the x's everywhere. But the more you work on these problems the easier it gets.

Answer: $|x|$.

Using the same equations, I could also find $f(f(x))$. To do so, again, I take the whole of what $f(x)$ equals (x^2) and plug it into the original function $f(x)=x^2$, replacing the x in each instance with x^2:

$$f(f(x))=(x^2)^2=x^4$$

Given $f(x)=2x+3$ and that $g(x)=x^2-4x$, which of the following is an expression for $g(f(x))$?

A. $2x^2-8x+3$ **B.** $4x^2-4x+9$ **C.** $4x^2+4x-3$

D. $2x^3-5x^2-12x$ **E.** x^2-2x+3

This question is essentially asking us to combine these two equations. Be sure to notice that $f(x)$ is on the inside and $g(x)$ is on the outside, so we will be plugging the whole of what function f equals $(2x+3)$ into the "input" value x of the function $g(x)=x^2-4x$. Everywhere we see an "x" in the function g, we will replace it with $2x+3$:

$$g(x)=x^2-4x$$
$$\nearrow \quad \uparrow \quad \nwarrow$$
$$2x+3 \quad 2x+3 \quad 2x+3$$

$$g(f(x))=g(2x+3)$$
$$=(2x+3)^2-4(2x+3)$$

Now we expand the square of $2x+3$ using the pattern from **Square of a Sum,** $(a+b)^2=a^2+2ab+b^2$ (See **Chapter 3: FOIL and Factoring** to review), setting $2x=a$ and $3=b$; then we distribute the negative four in the second half of the expression:

$$=(2x)^2+2(2x)(3)+(3)^2-(4)(2x)-4(3)$$
$$=4x^2+12x+9-8x-12$$
$$=4x^2+4x-3$$

Answer: **C.**

WEIRD SYMBOL PROBLEMS (SYMBOL DEFINED OPERATIONS)

By now, you know what most math symbols mean, whether addition $(+)$, multiplication $(\times)$, or factorial $(!)$ symbols. But did you know that the ACT® can define a new mathematical operation using a weird symbol you've never seen before?! These problems are essentially function problems. The elements on the left tell you what to plug in, and the second half of the equation sets up an expression that incorporates the variable(s) set up in the first half, telling you what to do with the "inputs" you're given in relation to the weird symbol (as denoted in the first half). These problems set up a "secret code" that you must plug values into. It's probably more confusing to describe than to actually do, so let's jump into an example.

If $x \,\&\, y = 3x + 4y^3$, what is $4 \,\&\, 3$?

The new operation "$\&$" has now been defined, and it indicates that you simply add 3 times the quantity on the left of the ampersand to 4 times the cube of the quantity on the right of the ampersand.

The question is a bit confusing because the numbers given for x and y are the same as the numbers in the problem, but just plug in one at a time and you'll be fine. First, let's figure out what $4 \,\&\, 3$ tells us:

$x = 4$ because 4 is to the left of the ampersand (just as x is in the example equation), and $y = 3$ because 3 is to the right of the ampersand (just as y is in the first given equation). Now we simply plug these values in for x and y:

$$x \,\&\, y = 3x + 4y^3$$

$$4 \,\&\, 3 = 3(4) + 4(3)^3$$

Thus:

$$\begin{aligned} 4 \,\&\, 3 &= 3(4) + 4(3^3) \\ &= 12 + 108 \\ &= 120 \end{aligned}$$

If $\circledast$ represents the operation defined by $a \circledast b = a + \frac{b}{a}$, then $(2 \circledast (3 \circledast 9)) = ?$

A. $\frac{70}{2}$ **B.** $\frac{13}{5}$ **C.** 6 **D.** 5 **E.** 2

When solving these types of problems, start from the innermost function and then work your way out. Take it one step at a time.

Plugging in the numbers of the innermost function:

$$(3 \circledast 9) = 3 + \frac{9}{3}$$
$$= 3 + 3$$
$$= 6$$

Plugging this into the outer function gives us:

$$(2 \circledast 6) = 2 + \frac{6}{2}$$
$$= 2 + 3$$
$$= 5$$

Answer: **D.**

INVERSE FUNCTIONS

Inverse functions are exactly what they sound like: functions that undo the effects of other functions. The inverse of a function, $f(x)$, is written $f^{-1}(x)$. To find an inverse, rewrite your equation, **swapping your x for your y and your y for your x**. For example:

Which of the following is the inverse of the function $f(x) = 2x$?

To make the problem easier to look at and manipulate, I can rewrite this to

$$y = 2x$$

To solve for $f^{-1}(x)$, simply **switch positions of your x and your y** (or your x and your $f(x)$). What was x becomes y, what was y becomes x:

$$x = 2y$$

and manipulate the equation to be in "$y =$" form if necessary. Here, that gives us:

$$y = \frac{x}{2}$$

We can then return to function notation, making our y value $f^{-1}(x)$:

$$f^{-1}(x) = \frac{x}{2}$$

Answer: $f^{-1}(x) = \frac{x}{2}$.

Let's look at a more complicated example:

Given $f(x)=\frac{3x-1}{7-2x}$, which of the following expressions is equal to $f^{-1}(x)$ for all real numbers x?

All this question wants is for us to find the inverse function for $f(x)$. The easiest way to do this is to simply switch y, or $f(x)$, with x, and then isolate y to get back into standard function notation.

Here, we've swapped out $f(x)$ with x and x with y $\quad x=\frac{3y-1}{7-2y}$

Now we multiply both sides to eliminate the denominator $\quad (7-2y)x=3y-1$

Simplify $\quad 7x-2yx=3y-1$

Place all terms with y on one side $\quad 7x+1=3y+2yx$

Factor out y $\quad 7x+1=(3+2x)y$

Divide to isolate the y $\quad \frac{7x+1}{2x+3}=y$

Don't forget to replace y with $f^{-1}(x)$! $\quad f^{-1}(x)=\frac{7x+1}{2x+3}$

Answer: $f^{-1}(x)=\frac{7x+1}{2x+3}$.

1. Given $f(x)=2x^2-3x+8$. what is the value of $f(-5)$?

A. 43
B. −27
C. 3
D. 13
E. 73

2. Given the function $m(t)=4t^2-7$, what is $m(-3)$?

A. −19
B. −31
C. −43
D. 17
E. 29

3. For the function $v(n)=3n^2-4n$, what is the value of $v(-6)$?

A. −624
B. −84
C. −6
D. 84
E. 132

4. A function $f(x)$ is defined as $f(x)=-9x^2$; what is $f(-4)$?

A. 144
B. −144
C. 72
D. −72
E. 7

5. What is the value of $h\left(\frac{1}{4}\right)$ when $h(x)=-16x^2+32x-9$?

A. −2
B. −5
C. −9
D. 0
E. −8

6. For the function $h(x)=3x^2-5x$, what is the value of $h(-4)$?

A. 28
B. 68
C. −28
D. 44
E. −172

7. What is the value of $g(-3)$ if $g(x)=3x^2-5x+11$?

A. −1
B. 53
C. 23
D. 44
E. −55

8. Given $f(x)=2x+3$ and $g(x)=x^2-1$, which of the following is an expression for $f(g(x))$?

A. $4x^2+12x+8$
B. $4x^2+12x+9$
C. x^2+2x+2
D. $2x^2+1$
E. $2x^2+5$

9. The function $f(x)=x^3+3x-2$. What is $f(x+h)$?

A. $x^3+h^3+3x+3h-2$
B. $h^3+3h^2x+3hx^2+3h+x^3+3x-2$
C. $h^3-3h^2x+3hx^2+3h-x^3+3x-2$
D. $h^3+3h^2x-3hx^2+3h+x^3-3x-2$
E. $h^3+3h^2x+3h^2x^2+3h+x^3+3x-2$

10. $f(x)=\begin{cases} |3x|+1, if\ x<-4 \\ |3x^3|+1, if\ x>-4 \end{cases}$

What is the value of $f(-3)$?

A. −80
B. −82
C. 80
D. 82
E. 193

11. There are 2 functions $f(x)$ and $g(x)$ such that $f(x)=\frac{4-x}{7+x}$ and $g(x)=2x^2-3x+1$. What is $f(g(3))$?

A. $\frac{-6}{17}$
B. $\frac{6}{17}$
C. $\frac{14}{17}$
D. $\frac{14}{3}$
E. $\frac{17}{5}$

12. If $f(x)=\frac{1}{x^2+5}$, what is $f(f(2))$?

A. $\frac{1}{9}$

B. $\frac{1}{86}$

C. $\frac{81}{406}$

D. $\frac{1}{76}$

E. $\frac{1}{21}$

13. Tables of values for the functions $f(x)$ and $g(x)$ are shown below. What is $f(g(7))$?

x	$f(x)$
−4	7
−1	−2
7	3
4	9

x	$g(x)$
−2	−1
3	8
5	2
7	4

A. 9
B. 7
C. 8
D. 4
E. 3

14. If $f(x)=2x+3$ and $g(x)=3x^2-1$, which of the following is the expression for $g(f(x))$?

A. $4x^2+12x+8$
B. $12x^2+36x+26$
C. $12x^2-36x+26$
D. $6x^2+36x-26$
E. $12x^2+18x+26$

15. Of the 5 functions below, each denoted by $h(x)$ and each involving a real number $a \geq 3$, which function yields the smallest value of $f(h(x))$, if $f(x)=\frac{1}{x}, x>1$?

A. $h(x)=ax$

B. $h(x)=\frac{a}{x}$

C. $h(x)=\frac{x}{a}$

D. $h(x)=x+a$

E. $h(x)=x^a$

16. We have 2 functions $f(x)=\sqrt[3]{x}$ and $g(x)=3x-d$. If $f(g(27))=4$, what is the value of d?

A. 13
B. 14
C. 15
D. 16
E. 17

17. If $P(x)=-x^4$, what is $P(P(x))$?

A. $-x^{16}$
B. x^8
C. x^{16}
D. x^{-8}
E. $-x^8$

18. The operation ℧ is defined as "add five to the square of the number on the left of ℧ and subtract the result from the number on the right. What is the value of 3 ℧(4℧5)?

A. 30
B. −2
C. 2
D. −30
E. −100

19. Given the operation $x ℌ y = xy^2+2y$, what is $(1ℌ2)ℌ3$?

A. 1200
B. −78
C. 78
D. 66
E. −1200

20. The function $f(x,y)=2x+9y$. What is $f(x,y)$ when $y=4$ and $x=y^{\frac{3}{2}}$?

A. 52
B. 34
C. 80
D. 26
E. 164

21. Given $f(x)=\sqrt[4]{2x+1}$, which of the following expressions is equal to $f^{-1}(x)$ for all real numbers x?

A. $\frac{1}{\sqrt[4]{2x+1}}$

B. $\frac{x^4+1}{2}$

C. $\frac{x^4-1}{2}$

D. $(2x+1)^{\frac{-1}{4}}$

E. $\sqrt[4]{2x+1}$

22. For each positive integer q, let $q\nabla$ be the product of all positive odd numbers less than or equal to q. For example, $9\nabla=(9)(7)(5)(3)(1) = 945$ and $10\nabla=(9)(7)(5)(3)(1)\rightarrow 945$. What is $\frac{13\nabla}{4\nabla}$?

A. 135135
B. 945
C. 45045
D. 154440
E. 10395

23. Which if the following pairs of functions, $f(x)$ and $g(x)$, form the composite function $g(f(x))=\sqrt{4x^3-9}$?

	$f(x)$	$g(x)$
A.	$4x$	$\sqrt{x^3-9}$
B.	x^3+9	$\sqrt{4x}$
C.	x^3	$\sqrt{4x-9}$
D.	$\sqrt{x^3-9}$	$4x$
E.	x^3+9	$\sqrt{4x-9}$

24. The graph of the functions $f(x)=x+1$ and $g(x)=-\frac{x^2}{2}-2x-3$ are shown in the standard (x,y) coordinate plane below. Which if the following is NOT true?

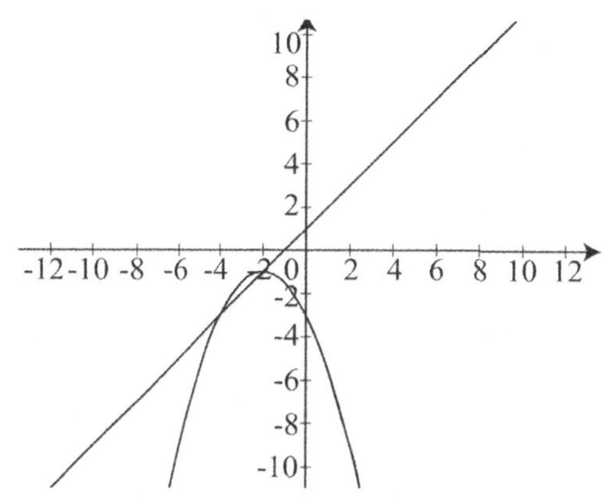

A. $f(-4)=g(-4)$
B. $g(-2)=f(-2)$
C. $f(g(2))=-8$
D. $-|f(x)|=f(x)$
E. $-|g(x)|=g(x)$

25. If $a(x)=x^3+2x^2-7$ and $b(x)=x^2-5y$ where y is some non-zero constant, which of the following expressions is equivalent to $(ab)(x)$?

A. $x^6+2x^4-15yx^4+50y^2x^2-125y^3-7$
B. $x^6+4x^5+4x^4-7x^3-28x^2-5y+49$
C. $x^5+2x^4-5yx^3-10yx^2-7x^2+35y$
D. x^3+3x^2-5y-7
E. x^3+x^2+5y-7

26. If $h(x)=x^4-2x^2+18$ and $g(x)=5x^4+2x^3-7x^2-19$ which of the following expressions is equivalent to $(h-g)(x)$?

A. $6x^4+2x^3-9x^2-1$
B. $-4x^4-2x^3+5x^2+37$
C. x^4-2x^2+37
D. $-5x^4-2x^3+7x^2+37$
E. $5x^8-14x^4+2x^3-352$

ANSWER KEY

1. E	**2. E**	**3. E**	**4. B**	**5. A**	**6. B**	**7. B**	**8. D**	**9. B**	**10. D**	**11. A**	**12. C**	**13. A**	**14. B**
15. E	**16. E**	**17. A**	**18. D**	**19. C**	**20. A**	**21. C**	**22. C**	**23. C**	**24. D**	**25. C**	**26. B**		

ANSWER EXPLANATIONS

1. **E.** Plugging in $x=-5$, we get $f(-5)=2(-5)^2-3(-5)+8$. Simplifying this gives us: $f(-5)=2(25)-(-15)+8 \rightarrow 50+15+8=73$. So, $f(-5)=73$.

2. **E.** Plugging in $t=-3$, we get $m(-3)=4(-3)^2-7$. Simplifying this gives us: $m(-3)=4(9)-7 \rightarrow 36-7=29$. So, $m(-3)=29$.

3. **E.** Plugging in $n=-6$, we get $v(-6)=3(-6)^2-4(-6)$. Simplifying this gives us $v(-6)=3(36)-(-24) \rightarrow 108+24=132$. So, $v(-6)=132$.

4. **B.** Plugging in $x=-4$, we get $f(-4)=-9(-4)^2$. Simplifying this gives us $f(-4)=-9(16) \rightarrow -144$. So, $f(-4)=-144$.

5. **A.** Plugging in $x=1/4$, we get $h\left(\frac{1}{4}\right)=-16\left(\frac{1}{4}\right)^2+32\left(\frac{1}{4}\right)-9$. Simplifying this gives us $h\left(\frac{1}{4}\right)=-16\left(\frac{1}{16}\right)+32\left(\frac{1}{4}\right)-9 \rightarrow -1+8-9 \rightarrow -2$. So, $h\left(\frac{1}{4}\right)=-2$.

6. **B.** Plugging in $x=-4$, we get $h(-4)=3(-4)^2-5(-4)$. Simplifying this gives us $h(-4)=3(16)-(-20) \rightarrow 48+20 \rightarrow 68$. So, $h(-4)=68$.

7. **B.** Plugging in $x=-3$, we get $g(-3)=3(-3)^2-5(-3)+11$. Simplifying this gives us $g(-3)=3(9)-(-15)+11 \rightarrow 27+15+11 \rightarrow 53$. So, $g(-3)=53$.

8. **D.** Plugging in $g(x)=x^2-1$ into the nested function $f(g(x))$, we get $f(g(x))=2(x^2-1)+3$. (Instead of plugging in a number, we are substituting in the entire expression x^2-1 to the x value in $f(x)=2x+3$). Simplifying this gives us $f(g(x))=2x^2-2+3 \rightarrow 2x^2+1$. So, $f(g(x))=2x^2+1$.

9. **B.** Plugging in $x+h$ for x in $f(x)$, we have $f(x+h)=(x+h)^3+3(x+h)-2 \rightarrow x^3+3x^2h+3xh^2+h^3+3x+3h-2$. We can rearrange this to be $h^3+3h^2x+3hx^2+3h+x^3+3x-2$. For more on expanding cubics, see page 32.

10. **D.** Comparing the number we plug in (-3) to the number we're comparing it to (-4) we know -3 is GREATER than -4. If this is confusing, draw out the number line. Because -3 is to the right of -4, it is greater than -4. Don't get confused by the fact that positive 4 is greater than 3: these are negative numbers! Now we know to plug into the second equation given. Plug in and simplify to find the answer. Since $-3>-4$, we use the second expression to evaluate $f(-3)$. So, $f(-3)=\left|3(-3)^3\right|+1 \rightarrow |3(-27)|+1 \rightarrow |-81|+1 \rightarrow 81+1=82$.

11. **A.** We first find $g(3)$ and then plug in that value to find $f(g(3))$. We plug in $x=3$ into $g(x)$ to get $g(3)=2(3)^3-3(3)+1 \rightarrow 2(9)-9+1 \rightarrow 10$. Now, to find $f(g(3))$, we plug in $x=g(3)=10$ for $f(x)$. This gives us $f(10)=\frac{4-10}{7+10} \rightarrow -\frac{6}{17}$.

12. **C.** We first find $f(2)$ and then plug in that value to find $f(f(2))$. So, solving for $f(2)$, we plug in $x=2$ for $f(x)$ to get $f(2)=\frac{1}{x^2+5} \rightarrow \frac{1}{2^2+5} \rightarrow \frac{1}{9}$. Now, to find $f(f(2))$, we plug in $x=f(2)=\frac{1}{9}$ for $f(x)$. This gives us $f\left(\frac{1}{9}\right)=\frac{1}{\left(\frac{1}{9}\right)^2+5} \rightarrow \frac{1}{\frac{1}{81}+5} \rightarrow \frac{1}{\frac{1}{81}+\frac{405}{81}} \rightarrow \frac{1}{\frac{406}{81}} \rightarrow \frac{81}{406}$.

13. **A.** We first find $g(7)$ and then plug in that value to find $f(g(7))$. Looking at the values given in the table for $g(x)$, we see that $g(7)=4$. Now, to find $f(g(7))$, we plug in $x=g(7)=4$ for $f(x)$ and see by the first table that $f(4)=9$.

14. **B.** To find $g(f(x))$, we replace the x 's in $g(x)$ with $f(x)=2x+3$. This gives us $g(f(x))=3(2x+3)^2-1$. Using FOIL to expand the polynomial, we get $3(4x^2+12x+9)-1$. Distributing the 3, we get $12x^2+36x+27-1=12x^2+36x+26$.

15. **E.** Since $f(x)=\frac{1}{x}$ for a positive x, the larger x is, the smaller $f(x)$ is. So, we are looking for the largest $h(a)$ value from our answer choices to yield the smallest value of $f(h(x))$. Since $a\geq 3$, x^a would be the greatest value because the exponent increases x most rapidly.

16. **E.** Plugging in $x=27$ for $g(x)=3x-d$, we get $g(27)=3(27)-d\rightarrow 81-d$. Then, plugging in $x=81-d$ for $f(x)=\sqrt[3]{x}$, we get $f(g(27))=f(81-d)\rightarrow\sqrt[3]{81-d}$ we are given that this value is equal to 4. So, we now want to solve the equation $4=\sqrt[3]{81-d}$ for d. Cubing both sides, we get $64=81-d$. Subtracting 81 on both sides, we get $-17=-d$. So, $d=17$.

17. **A.** Plugging in $P(x)=-x^4$ for x in $P(x)$, we get $P(P(x))=-(-x^4)^4\rightarrow-(x^{16})\rightarrow -x^{16}$.

18. **D.** We first solve for the value inside the parentheses. $(4Ʊ5)$ means $5-(4^2+5)=5-(16+5)\rightarrow 5-21\rightarrow -16$. Taking this value and plugging it back in we have $3Ʊ(-16)$ or $-16-(3^2+5)=-16-(9+5)\rightarrow -16-14\rightarrow -30$.

19. **C.** We first solve for the value inside the parentheses. $(1ℌ2)$ means we plug in $x=1$ and $y=2$ for xy^2+2y. This gives us $(1(2^2))+2(2)=8$. Then, we plug in $1ℌ2=8$ to solve for $(1ℌ2)ℌ3=8ℌ3$. This means we plug $x=8$ and $y=3$ in for xy^2+2y. This gives us $8(3)^2+2(3)\rightarrow 8(9)+6\rightarrow 72+6=78$.

20. **A.** Plugging in $y=4$ and $x=y^{\frac{3}{2}}\rightarrow 4^{\frac{3}{2}}=8$ we get $f(8,4)=2(4)+9(4)\rightarrow 16+36=52$.

21. **C.** To find $f^{-1}(x)$ we switch the x and y values in the function $y=\sqrt[4]{2x+1}$ and solve for y. This gives us $x=\sqrt[4]{2y+1}$. Taking both sides to the power of 4, we get $x^4=2y+1$. Subtracting 1 and dividing both sides by 2, we get $\frac{x^4-1}{2}=y$.

22. **C.** $\frac{13\nabla}{4\nabla}=\frac{(13)(11)(9)(7)(5)(3)(1)}{(3)(1)}$. Canceling out common factors in the numerator and denominator, we get $\frac{(13)(11)(9)(7)(5)(\cancel{3})(\cancel{1})}{(\cancel{3})(\cancel{1})}=(13)(11)(9)(7)(5)\rightarrow 45045$.

23. **C.** We are told to find $g(f(x))$, which is not the same as $f(g(x))$. That rules out answer choice D, because clearly it is $f(g(x))$. We want the square root to encapsulate the entire expression, and choice D would leave the four to the left of the root sign. We need the square root in the function $g(x)$ which is applied last so that we plug in somthing under that radical. Choice A looks tempting, but when we plug it into $g(x)$ we must also cube the four, creating a coefficient of 16, which is not in the originally given expression. In choice B, when we plug in, the four distributes and multiplies by 9, creating a 36 nowhere in the orignally given expression. With choice C, we simply plug the x^3 in for x to get: $\sqrt{4(x^3)-9}$. Thus choice C is correct. Choice E again creates a situation in which the four distributes in a way so that the negative nine goes away and is replaced by a number not in our original expression (positive 27).

24. **D.** Don't forget we're looking for the answer choice that is NOT true, not the one that IS true. We can start by working through the answers one at a time. Choice A simply asks whether the same x-input into each function will give us the same y-output. This is true and we can eyeball that for the moment. Both graphs intersect at x=-4. Choice B also appears true. At x=-2 both graphs appear to intersect again. Here, I could test A & B algebraically to confirm (by plugging in each x value

into $f(x)$ and $g(x)$) but I want to save time so let's look at the rest first. To check choice C, I find the y-value of graph $g(x)$ when x=2. I can do this algebraically by plugging in to be 100% or if I am low on time I can eyeball this. Either way I get -9 (maybe -10 if I eyeball it). I now plug -9 into $f(x)$, again, working algebraically if I have time and see that it equals -8. Great. That works. Not the answer. Choice D is essentially saying that every y-value of the graph of $f(x)$ is negative. Remember, absolute value makes everything positive, so $|f(x)|$ will always be positive and by extension, $-|f(x)|$ will always be negative. Thus if the original function equals $-|f(x)|$ then it simply means every y-value in the function is negative.The graph of $f(x)$ crosses the x-axis, which makes all the difference: its y-values are NOT always negative. If it were to exist only below the x-axis as $g(x)$ does, $-|f(x)|=f(x)$ would be true (thus answer choice E is true; $-|g(x)|=g(x)$ because that graph is always negative for y). But here, when we calculate $-|f(x)|$, everytime the original graph was negative for y, we'll have the same value, but every time the original graph was positive, we will flip its value to become negative. We can also think about this in steps of graph translation. If we know were to graph the absolute value of $f(x)$, we would get a graph that looked like a "v", since all the points must be positive. We could take the graph of $f(x)=x+1$, and whenever it became negative, flip the line across the x-axis so that instead it is positive. If we were to reflect this graph over the x-axis to graph $-|f(x)|$ the graph would still be shaped like a "v", but it would be upside down. This is clearly not the same as the line we see graphed in the question. Thus choice D is correct, because it is NOT true.

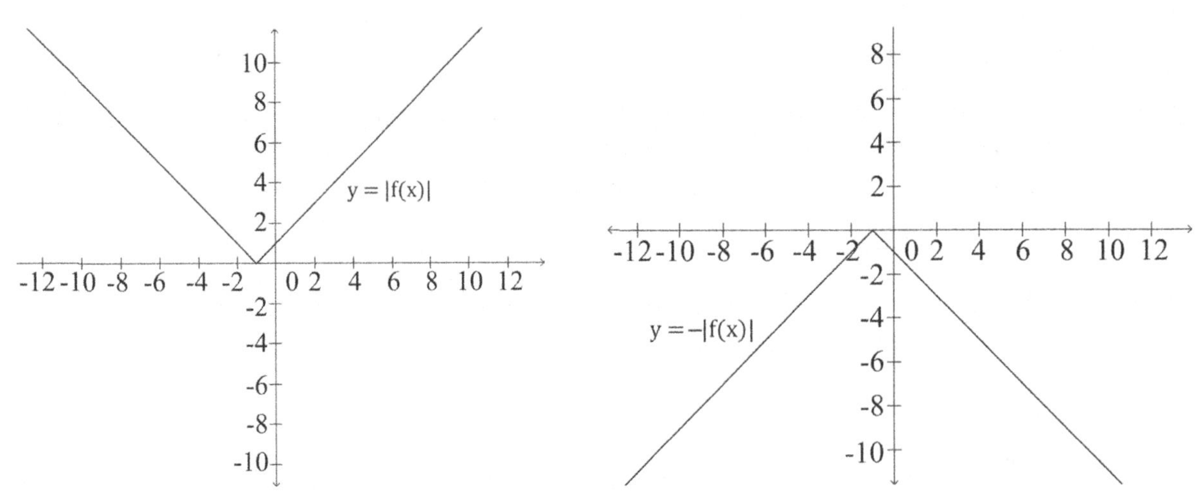

25. **C.** According to function notation, $(ab)(x)$ means $a(x)\times b(x)$. Therefore, all we have to do is multiply the two polynomials together and solve:

$$\left(x^3+2x^2-7\right)\left(x^2-5y\right)=$$
$$x^2x^3+2x^2x^2-7x^2-5yx^3-5y\left(2x^2\right)-5y(-7)=$$
$$x^5+2x^4-7x^2-5yx^3-10yx^2+35y=$$
$$x^5+2x^4-5yx^3-10yx^2-7x^2+35y$$

26. **B.** According to function notation, $(h-g)(x)$ means $h(x)-g(x)$. Therefore, all we have to do is subtract the two polynomials and solve:

$$\left(x^4-2x^2+18\right)-\left(5x^4+2x^3-7x^2-19\right)\rightarrow x^4-2x^2+18-5x^4-2x^3+7x^2+19\rightarrow -4x^2-2x^3+5x^2+37.$$

CHAPTER

17 QUADRATICS AND POLYNOMIALS

SKILLS TO KNOW

- Polynomial Tips & Terminology
- Simplifying Polynomial Expressions
- Quadratic Equations:
    - Standard, Factored Zeros, and Vertex Equation Forms
    - How to find the Vertex of any Quadratic
    - Using the Discriminant to Find the Number of Solutions
- Polynomial Long Division
- How to solve for unknowns (a, k or h) when given a polynomial equation, vertex, and/or point(s)

NOTE: Many problems that involve polynomials and quadratics are also coordinate geometry related. Find more in-depth coverage of related ideas in the **Coordinate Geometry, Graph Behavior,** and **Translations and Reflections** chapters (**13, 22 & 23**) in this book.

Additionally, this chapter builds on the knowledge presented in the chapter **FOIL and Factoring** in Part One of this book. Think of this chapter as that chapter on overdrive!

POLYNOMIAL TIPS & TERMINOLOGY

Often with polynomials, a problem looks harder than it is. Many of these problems throw around fancy sounding vocabulary that really isn't very complex, but still intimidates some students. A few things to remember:

1. **Many polynomial "terms" mean the same thing.** If we know a **"solution"** or a **"root"** of a quadratic or polynomial, it is the same as a **"zero."** A solution to a quadratic or polynomial is essentially an **"x-intercept."** Remember an x-intercept occurs when x is a number and y, or the entire expression (or equation in the form $f(x)=$ or $y=$), is equal to zero. In a coordinate plane, these "solutions" are the points at which the line crosses the x-axis and $y=0$.

2. **If we know a "solution" "x-intercept" or "zero," we instantly know one "factor" of the polynomial.** Because of the Zero Product Property, we know that if a solution is n, then $(x-n)$ must be a factor of that polynomial. So if a solution is -3, then $(x+3)$ must be a factor of the polynomial. If a factor is $(x-9)$, then $x=9$ must be a solution.

3. **The signs of solutions are the OPPOSITE of the signs of numbers in factors.**
For example, if a polynomial $f(x)=(x-7)(x+9)$ has a factor of $(x-7)$ the sign is negative. But the related SOLUTION or root would be $x=7$, which is positive. That's because to solve for the value of x, we need to set the factored piece equal to zero and solve. $x-7=0$ simplifies to $x=7$. Many students get confused on this point and jump the gun, selecting an answer choice with the wrong sign before carefully thinking through what they need.

4. If given a graph of a polynomial, **every time a line crosses the horizontal axis (the x-axis)**, it represents a **"distinct, real solution."** For each distinct, real solution, we will see a factor in the factored form of the polynomial. Remember the word **"distinct"** just means unique.

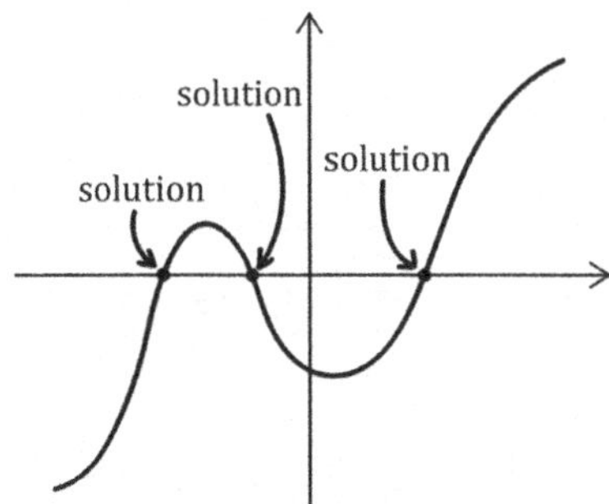

5. If a factor occurs twice in the factorization of a polynomial, i.e. $(x-2)^2$, it is called a **double root.** Factors that occur more than once (and create equal roots or **"multiple roots"**) are not **"distinct"** factors. Multiple roots, in graphs, create a "bounce" or direction change at the zero when they occur in even quantities and an "inflection" when they occur in odd ones.

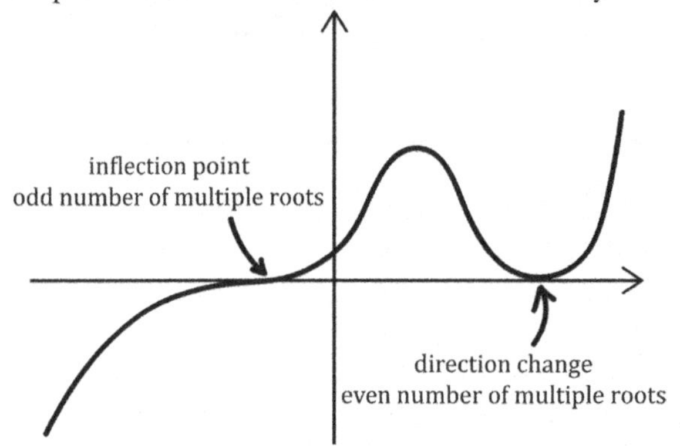

SIMPLIFYING EXPRESSIONS

On the ACT®, you'll be expected to simplify expressions that involve polynomials. We cover the basics of simplifying expressions in Part One (Algebra Core) in the chapters **FOIL and Factoring** and **Basic Algebra.** Because this basic skill is necessary for ANY problems involving polynomials, you may want to review those chapters if you find yourself making careless mistakes in the problem set at the end of this chapter.

We'll discuss one quick concept here, though, specific to polynomials

Remember that you can only combine **like terms. Like terms** have the same polynomial variable(s) and respective degree(s).

For example:
x^3, $2x^3$, and $5x^3$ are like terms. Here, we could add all three together to create a single sum.

$$x^3 + 2x^3 + 5x^3 = 8x^3$$

Similarly $2xy + 6xy$ are like terms:

$$2xy + 6xy = 8xy$$

x^3, x^2 and $2x$ are NOT like terms. If we add them together, they do not create a single term.

$$x^3 + x^2 + 2x$$

The "degree" or exponent on the variable(s) must match in order for elements to be like terms.

Because large polynomial expressions with terms such as x^7 or $8x^3y^7$ would never occur in Algebra 1, we have saved the more complex problems that may involve such large terms for the end of this chapter. Still, the idea of simplifying is pretty much the same as what you did back in middle school (and what we covered in Part One): make sure terms have the same variable(s) and same degree(s) before you combine them.

What polynomial must be added to $4x^3 + x - 12$ so that the sum is $x^3 - 4x - 2$?

A. $-3x^3 - 5x - 10$ **B.** $5x - 3x^3 - 10$ **C.** $3x - 5x^3 + 10$

D. $-3x^3 - 5x + 10$ **E.** $3x^3 - 5x - 10$

This problem requires only basic algebra. We can visualize the problem better by setting up the equation vertically, lining up each term with a "blank." **The vertical columns are all representative of a single like term** (the first column the x^3 terms, the center column the x terms, and the final column the integer or constant terms.)

$$\begin{array}{r} 4x^3 + x - 12 \\ + \_\_\_ + \_ + \_\_ \\ \hline x^3 - 4x - 2 \end{array}$$

We can break this down, now, one piece at a time.

$$4x^3 \text{ plus what equals } x^3?$$

I can see that subtracting $3x^3$ would give me that answer, so I begin by writing in $-3x^3$:

$$\begin{array}{r} 4x^3 + x - 12 \\ -3x^3 + \_ + \_\_ \\ \hline x^3 - 4x - 2 \end{array}$$

Now I move to the 2nd term: x plus what equals $-4x$? $-5x$ is clearly what I would add to get to $-4x$, so I place that in the 2nd blank.

$$\begin{array}{r} 4x^3 + x - 12 \\ -3x^3 - 5x + \_ \\ \hline x^3 - 4x - 2 \end{array}$$

Now for the final term: -12 plus what is -2? I can see that I add 10 to -12 to get -2. So I fill that in:

$$\begin{array}{r} 4x^3+x-12 \\ -3x^3-5x+10 \\ \hline x^3-4x-2 \end{array}$$

Answer: $-3x^3-5x+10$.

We could also solve this by creating a variable for our answer, n, and making an algebraic equation:

$$\left(4x^3+x-12\right)+n=\left(x^3-4x-2\right)$$

After isolating n we get:

$$n=\left(x^3-4x-2\right)-\left(4x^3+x-12\right)$$

Then we distribute our negative sign:

$$n=\left(x^3-4x-2\right)-4x^3-x+12$$

We use the commutative property to place like terms together:

$$n=x^3-4x^3-4x-x-2+12$$

And finally simplify:

$$n=-3x^3-5x+10$$

QUADRATICS

The new SAT® fiercely tests your knowledge of parabolas and quadratic equations. If you are also studying for the SAT®, be sure to spend extra time on this section. These questions appear on the ACT® as well, but less often. Still, you should know all **the basic forms** of the quadratic equation.

VERTEX FORM

The Vertex Form of a parabola is: $f(x)=a(x-h)^2+k$

- The vertex of the parabola in this form is (h,k).
- When a is positive, the parabola opens upwards, and the minimum is (h,k).
- When a is negative, the parabola opens downwards, and the maximum is (h,k).

Which of the following equations of a parabola has a vertex at $(2,-4)$?

A. $f(x)=(x-2)^2+4$ **B.** $f(x)=(x+2)^2-4$ **C.** $f(x)=(x-2)^2-4$

D. $f(x)=(x+2)^2+4$ **E.** $f(x)=(2x)^2-4$

Because all of the answer choices are in vertex form, our job is easy. $(2,-4)$ is equivalent to (h,k). $h=2$ and $k=-4$. We're looking for $(x-h)$ or $(x-2)$ in the parentheses, and $+k$ or -4 to the right of the parentheses. This gives us the function $f(x)=(x-2)^2-4$, which matches answer choice C.

Answer: **C.**

FACTORED FORM

A polynomial is in factored form when it is expressed as a product of two or more monomials, polynomials or constants.

- The factored form of a polynomial usually takes the form: $f(x)=a(x-n)(x-m)$
- When a is positive, the parabola opens upwards.
- When a is negative, the parabola opens downwards.

For example:

$$f(x)=-2(x-4)(x+9)$$

is in factored form.

Each factored element represents a solution or a "zero" of the polynomial, which can be found by setting any factored element equal to zero. In the example above, $x-4=0$ would produce a solution of $x=4$, while $x+9=0$ produces a solution of $x=-9$.

Note: **when quadratics have no real solutions, they cannot be expressed in this form unless complex numbers are used.** For example: $y=(x-(2i))(x+(2i))$.

In the case of quadratics, **the x-value of the vertex is always the average of the two zeros**, so factored form can be used to easily find the x-value of the vertex or axis of symmetry in a quadratic.

STANDARD FORM

The standard form of a parabola is: $f(x)=ax^2+bx+c$

- $-\frac{b}{2a}$ is the x-value of the vertex. The y-value of the vertex can be found by plugging in this value for x and solving for y (or $f(x)$).
- The vertex is always either the maximum or the minimum of the graph.*
- When a is positive, the parabola opens upwards.
- When a is negative, the parabola opens downwards.
- The sum of the two roots is $-\frac{b}{a}$ (not necessary to know)
- The product of the two roots is $\frac{c}{a}$ (not necessary to know)

*This is a great tip for those word problems when you throw a ball in the air and want to know when it reaches its highest point. See **Chapter 10: Function as a Model** for more of these.

To solve a quadratic equation using the **quadratic formula**, you must first have an equation in standard form. For this reason, recognizing this form is essential on the ACT®!

THE QUADRATIC FORMULA

Earlier in this book, we covered how to solve a quadratic by factoring in the chapter on **FOIL and Factoring**. Another way to solve quadratic equations is the quadratic formula. Most students learn the basics of this formula early on in algebra, but the ACT® tends to invoke such knowledge in more challenging or creative ways, so we've saved covering it until now. Additionally, factoring tends to work on easier problems, but it doesn't always work on hard ones.

THE GENERAL QUADRATIC EQUATION

For a quadratic equation of the form:

$$x = \frac{-b \pm \sqrt{b^2 - 4ac}}{2a}$$

Where x represents an unknown, and a, b, and c are constants with a not equal to 0.

You should have this memorized. However, if you have a TI-83 or TI-84, I also recommend programming this into your calculator. You can find more information on how to do so on our website: SupertutorTV.com/BookOwners

What are the two roots of the equation $-2x^2 + 5x - 1 = 0$?

A. 2 and −1 **B.** −2 and 1 **C.** $\frac{-5 \pm \sqrt{33}}{-4}$ **D.** $\frac{-5 \pm \sqrt{17}}{-4}$ **E.** $\frac{-5 \pm \sqrt{17}}{4}$

Before using the quadratic equation, let's factor out a -1 to make our lives a little easier. All I need to do is multiply each term by negative 1:

$$2x^2 - 5x + 1 = 0$$

From this equation we have $a = 2$, $b = -5$, and $c = 1$. Plugging these values into the quadratic equation, we get:

$$x = \frac{-(-5) \pm \sqrt{(-5)^2 - 4(2)(1)}}{2(2)}$$

$$x = \frac{5 \pm \sqrt{25 - 8}}{4}$$

$$x = \frac{5 \pm \sqrt{17}}{4}$$

Since we factored out a -1 earlier, our answer is in a slightly different form from available answers, so we can multiply the numerator and denominator by -1:

$$x = \frac{5 \pm \sqrt{17}}{4}\left(\frac{-1}{-1}\right) = \frac{-5 \pm \sqrt{17}}{-4}$$

Answer: **D.**

For what integer k are both solutions of the equation $x^2 + kx + 13$ negative integers?

A. 14 **B.** −14 **C.** −12 **D.** 12 **E.** 1

First, we could approach this by thinking about factors. To have two real solutions, there must be some product $(x+n)(x+m)$ that creates this function, and we know it must be factorable if the answer choices are integers. Whenever you see the word integer, think FACTOR and GUESS AND CHECK. These two ideas often come into play when a problem specifies something is an integer. We know both n and m are positive because we want our solutions to be negative (remember the five tips at the beginning of the chapter: signs in factors are not the same as signs of the solutions). Off hand, $x+1$ and $x+13$, two values whose product is 13, but which produce negative integer solutions, appear to work. The middle term of the product of $(x+1)$ and $(x+13)$ would be $x+13x$ or $14x$. That would make $k=14$.

On looking through the answer choices, we see that this must be correct. Though we can likely get this correct by guessing without the quadratic equation, we can prove it is true with the quadratic equation. Also, that explanation is complex, while the quadratic formula is straightforward:

$$= \frac{-14 \pm \sqrt{14^2 - 4(1)13}}{2(1)}$$

$$= \frac{-14 \pm \sqrt{196-52}}{2}$$

$$= \frac{-14 \pm \sqrt{144}}{2}$$

Because the square root of 144 is 12, this simplifies to:

$$= \frac{-14 \pm 12}{2}$$

$$= -\frac{2}{2} \text{ and } -\frac{26}{2}$$

$$= -1 \text{ and } -13$$

We could also use the quadratic formula to *disprove* answer choices on this question. From factoring I'm pretty sure $k=14$, but I might not know why k can't be 12 or -12. With the quadratic formula, I have a tool to check:

$$= \frac{-12 \pm \sqrt{12^2 - 4(1)13}}{2(1)}$$

$$= \frac{-14 \pm \sqrt{144 - 52}}{2}$$

$$= \frac{-14 \pm \sqrt{92}}{2}$$

I know root 92 is not an integer, so this cannot create an integer answer. Alternatively, with this programmed in my calculator, I can whip through the answer options even more quickly to see if they work.

Did I *need* the quadratic equation here? Maybe not. Do I always have time to double-check my answers? No. But it's a wonderful backup plan when you can't see the factors to an equation or a problem involves a quadratic and you want to be 100% sure. One of the reasons we didn't teach this earlier is that it isn't necessary on most of the problems you could use it on. Still, it's a great technique to understand if you're looking for an elusive near perfect or perfect score on the math. It could be the back-up plan that helps you unpack some of the most challenging problems on the test.

THE DISCRIMINANT

One of the most powerful parts of the quadratic formula is the discriminant, or the portion of the quadratic formula that falls under the radical sign: $b^2 - 4ac$.

We can use the discriminant to determine how many (if any) solutions a quadratic equation has.

FINDING SOLUTIONS USING THE DISCRIMINANT

Given that $f(x) = ax^2 + bx + c$ the discriminant is defined as $b^2 - 4ac$ (the argument of the root in the quadratic equation) :

1. When this value is positive, there are two real roots
2. When this value is 0, there is one real root
3. When this value is negative, there are no real roots (but there are two imaginary roots)

Remember, **Roots**, **Zeros**, and **Solutions** refer to the places where the graph **intercepts the x-axis**, or $y = 0$.

Let's unpack why this works. For rule #1, remember that the quadratic formula always has plus or minus before the radical. The plus and minus are the source of the two different solutions, right? So it makes sense that when what follows exists (i.e. there is a real, positive number under the radical), then two real solutions are created.

For rule #2, we can calculate the square root of 0, but "plus or minus 0" doesn't mean much. Whether we add zero or subtract it, we get the same number. Clearly, having a determinant equal to zero creates a single solution, or a double root.

For rule #3, if the number under the radical is negative, we cannot calculate a real solution because the square root of a negative number is by nature an imaginary number. As a result, negative discriminants indicate no real solutions.

Though in many cases we could figure out the number of solutions using the entire quadratic formula or in some cases by factoring, and essentially solving down to the solutions themselves, checking the discriminate can save time and also offers a strategy for more complex problems.

For what non-zero whole number k does the quadratic $x^2+kx+2k=0$ have exactly 1 real solution for x?

For this problem we can use the discriminant to find the solution. We know that **one real solution** occurs when the **discriminant is equal to zero**, so we set $b^2-4ac=0$ and solve.

$$b^2-4ac=0$$

$$k^2-4(2k)=0$$

$$k^2-8k=0$$

$$k(k-8)=0$$

Now we know $k=0$ or $k-8=0,$ but the problem asks for a **non-zero whole number**, so we can disregard $k=0$.

$$k=8$$

Answer: 8.

LONG DIVISION

Some of you might solve polynomial division problems using Synthetic Division. If so, great! That works for typical polynomial division problems! In this book, we focus on traditional polynomial long division because I find it more intuitive, easier to remember, and more useful for complex problems involving unknown constants.

If $x=3$ is a zero of the expression $3x^3-11x^2-2x+24$, which of the following must also be a factor?

When we're given a "zero" we know we have a factor at $(x-n)$, where n is that zero. In this case, we know that $(x-3)$ must be a factor of this expression. Unlike quadratics, third degree polynomial factors can't usually be deduced efficiently by sight and creative guess and check. Instead, use long division.

We now can set up our long division problem:

$$x-3\overline{)3x^3-11x^2-2x+24}$$

Take the division one step at a time, one term at a time. For the first step we only need to pay attention to the terms of highest degree in both the divisor and dividend, x and $3x^3$, respectively.

How many times does x go into $3x^3$? $3x^2$ times, since $x \times 3x^2 = 3x^3$. We write this term on the top line as the first part of our final answer:

$$\begin{array}{r|l} & 3x^2 \\ \hline x-3 & 3x^3 - 11x^2 - 2x + 24 \end{array}$$

Now, just like in ordinary long division, we multiply this first term by the divisor, write the product as shown below, and subtract from the dividend to get our initial remainder. We subtract the $-9x^2$ from the $-11x^2$ to get $-2x^2$.

$$\begin{array}{r|l} & 3x^2 \\ \hline x-3 & 3x^3 - 11x^2 - 2x + 24 \\ & \underline{-(3x^3 - 9x^2)} \\ & \quad\quad -2x^2 \end{array}$$

Now we bring down the next term of lower degree ($-2x$). We ask a similar question: how many times does x go into $-2x^2$? $-2x$ times. We place the $-2x$ above the division bar, multiply $-2x$ times $(x-3)$, and write that product underneath in our work below. We then subtract to simplify (don't forget to change the sign on the 2nd item $+6x$!).

$$\begin{array}{r|l} & 3x^2 - 2x \\ \hline x-3 & 3x^3 - 11x^2 - 2x + 24 \\ & \underline{-(3x^3 - 9x^2)} \\ & \quad\quad -2x^2 - 2x \\ & \quad\quad \underline{-(-2x^2 + 6x)} \\ & \quad\quad\quad\quad -8x \end{array}$$

Now we bring down the 24 to meet the $-8x$ and continue. x goes into $-8x$ -8 times, so we write -8 above the division line in the top row. Then we multiply -8 times $(x-3)$ to get $-8x+24$. Now we write $-8x+24$ underneath the rest of our work and once again subtract to at last find our remainder. Because $24-24$ is 0, our remainder is 0.

$$\begin{array}{r|l} & 3x^2 - 2x - 8 \\ \hline x-3 & 3x^3 - 11x^2 - 2x + 24 \\ & \underline{-(3x^3 - 9x^2)} \\ & \quad\quad -2x^2 - 2x \\ & \quad\quad \underline{-(-2x^2 + 6x)} \\ & \quad\quad\quad\quad -8x + 24 \\ & \quad\quad\quad\quad \underline{-(-8x + 24)} \\ & \quad\quad\quad\quad\quad\quad 0 \end{array}$$

If you are left with a constant, when doing long division, place the constant remainder over the divisor, add in this fraction to the rest of your work and you are done! Note in this problem a remainder would indicate an error as you are told this is a factor. Factors should never leave a remainder in long division.

Answer: $3x^2 - 2x - 8$.

One of the roots of $2x^4 + 8x^3 - 2x^2 - 8x = 0$ is -4. What are the other roots?

We are told that -4 is a root, therefore the expression must be divisible by $(x+4)$. If we divide the expression by $(x+4)$ using polynomial long division:

$$x+4 \overline{)2x^4 + 8x^3 - 2x^2 - 8x}$$

$$\begin{array}{r} 2x^3 \\ x+4 \overline{)2x^4 + 8x^3 - 2x^2 - 8x} \\ \underline{-2x^4 + 8x^3} \\ 0 \end{array}$$

$$\begin{array}{r} 2x^3 + 0x^2 - 2x \\ x+4 \overline{)2x^4 + 8x^3 - 2x^2 - 8x} \\ -\underline{2x^4 + 8x^3} \\ 0 - 2x^2 - 8x \\ \underline{-0 - 2x^2 - 8x} \\ 0 \end{array}$$

We get $\frac{2x^4 + 8x^3 - 2x^2 - 8x}{x+4} = 2x^3 - 2x = 0$. Now we just have to find the zeros of $2x^3 - 2x = 0$.

We can factor out a two.

$$2\left(x^3 - x\right) = 0$$

Then divide both sides by two.

$$x^3 - x = 0$$

We can now factor out an x:

$$x\left(x^2 - 1\right)$$

And then apply the pattern from the Difference of Squares (See **FOIL and Factoring, Chapter 3**).

$$= x(x+1)(x-1) = 0$$

Using the Zero Product Property, we have three equations we can form:

$$x = 0 \text{ or } x + 1 = 0 \text{ or } x - 1 = 0$$

Solving out these equations we find the solution set for $x = \{0, -1, 1\}$.

SOLVE FOR A, K, OR H (PLUGGING IN)

One of the most basic algebraic skills you'll need on the ACT® is plugging in, which is covered in our Algebra Core (Part One) chapter on Basic Algebra.

But sometimes the same principle plays out in more complex quadratic and polynomial problems. Remember the steps you take when finding a linear equation using slope intercept form, given the slope and a point. To solve those problems, you plug in the (x,y) point you have and the slope and solve for "b" to find the slope-intercept form of the equation. Here we will do the same thing: **plug in what we know and solve for what we don't know**. Often what we don't know will be some random letter (such as a, h, or k), and what we know is an (x,y) pair or even a factor $(x-n)$, which in turn implies a solution $(n,0)$.

This principle can sometimes save a huge amount of time over long division. I.e., sometimes a problem could be solved either way, but this way is often faster.

The solution set for x of the equation $x^2 - \frac{k}{2}x + 9 = 0$ is $\{3\}$. What does k equal?

A. 9 **B.** 12 **C.** –12 **D.** 3 **E.** -3

This question is simply asking you to solve for a variable, k. We know x. Don't be thrown by fancy wording: "solution set" just means that is what x equals.

Here, we'll plug in what we know (the given value of x) to most quickly find the answer:

$$(3)^2 - \frac{k}{2}(3) + 9 = 0$$

Now, simplify:

$$9 - \frac{3}{2}k + 9 = 0$$

$$18 - \frac{3}{2}k = 0$$

$$18 = \frac{3}{2}k$$

$$k = 18\left(\frac{2}{3}\right)$$

$$= 12$$

We alternatively could solve this problem with long division, knowing that $(x-3)$ is a factor of the polynomial given the solution $x=3$, but doing so would be time consuming. Likewise, we could set up $(x-3)(x-n)$ and solve for n, setting that FOIL product equal to what we know, and matching up pieces to form equations as necessary. Again, time consuming, not to mention confusing. Our best bet here is plugging in.

Answer: **B.**

What is the value of k if $(x-5)$ is a factor of $x^3-3x^2-3kx-5$?

To solve this problem we could make use of polynomial long division to solve for k. As we are given a factor, this is probably the first method many students would turn to. However, there's a faster way: plugging in.

Because we know $x-5$ is a factor, we also know $x=5$ is a zero, or a value of x that makes the expression equal zero (remember the tips at the beginning of the chapter! A solution gives you a factor, and a factor gives you a solution!). Knowing this, I can set this expression equal to zero, plug in 5 (the solution we know given the factor we know), and solve for k.

$$x^3-3x^2-3kx-5=0$$

$$5^3-3(5)^2-3k(5)-5=0$$

$$125-3(25)-15k-5=0$$

$$125-75-5=15k$$

$$45=15k$$

$$3=k$$

Now back to polynomial division. What if you wanted to use that method?

We could divide the long polynomial by the given factor to get:

$$\frac{x^3-3x^2-3kx-5}{x-5}=x^2+2x+\frac{(-3k+10)x-5}{x-5}$$

In order for $(x-5)$ to be a perfect factor, it needs to divide evenly into the remainder of the expression, so the numerator of the fraction $\frac{(-3k+10)x-5}{x-5}$ must be divisible evenly by the denominator $(x-5)$. The only numerator that we can easily attain that is divisible by $(x-5)$ is $x-5$. In other words, the x must have a coefficient of 1. Try any other value of $-3k+10$ and you'll see it makes nothing cancel. We set the expression $-3k+10$ equal to 1 and solve for k to get $k=3$. As you can see, the first method may not be as obvious, but it is faster, not to mention less confusing!

Answer: $k=3$.

DEGREES OF POLYNOMIALS

The degree of the polynomial refers to the highest power that exists within the equation.

- Each factor, each zero, and each time a line touches the x-axis of a graph indicates a possible solution. For each possible solution, we can count at least one "degree" in our polynomial, because the maximum number of solutions for any polynomial is the same as its highest degree. For example, a 2nd degree (quadratic) polynomial, such as x^2+2x-7, can have at most 2 distinct, real solutions. A 5th degree polynomial, such as $x^5-5x^4+3x^2-625$, can have at most 5 real solutions.

- We can always have a higher degree linear polynomial than the number of solutions because of imaginary solutions and multiple roots.

NOTE: For more information on this type of polynomial problem see **Chapter 22: Graph Behavior** in this book, as we cover these problems more extensively there.

What is the minimum degree possible for a polynomial function with factors $(x-3)$, $(x-5)$ and a double root at $x=-1$?

A. 1 **B.** 2 **C.** 3 **D.** 4 **E.** 5

Because each factor indicates at least one degree, our two factors imply two degrees. Our "double root" implies that we have $(x+1)^2$ as part of the factorization of this function. Thus for those two factors we add another two degrees for a total of 4. At minimum, this polynomial would be a 4th degree polynomial.

Answer: **D**.

A polynomial $f(n)$ has p non-zero terms. If $f(n)$ is the product of function $g(n)$, which has r non-zero terms, and function $h(n)$, which has q non-zero terms, $p>r$ and $p>q$, which of the following could be equal to q?

A. 1 **B.** $p+r$ **C.** $\frac{p}{r}$ **D.** $\frac{r}{p}$ **E.** $p+1$

HARD! Skip this problem if you're not aiming for a 34+. If you are an ACT course subscriber, I recommend following along with the video.

First, to understand this problem, you need to make it real. What are terms again? Oh yeah, those are the pieces of a polynomial expression or equation. For example, $3x^4+2x^2+7$ is a polynomial with three terms. The first term is $3x^4$ the second term is $2x^2$ and the third is 7.
Terms aside, we are dealing with three polynomials. We know $f(n)=g(n)$ times $h(n)$. We thus would multiply every term in $g(n)$ times every term in $h(n)$ to get every term in $f(n)$.
For example, if we let $g(n)$ equal $3x^4+2x^2+7$, an expression with three terms, and $h(n)$ equal $-8x$, an expression with one term, then distribute the $-8x$ across each term of $g(n)$ we get $f(n)$.

$$f(n)=-8x(3x^4+2x^2+7)=-8x(3x^4)-8x(2x^2)+(-8x)7$$

Here, $r=3$ and $q=1$. When we distribute, because we have three terms in our final function, we would figure out that $p=3$. From here we can see if the inequalities are true or not. (Hint: they're not, because $p=r$, so this scenario can't be true given that $p>r$) Cool?

Let's analyze each answer. I'm going to use more hypothetical language, but feel free to make up a bunch of "real" terms like I did above for each scenario to work out what happens.

Answer A: We know that neither r nor q can equal 1. If $r=1$ than q would equal p, because there would be the same number of terms in $h(n)$ as in $f(n)$. Think about it—let k equal the one term that r represents:

$$k(x+1) = kx + k$$

As in the first example, where $r=3$ and $q=1$, we see how a single term ($-8x$ in the previous example or in this example k) does not create any more terms when you multiply by it, i.e. when this single term expression is multiplied by an expression with three terms, the product of the single term and the three term expression also has three terms. It's just like multiplying by a scalar. Thus if $g(n)$ has a single term, $f(n)$ will have the same number of terms as $h(n)$. Likewise, if $h(n)$ has a single term, $f(n)$ will have the same number of terms as $g(n)$. Given that $p>r$ and $p>q$, $f(n)$ must have more terms than either of the other functions, so q cannot equal 1.

Answer B: If $q=p+r$ then we should be able to substitute this value into our inequality given by the problem and that inequality must be true. Let's plug this value for q into the inequality $p>q$ and see what happens: $p>p+r$. Now I can use algebra and subtract from both sides to get: $0>r$. However, the problem defines r as the number of terms, and we can't have a negative number of terms. Thus this answer choice doesn't make sense.

By extension, **Answer E** also doesn't work. We can't add one to p and expect it to be a value less than p itself. You could also run the algebra above again with the number one instead of the variable r and you'll see once more we get an impossible inequality.

Answer C: We'll make up a couple of polynomials where $r=$ two terms and $q=$ three terms. If $p/r=q$ then $qr=p$. Thus, we want the product of the two functions to have six terms.

$$(x^2+5)(x^2+x+1)$$

$$x^4+x^3+x^2+5x^2+5x+5$$

Though this looks like six, the x^2 terms will combine and we will get 5 terms in total, so p would be 5, which divided by 2 is not 3. However, we can see that if the x^2 terms didn't combine, we could have had 6 terms, and that would make $p=6$, which divided by 2 is 3. Let's make up another example and see if we can get zero overlapping terms so nothing "disappears" as we simplify.

$$(x^{72}+5x^{13})(x^2+x+1)$$

$$x^{144}+x^{73}+x^{72}+5x^{26}+5x^{14}+5x^{13}$$

Six equals 2×3. Thus we found an example that makes C true: $\frac{6}{2}=3$. Because the problem asks for something that "could" be true, we only need one example to confirm the answer.

Answer D doesn't make sense. Because these values are all positive integers, if I take the larger one and put it on the bottom and a smaller one and put it on the top, we get a fraction less than one. That can't be the number of terms.

Answer: **C.**

This question is a hybrid of a polynomials problem and a "Properties of Numbers" problem. If you struggled with this, check out **Chapter 2: "Properties of Numbers"** in Book 2 (Numbers, Trig, Stats and Geometry).

Answer: **C.**

1. The sum of $\left(-3x^2+4x-8\right)$ and which of the following polynomials is $\left(2x^2-7x+10\right)$?

   A. $5x^2-11x+18$
   B. $-5x^2+11x-18$
   C. $5x^2+18$
   D. $5x^2-18$
   E. $5x^2-11x$

2. What is the minimum degree possible for the polynomial function whose graph is shown in the standard (x,y) plane below?

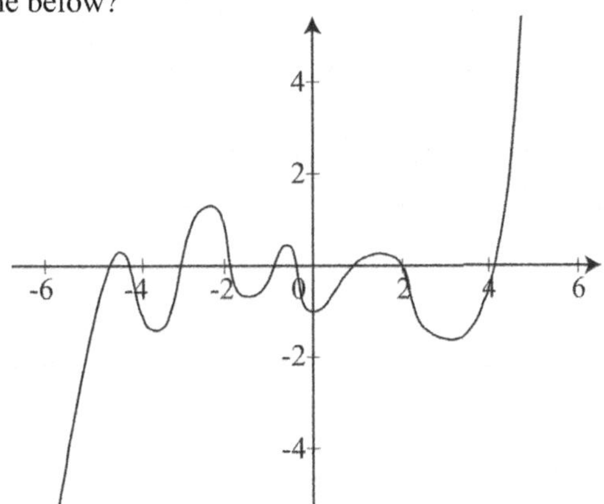

   A. 5
   B. 6
   C. 7
   D. 8
   E. 9

3. What is the solution set of the equation $-2x^2+7=0$?

   A. $\left\{-\sqrt{\frac{7}{2}},\sqrt{\frac{7}{2}}\right\}$
   B. $\left\{-\sqrt{3},\sqrt{3}\right\}$
   C. $\left\{-\frac{4}{2},\frac{4}{2}\right\}$
   D. $\{-3,3\}$
   E. $\left\{-\sqrt{5},\sqrt{5}\right\}$

4. The graph of $y=-3x^2+5$ passes through the point $(3,4a)$ in the standard (x,y) coordinate plane. What is the value of a?

   A. 32
   B. −22
   C. 8
   D. −5.5
   E. −8

5. For what nonzero whole number k does the quadratic equation $x^2+4kx+k^3$ have exactly 1 real solution for x?

   A. 2
   B. 4
   C. 8
   D. 16
   E. 1

6. Which of the following is the set of real solutions for the equation $5x+12=2(4x+6)$?

   A. The empty set
   B. The set of all real numbers
   C. $\{0,5\}$
   D. $\left\{\frac{5}{8}\right\}$
   E. $\{0\}$

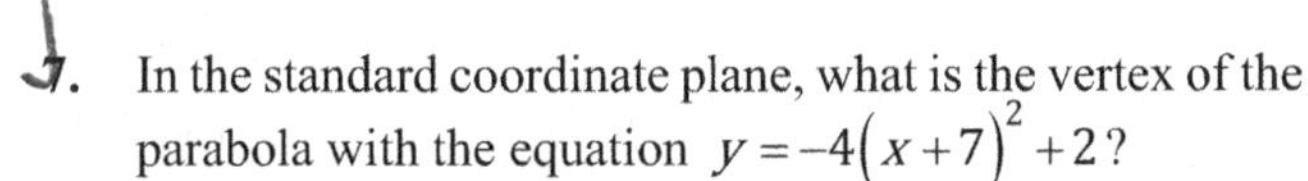

7. In the standard coordinate plane, what is the vertex of the parabola with the equation $y=-4(x+7)^2+2$?

   A. $(-7,-2)$
   B. $(7,2)$
   C. $(7,-2)$
   D. $(-7,2)$
   E. $(-14,2)$

8. The graph of the parabola with the equation $y=-x^2-4x+7$ is shown in the standard (x,y) coordinate plane below. Which of the following graphs is the graph of the given equation rotated 90° counterclockwise about the origin?

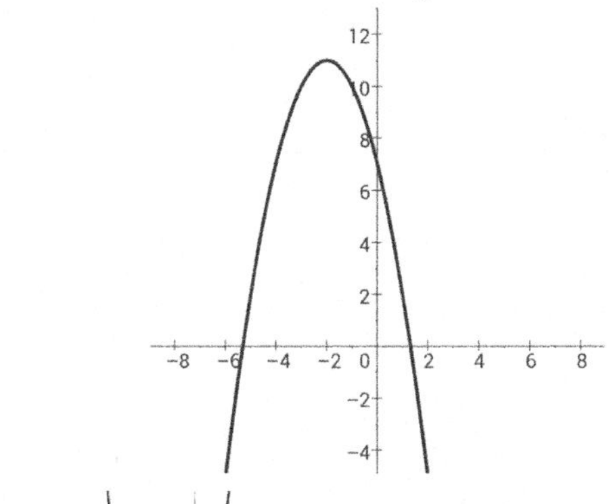

A.

B. 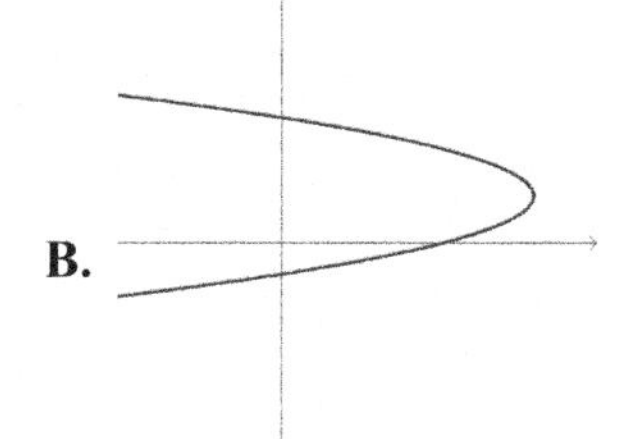

C.

D. 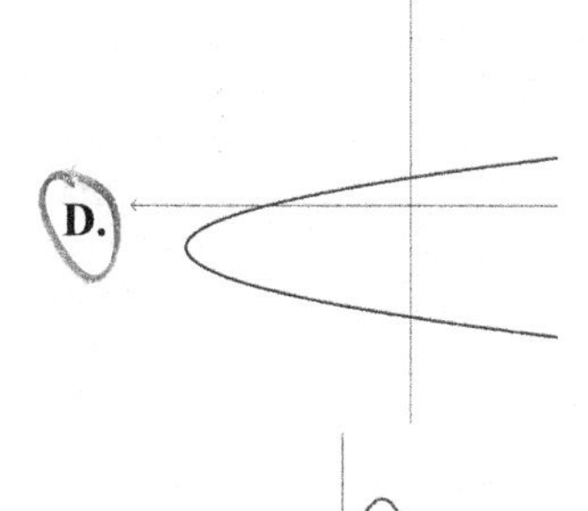

E.

9. The graph of which of the following equations is the parabola shown in the standard (x,y) coordinate plane below?

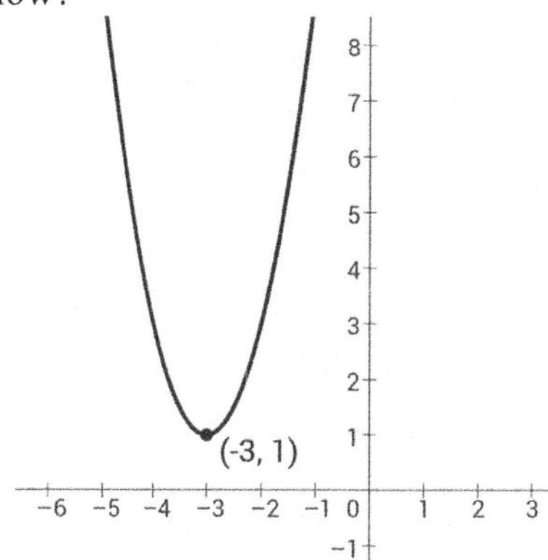

A. $y-1=(x+3)^2$

B. $y-1=2(x+3)^2$

C. $y+1=2(x-3)^2$

D. $y-1=\frac{1}{2}(x+3)^2$

E. $y+1=\frac{1}{2}(x-3)^2$

10. Using the quadratic formula, what are the two roots for the equation $7x^2-3x=17$?

A. $\frac{3\pm\sqrt{485}}{14}$

B. $\frac{3\pm\sqrt{-467}}{14}$

C. $\frac{5}{7}$ and 3

D. $-\frac{5}{7}$ and −3

E. 7

11. For what integer k are both solutions of the equation $x^2+kx+19=0$ negative integers?

A. 20
B. 19
C. 1
D. −19
E. −20

12. The solution set for x of the equation $x^2+mx-4=0$ is $\{-4,1\}$. What does m equal?

A. 1
B. 4
C. −4
D. 3
E. −3

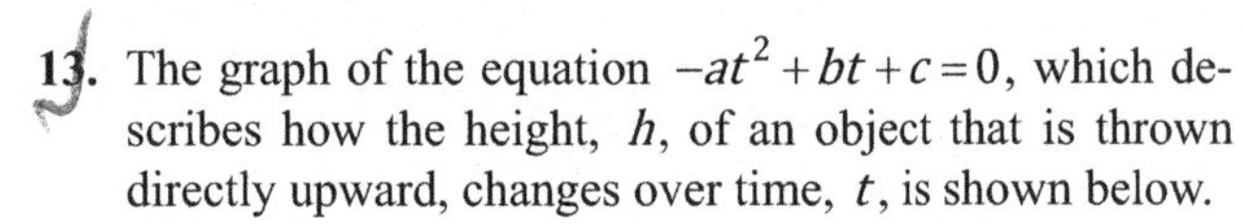

13. The graph of the equation $-at^2+bt+c=0$, which describes how the height, h, of an object that is thrown directly upward, changes over time, t, is shown below.

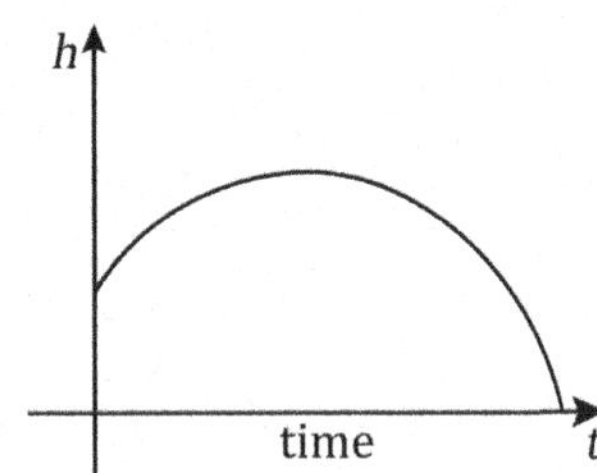

If you alter only this equations a term, the leading coefficient, the alteration has an effect on which of the following? (Also see, Graph Behavior Ch 22)

I. The t-intercept
II. The h-intercept
III. The maximum value of h

A. I only
B. II only
C. III only
D. I and II only
E. I and III only

14. Which of the following equations shows a correct use of the quadratic formula to solve $2x^2+8x-1=0$?

A. $x=\dfrac{8\pm\sqrt{64-4(2)(-1)}}{2(2)}$

B. $x=\dfrac{-8\pm\sqrt{64-4(2)(1)}}{2(2)}$

C. $x=\dfrac{-8\pm\sqrt{64-4(2)(-1)}}{2(2)}$

D. $x=\dfrac{8\pm\sqrt{64-4(2)(-1)}}{2}$

E. $x=\dfrac{-8\pm\sqrt{64-4(2)(1)}}{2}$

15. In the equation $x^2+2mx-\left(\frac{1}{2}\right)n=0$, m and n are integers. The *only* possible value for x is −2. What is the value of n?

A. −2
B. −8
C. 2
D. 8
E. 4

16. $7w^3+65w-w^3-20-35w+2$ is equivalent to:

A. $8w^3+30w-22$
B. $6w^3+30w-22$
C. $6w^3+30w-18$
D. $18w^3$
E. $18w$

17. What polynomial must be added to $2x^3-4x-2$ so that the sum is $-x^3-4$?

A. $4x-x^3+2$
B. $-3x^3+4x-2$
C. $-x^3-4x-2$
D. $-3x^3-4x-2$
E. $-3x^3-4x+2$

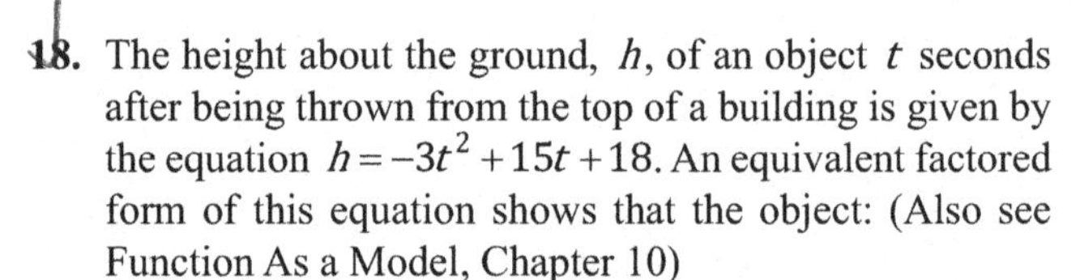

18. The height about the ground, h, of an object t seconds after being thrown from the top of a building is given by the equation $h=-3t^2+15t+18$. An equivalent factored form of this equation shows that the object: (Also see Function As a Model, Chapter 10)

A. Starts at a point 6 units off the ground
B. Reaches the ground in 6 seconds
C. Reaches the ground in 1 second
D. Reaches a maximum in 18 seconds
E. Reaches a maximum in 1 second

19. Which of the following expressions is equivalent to $(3x^3+5)-(2x^2-6x+7)+(7x-5)-(5x^2+3x+3x+2x)$?

A. $3x^3-7x^2+5x-7$
B. $3x^3-10x^2+9x+7$
C. $-7x^2+5x-7$
D. $3x^3-7$
E. $3x^3+5x-7$

20. $(x+4y-2z)-(-3x+2y+5z)$ is equivalent to:

A. $-2x+8y+3z$
B. $-3x+2y-7z$
C. $4x+2y-7z$
D. $-3x+8y+3z$
E. $4x+2y+3z$

21. If $f(x)=4x^3-64x$, which of the following correctly describes the zeroes of the polynomial? (Zeroes are the values where $f(x)=0$.)

A. 2 different rational zeroes
B. No real zeroes
C. Only 1 rational zero
D. 3 different rational zeroes
E. 1 number is a double zero

22. One of the roots of $4x^3-18x^2+32x-24=0$ is 2. What are the other roots?

A. $\frac{5}{2}\pm i\sqrt{23}$
B. $\frac{5}{4}\pm\sqrt{23}$
C. $\frac{5}{4}\pm\frac{i\sqrt{23}}{4}$
D. $\frac{5}{2}\pm\frac{i\sqrt{23}}{2}$
E. $5\pm i\sqrt{23}$

23. What is the value of c if $x+1$ is a factor of $x^3+2x^2-cx-20$?

A. 19
B. 18
C. 17
D. 16
E. 15

24. What is the equivalent of $(n+4)^3$?

A. n^3+64
B. $n^3+6n^2+24n+32$
C. $n^3+12n^2+48n+64$
D. $n^3+12n^2+48n+32$
E. $n^3+24n^2+48n+64$

25. Consider the equation $y=-(x+2)^2-4$, where x and y are both real numbers. The table below gives the values of for selected values of x.

x	y
−11	−85
−9	−53
−7	−29
−5	−13
−3	−5
−1	−5
1	−13

For the equation above, which of the following values of x gives the greatest value of y?

A. −8
B. −6
C. −4
D. −2
E. 0

26. Which of the following values is a zero of $f(x)=3x^4+8x^3+4x^2$?

A. $\frac{2}{3}$
B. $\frac{3}{2}$
C. −2
D. 3
E. −3

27. Which of the following expressions is equivalent to $4x^4+16x^2+12$?

A. $(x^2+4)(x^2-3)$
B. $4(x+1)(x+3)$
C. $4(x+4)(x-3)$
D. $4(x^2+3)(x^2+4)$
E. $4(x^2+1)(x^2+3)$

28. Which of the following is NOT a factor of $a^7 - 81a^3$?

A. a
B. a^2
C. $a+3$
D. $a-3$
E. a^2+3

29. The function $f(x)$ is a cubic polynomial that has the value of 0 when x is $0, -3$, and 4. If $f(1) = -6$, which of the following is an expression for $f(x)$?

A. $x(x-3)(x+4)$
B. $x(x+3)(x-4)$
C. $2x(x+3)(x-4)$
D. $\frac{x}{2}(x+3)(x-4)$
E. $x^2(x-3)(x+4)$

30. $f(x)$ is a quartic (fourth order) polynomial that has zeroes at $x = 2, 6, -4, -9$. If $f(3) = 63$, which of the following is an expression for $f(x)$?

A. $(x-2)(x-6)(x+4)(x+9)$
B. $\frac{1}{4}(x-2)(x-6)(x+4)(x+9)$
C. $-\frac{1}{4}(x-2)(x-6)(x+4)(x+9)$
D. $\frac{1}{4}(x+2)(x-6)(x+4)(x+9)$
E. $-\frac{1}{4}(x+2)(x+6)(x-4)(x-9)$

31. Which of the following equations, when graphed in the standard (x, y) coordinate plane, would cross the x-axis at $x = -13$ and at $x = 7$?

A. $y = -5(x+13)(x-7)$
B. $y = -6(x+13)(x+7)$
C. $y = 2(x-13)(x-7)$
D. $y = 3(x-13)(x+7)$
E. $y = -3(x-13)(x-7)$

32. A parabola with vertex $(5, -9)$ and axis of symmetry at $x = 5$ crosses the x-axis at $(5 - \sqrt{35}, 0)$. At what other point, if any, does the parabola cross the x-axis?

A. $(5 + \sqrt{35}, 0)$
B. $(-5 - \sqrt{35}, 0)$
C. $(-5 + \sqrt{35}, 0)$
D. No other point
E. Cannot be determined from given information

33. A polynomial function $F(x)$ has degree a. If the graph of $F(x)$ touches or crosses the x-axis exactly 5 times, which of the following cannot be the value of a?

A. 8
B. 7
C. 6
D. 5
E. 4

ANSWER KEY

1. A	2. E	3. A	4. D	5. B	6. E	7. D	8. D	9. B	10. A	11. A	12. D	13. E	14. C
15. B	16. C	17. B	18. B	19. A	20. C	21. D	22. C	23. A	24. C	25. D	26. C	27. E	28. E
29. D	30. C	31. A	32. A	33. E									

ANSWER EXPLANATIONS

1. **A.** We wish to solve the equation $-3x^2+4x-8+Y=2x^2-7x+10$ for the polynomial Y. So, subtracting $-3x^2+4x-8$ on both sides, we get $Y=2x^2-7x+10-\left(-3x^2+4x-8\right)$. Distributing the negative sign, we get $Y=2x^2-7x+10+3x^2-4x+8$. Now, combining like terms, we get $Y=5x^2-11x+18$.

2. **E**. For a polynomial with n turning points (whenever the slope of the graph changes signs, the minimum degree of the polynomial is $n+1$. The graph has 8 turning points, so the minimum degree of the polynomial is $8+1=9$. Alternatively, the graph has 9 "zeros" or x-intercepts. Each x-intercept must represent at LEAST one root, so we know the function must have at least 9 degrees, one for each root or solution. Because 9 is the largest number available, we can reliably choose it based on this evidence, as well.

3. **A.** Using the quadratic formula with $a=-2,$, $b=0$, and $c=7,$, we have:
$$x=\frac{0\pm\sqrt{-4(-2)7}}{2(-2)}=\pm\frac{\sqrt{56}}{-4}=\pm\frac{2\sqrt{14}}{4}=\pm\frac{\sqrt{14}}{2}=\pm\frac{\sqrt{14}}{\sqrt{4}}=\pm\sqrt{\frac{14}{4}}=\pm\sqrt{\frac{7}{2}}.$$

4. **D.** Plugging in $x=3$, we get $y=-3(3)^2+5=-3(9)+5=-27+5=-22$. So, we can equate $4a=-22\rightarrow a=-\frac{22}{4}=-5.5$

5. **B.** If the polynomial only has one solution, it means that it is a perfect square that can be factored into $(x+a)(x+a)$. So, we set $x^2+4kx+k^3=(x+a)(x+a)=x^2+2ax+a^2$. This means that $2ax=4kx$ and $a^2=k^3$. Simplifying the first equation, we get $a=2k$. Plugging in this value for a in $a^2=k^3$, we get $(2k)^2=k^3\rightarrow 4k^2=k^3\rightarrow k=4$.

6. **E.** Distributing the 2 on the right hand side of the equation, we get $5x+12=8x+12$. Subtracting 12 on both sides, we get $5x=8x$. This is only true if $x=0$. This problem is Basic Algebra...maybe it should be in Chapter 1.

7. **D.** The equation of a parabola is in the form $y=a(x-h)^2+k$ where (h,k) is the vertex of the parabola. So, the parabola with equation $y=-4(x+7)^2+2$ has vertex $(-7,2)$.

8. **D.** Although we could attempt to figure out what the original parabola looked like and then try to match specific points to a graph, since we know that the original parabola was downward facing $(-x^2)$, we know that rotating the graph $90°$ would produce a parabola that opens to the right. Only one answer choice has a rightwards-opening parabola.

9. **B.** The parabola shown has a vertex at $(-3,1)$, so these are our (h, k) values in the vertex form of the parabola: $y=a(x-h)^2+k$ where (h, k) is the vertex and a is a constant. I notice all the answer choices have the "*k*" value next to the *y*, so I subtract *k* from both sides to get: $y-k=a(x-h)^2$. Now I plug in $(-3,1)$ for (h, k) to get $y-1=a(x--3)^2\rightarrow y-1=a(x+3)^2$. This matches answer choices A, B, and D. Now to see which is correct, we can pluck a point and plug it into our working equation to solve for the value of *a*. Let's try $x=-2$. Tracing this point on the graph it appears to coincide with the parabola around $y=3$, so we'll plug those into $y-1=a(x+3)^2$ to get $3-1=a(-2+3)^2\rightarrow 2=a(1)^2\rightarrow 2=a$. Now we plug *a* back into the original equation to get: $y-1=2(x+3)^2$, (B).

10. **A.** We first subtract 17 on both sides to bring everything to the left side of the equation. We get $7x^2-3x-17=0$. Now, we plug in $a=7$, $b=-3$, and $c=-17$ into the quadratic equation to get:
$$x=\frac{-(-3)\pm\sqrt{(-3)^2-4(7)(-17)}}{2(7)}=\frac{3\pm\sqrt{9+476}}{14}=\frac{3\pm\sqrt{485}}{14}$$

11. A. Let's plug in the values $a=1$, $b=k$ and into the quadratic equation and set equal to x. We get $x=\frac{-k\pm\sqrt{k^2-4(1)(19)}}{2(1)}=\frac{-k\pm\sqrt{k^2-76}}{2}$. If both solutions for x are negative, then we know that $\frac{-k+\sqrt{k^2-76}}{2}$ is negative and so is $\frac{-k-\sqrt{k^2-76}}{2}$. For the first solution to be negative, k must be positive because if k were negative, the negative sign in front of it would cancel it out to make a positive. We'd then add that positive to a positive (all radicals are positive) over a positive, and the first solution for x would be a positive value. We can thus narrow our options to choices (A), (B), and (C). We also know that in order for either solution to be an integer, $\sqrt{k^2-76}$ must be an integer. If we plug in the positive answer choices 20,19 and 1 for k, only $k=20$ gives an integer solution. $\sqrt{20^2-76}=18$ while $\sqrt{19^2-76}=16.88$ and $\sqrt{1^2-76}=$ undefined. Thus $k=20$; (A) is correct.

12. D. The question says that -4 and 1 are solutions to a quadratic equation, so one option is to work backwards. If those are solutions, reverse the signs and plug into factored form: $(x-1)(x+4)=0$. FOIL to get $x^2+3x-4=0$. Comparing the equation given with the one we found, we see that $m=3$. A second method is described on page 223.

13. E. One way to solve this is to graph our own made up, arbitrary examples on a calculator and see for ourselves. First, let's swap out the "h" for "y" and the "t" for "x." Try graphing $-x^2+5x+2$ and $-\frac{1}{2}x^2+5x+2$.

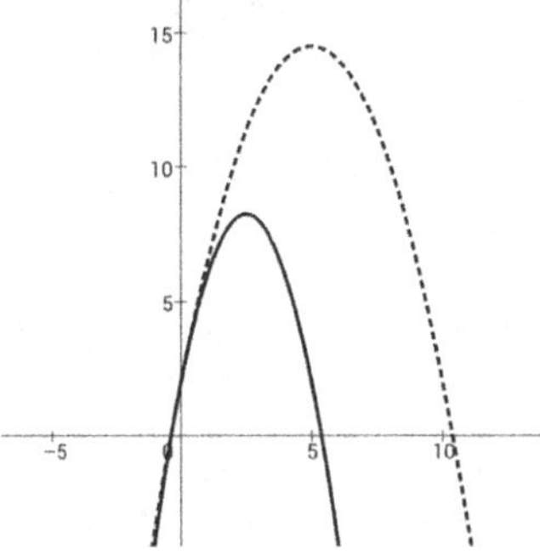

From the graphs we can see that the x-intercept (t in the question, option I) and the vertex have moved, but the y-intercept (h in the question, option II) hasn't. If we don't have a graphing calculator or don't have time to graph them, remember that the leading coefficient of a parabola can tell us only a couple things about the parabola: its sign indicates what direction the parabola is facing, its magnitude tells how 'fat' or 'skinny' the parabola is, and it in part determines the x-value of the vertex, whose formula is $\frac{-b}{2a}$. Thus, altering the a term would potentially change both the x and y value of the vertex, and the y value is the maximum height (option III). However, the only way to change the y-intercept would be to change the c value, so it makes sense that the y-intercept would stay the same, as c is not affected when changing the a value. Thus options I and III are altered, choice E. See Chapter 22 for more on the above ideas.

14. C. The quadratic formula is $x=\frac{-b\pm\sqrt{b^2-4ac}}{2a}$, so plugging in 2 for a, 8 for b, and -1 for c, we get $\frac{-8\pm\sqrt{64-4(2)(-1)}}{2(2)}$, answer (C).

15. B. We cannot solve this by plugging in -2 and solving for n as we have a second variable m. Instead, work backwards. Think about what the question means by "the *only* possible value for x is -2". Putting together the fact that it must be a polynomial of degree two, which means that if factored there would be two expressions with x, and that both must show that $x=-2$, it makes sense to assume that $(x+2)(x+2)=0$. Simplify to get x^2+4x+4. Compare this with the expression given, and we realize that $2mx=4$ and $-\frac{1}{2}n=4$. Solving for n in the second equation gives us -8.

16. C. Add like terms:

$$7w^3 + 65w - w^3 - 20 - 35w + 2$$
$$= 7w^3 - w^3 + 65w - 35w - 20 + 2$$
$$= 6w^3 + 30w - 18$$

17. B. If $2x^3 - 4x - 2$ plus some random expression (let's call it *N*) equals $-x^3 - 4$, then subtract the first from the second:

$$\left(2x^3 - 4x - 2\right) + N = \left(-x^3 - 4\right)$$
$$N = \left(-x^3 - 4\right) - \left(2x^3 - 4x - 2\right)$$
$$N = -x^3 - 2x^3 - \left(-4x\right) - 4 - \left(-2\right)$$
$$N = -3x^3 + 4x - 2$$

which is answer (B).

18. B. Looking at the coefficients, realize that we can pull out -3, which gives us $h(t) = -3\left(t^2 - 5t - 6\right)$, which is much easier to factor. $h(t) = -3\left(t^2 - 5t - 6\right)$ factored becomes $h(t) = -3(t-6)(t+1)$. Since the graph relates height and time, we know that $h = 0$ at $t = 6$ seconds and $t = -1$ seconds. There is no such thing as negative time, so we know that the ball is dropped from a height of 18 feet (our y-intercept found from the original form) and hits the ground at $t = 6$. Looking at our answer choices, hitting the ground after 6 seconds is the only correct statement.

19. A. Distributing out the negative signs, we write the expression as $3x^3 + 5 - 2x^2 + 6x - 7 + 7x - 5 - 5x^2 - 3x - 3x - 2x$. Adding like terms, we get $3x^3 - 2x^2 - 5x^2 + 6x + 7x - 3x - 3x - 2x + 5 - 7 - 5 = 3x^3 - 7x^2 + 5x - 7$.

20. C. Distributing out the negative sign, we get $x + 4y - 2z + 3x - 2y - 5z$. Adding like terms, we get $x + 3x + 4y - 2y - 2z - 5z = 4x + 2y - 7z$.

21. D. Factoring out $4x$, we get $f(x) = 4x\left(x^2 - 16\right)$. $x^2 - 16$ is a difference of squares, so we can rewrite the equation as $f(x) = 4x(x+4)(x-4)$. This gives us three different rational zeros. Namely, $0, 4$, and -4.

22. C. We can first factor out 2 from the equation because every term in the equation is divisible by 2. We get $2\left(2x^3 - 9x^2 + 16x - 12\right) = 0$. Knowing that 2 is a root, we know that when $x = 2$, the equation equals zero. Thus, it must have the factor $(x-2)$. Using long division to factor out $(x-2)$, we get $2(x-2)\left(2x^2 - 5x + 6\right) = 0$. Now, we find the remaining two roots by using the quadratic formula on $2x^2 - 5x + 6$. Plugging in $a = 2$, $b = -5$, and $c = 6$, we get $x = \frac{-(-5) \pm \sqrt{(-5)^2 - 4(2)(6)}}{2(2)} = \frac{5 + \sqrt{25 - 48}}{4} = \frac{5}{4} \pm \frac{\sqrt{-23}}{4} = \frac{5}{4} \pm \frac{i\sqrt{23}}{4}$.

23. A. Plug in for the fastest solution! From the factor given, we know -1 is a solution or zero. Plug in -1 for x, set the expression equal to zero, and solve for c. $(-1)^3 + 2(-1)^2 - c(-1) - 20 = 0$ which simplifies to $-1 + 2 + c - 20 = 0$ and $c = 19$. Alternatively: Using long division to factor out $(x+1)$ from $x^3 + 2x^2 - cx - 20$, we get $(x+1)\left(x^2 + x - (c+1)\right)$ where the constant term is equal to 20. $c + 1 = 20$ so $c = 19$.

24. C. Foiling out $(n+4)^3$, we get $(n+4)^3 = (n+4)\left(n^2 + 8n + 16\right) = \left(n^3 + 4n^2 + 8n^2 + 32n + 16n + 64\right)$ $= n^3 + 12n^2 + 48n + 64$

25. D. Without even looking at the table, we see that the equation is in vertex form. From the equation we see that it is a downward facing parabola (because the leading term is negative) with a vertex at $(-2, -4)$. Because the parabola is facing down, we know the vertex has the greatest y-value, so the answer is -2. Alternatively, looking at the table we see that the y-values increase while $x < -3$, and decrease when $x > -1$, so the greatest point must be in between those two numbers, and -2 is the only answer that fulfills that condition.

26. **C.** First factor out an x^2: $f(x)=3x^4+8x^3+4x^2 \to f(x)=x^2(3x^2+8x+4)$. Now we can factor by reverse foiling: $f(x)=x^2(3x+2)(x+2)$. Alternatively, plug this into your graphing calculator and trace to find the zeros, or solve the quadratic portion of the polynomial down by using the quadratic equation program. Our zeros are: 0, $\frac{-2}{3}$, and -2, and -2 is the only correct answer given.

27. **E.** Notice that we can pull out a constant of 4: $4x^4+16x^2+12 \to 4(x^4+4x^2+3)$. We don't know how to factor polynomials to the fourth degree easily, but we can substitute. If we let $w=x^2$, our expression becomes $4(w^2+4w+3)$, which factors easily into $4(w+1)(w+3)$. When we plug back in x^2 for w, we get $4(x^2+1)(x^2+3)$.

28. **E.** Factor out a^3 and then look at the problem as the difference between squares special pattern: $a^7-81a^3 \to a^3(a^4-3^4) \to a^3\left((a^2)^2-(3^2)^2\right) \to a^3(a^2-3^2)(a^2+3^2) \to a^3(a-3)(a+3)(a^2+9)$ Thus, a, a^2, a^3, $(a+3)$, $(a-3)$, (a^2-9), and (a^2+9) are all factors, but a^2+3 is not. You could also backsolve this problem, knowing that each factor determines a zero. Set each answer choice equal to zero and solve for the solution each indicates, then plug in each value and see which one(s) equal zero when plugged in until you find the one that doesn't.

29. **D.** If our zeros are at 0, -3, and 4, then we can say that $f(x)=x(x+3)(x-4)$. When we plug in 1 to test, $f(1)=1(1+3)(1-4)=-12 \neq -6$. In order to satisfy the condition that says that $f(1)=-6$, we look at what our $f(1)$ currently equals and adapt the equation accordingly. In order to get $f(1)=-6$, we must divide our current $f(1)$ by 2, so our equation becomes $\frac{x}{2}(x+3)(x-4)$.

30. **C.** Since $2, 6, -4$, and -9 are zeros for the polynomial, we know that it can be written in the form $(x-2)(x-6)(x+4)(x+9)k=0$ for some constant k. We are given that $f(3)=63$, so plugging in 3, we have $(3-2)(3-6)(3+4)(3+9)k=63 \to 1(-3)(7)(12)k=63 \to -252k=63 \to k=-\frac{63}{252}=-\frac{1}{4}$. So, the polynomial is $-\frac{1}{4}(x-2)(x-6)(x+4)(x+9)$.

31. **A.** The equation that crosses the x-axis at $x=-13$ and $x=7$ has roots that are equal to zero at both of those points, respectively. Thus, it must have $(x+13)$ and $(x-7)$ in its factorization. The only answer choice that has both is A.

32. **A.** Since the parabola has a vertical axis of symmetry, its second crossing must be as equidistant from the axis of symmetry to the first crossing, but on the right instead of the left. Thus, it must cross the x-axis a second time at $(5+\sqrt{35},0)$.

33. **E.** A 4th degree polynomial function can only cross the x-axis a maximum of 4 times. This is because the term with the highest degree will be some term containing x^4. Thus we cannot have five x-intercepts. A 4th degree polynomial cannot cross or touch the x-axis five times.

CHAPTER

18 COMPLEX NUMBERS

SKILLS TO KNOW

- Definition of an imaginary number, complex number, and i
- Solving equations and simplifying expressions which include i
- Adding and subtracting imaginary and complex numbers
- Multiplying by a complex conjugate (rationalization)
- Using the complex plane

NOTE: The problems in this chapter require a solid knowledge of **FOIL and Factoring, Chapter 3**, and **Exponents and Radicals, Chapter 15.** Problems involving multiples of i are a special case of "Remainder Problems." See **Book 2, Chapter 6: Sequences & Series** for in-depth coverage of Remainder Problems.

DEFINING IMAGINARY NUMBERS

What is i ?

i equals $\sqrt{-1}$. Because this number has no place on a number line, it is called an "imaginary number." This does not mean, however, that i does not exist or that it has no use in mathematics.

Because $\sqrt{-1}=i$, radicals of negative numbers can be rewritten without the negative sign by placing an i outside the radical: $\sqrt{-x}=i\sqrt{x}$. Example: $\sqrt{-5}=\sqrt{-1}\sqrt{5}=i\sqrt{5}$. We essentially "factor out" the negative part of the root and replace it with i.

What is a complex number?

A complex number is a number that can be expressed as a combination of a real and an imaginary number in the form of $a+bi$ where a and b are real numbers. For example, $5+2i$ is a complex number with imaginary parts. A real "part" of a complex number cannot be combined with its imaginary part. $5+2i$, for example, is fully simplified. You cannot combine the 5 with the $2i$ portion. But the catch is, the set of complex numbers includes **all numbers real and imaginary**. If b is zero, then the imaginary part of the number does not exist, but the number is still complex as it can be expressed in $a+bi$ form. All real numbers can be expressed in $a+bi$ notation when $b=0$. For example, 5 equals $5+0i$.

$a+bi$ MODE on a TI-84

Many graphing calculators have an $a+bi$ mode (on a TI-84 hit MODE then select $a+bi$). **In this mode, your calculator can AUTOMATICALLY do many of the problems in this chapter.** Hit 2nd then the "." button (bottom row center) to access the "i" in this mode. Be careful, though, as if you take i to a power, the accuracy diminishes as the exponent gets bigger. New calculators can take i to the power of 100 accurately. Some calculators will print ugly scientific notation (such as -3E-13) before the i. If your answer has this ugly type in it, just IGNORE the scientific notation part or set it equal to zero and look at the value of i $(+i,-i)$ or of 1 ($+1$ or -1). Still, it's a good idea to understand the ideas behind complex numbers if you have time, so we'll use other methods below, too.

MULTIPLES OF i

A pattern emerges when we take i to a power.

i^1	$\sqrt{-1}$
i^2	-1
i^3	$-i$
i^4	1
i^5 (equivalent to i^1)	$\sqrt{-1}$
etc.	etc.

Because of this pattern, we can simplify many expressions that include i to a power. To find i to the x power, first divide x by four and find the remainder. Then take i to the power of that remainder.

Formally, $i^n = i^{n \bmod (4)}$. The function **"mod"** simply means to find the remainder when n is divided by 4. Your graphing calculator may have this function built in. On a TI-84 it is called "remainder." You can find it by hitting **MATH** then **NUM** then selecting **0** for **"remainder("**

Which of the following is equal to i^{53}?

A. i **B.** i^2 **C.** i^3 **D.** i^4 **E.** -1

If your calculator has $a+bi$ mode and plays nice with large exponents (see tip on page 235), you can solve this one fast: in $a+bi$ mode type i ("2ND" then ".")^{53}, which equals i. But if your calculator outputs scientific notation garble (i.e. -3E-13+i), ignore the parts that aren't i or 1 to get i (answer choice A). I realize the garble is confusing, and thus you may want alternate methods.

Let's try the remainder method: first find the remainder of 53 divided by 4 Using your calculator, "remainder$(53,4)$" is 1. Now take i to the 1st power and you get i, answer choice A. When the exponent is greater than one, I'll often write out the chart below to help me figure out the answer.

i^1	$\sqrt{-1}$
i^2	-1
i^3	$-i$
i^4	1

We can also solve just using this pattern. What happens when we continue it? Anything to a power that is a multiple of 4 will equal 1. For example, if we continue this pattern, multiplying i to the 8th, 12th, 52nd and every other exponent divisible by 4, we see each fourth value in the pattern equals one. Now, 53 is not divisible by 4, but we can find the closest number that is, 52. From the pattern, we conclude $i^{52} = i^4$, or 1 Then I can count one forward in the pattern to see what value corresponds to 53, i.e. the 53rd power of i, which equals i^1, i.e. i or $\sqrt{-1}$: Answer **A.**

We could also solve this problem more algebraically with the same idea in mind; because 4 goes into 52 evenly (13 times), we can rewrite our expression, using the exponent rules to factor the exponent:

$$i^{53}$$

Applying the Product of Powers: $a^b a^c = a^{b+c}$

Now we apply Power of a Power: $\left(a^b\right)^c = a^{bc}$ $= i^{52} \times i^1$

Now we replace i^4 with its value, 1: $= \left(i^4\right)^{13} \times i$

$= 1^{13} \times i$ or i

Answer: **A.**

For the complex number i such that $i^2 = -1$, what is the value of $i^4 + 2i^3 + i$?

A. $-1-i$ **B.** $-1+i$ **C.** $1-i$ **D.** $1+3i$ **E.** $-1+3i$

We can just take this problem one part at a time. In general, our goal is to reduce down any pieces by extracting out even powers of i, as these have real number values since $i^2 = -1$ and $i^4 = 1$. The first monomial in this expression, thus, can be rewritten as $\left(i^2\right)^2 = (-1)^2 = 1$, or you use your calcuator (see pg 235) or can memorize that the fourth power of i is positive 1. In any case, we have:

$$1 + 2i^3 + i$$

$2i^3$ is the same as $2i\left(i^2\right) = 2i(-1) = -2i$, leaving us with:

$$1 - 2i + i$$

We combine the like terms (the "i" terms) to get our final answer:

$$1 - i$$

Answer: **C.**

ADDING COMPLEX NUMBERS

As we mentioned above, real number parts and imaginary number parts cannot be added together. However, complex numbers can be added together by combining all the real terms and all the imaginary terms. Again in $a+bi$ mode, you can make your calcluator do all this for you. However, given the radicals you would then need to backsolve a bit as your calculator will simplify everything down.

What is the sum of $\sqrt{-27}$ and $\sqrt{-48}$?

To simplify radicals, separate the root expression into the product of two (or more) factors, with each factor under its own square root symbol.

First we simplify $\sqrt{-27}$:

$$\sqrt{-27} = \sqrt{-1 \times 9 \times 3}$$
$$= \sqrt{9} \times \sqrt{-1} \times \sqrt{3}$$
$$= 3i\sqrt{3}$$

Next we simplify $\sqrt{-48}$:

$$\sqrt{-48} = \sqrt{-1 \times 16 \times 3}$$
$$= \sqrt{16} \times \sqrt{-1} \times \sqrt{3}$$
$$= 4i\sqrt{3}$$

Adding these together:

$$3i\sqrt{3} + 4i\sqrt{3} = 7i\sqrt{3}$$

MULTIPLYING BY A COMPLEX CONJUGATE

Answers with imaginary numbers in the denominator are not "simplified." To simplify a number with the imaginary parts in the denominator, multiply both the numerator and denominator by **"the complex conjugate."** Sometimes this is called **"rationalizing"** the denominator. Multiplying a complex number by its complex conjugate will leave you with a real number in the denominator.

RATIONALIZING A COMPLEX NUMBER

For any complex number $(a+bi)$, its complex conjugate is $(a-bi)$. To rationalize any complex number, multiply by its conjugate to clear the imaginary part:

$$(a+bi)(a-bi)=\mathrm{a}^2+b^2$$

How do we find the complex conjugate? If our denominator is $a+bi$, with a being the real part and bi being the complex part, then our complex conjugate is simply $a-bi$. Just flip the sign on the imaginary part! When we multiply the numerator and denominator by this term and FOIL the original term and its complex conjugate, we eliminate the i's in the denominator.

Remember the special product called **"the difference of squares"** or the **"product of a sum and a difference"?**

$$(a+b)(a-b)=a^2-b^2$$

That is the pattern the above process derives from. The goal is to "square" the i term so that it becomes real. Conjugates eliminate the i:

$$\begin{aligned}(a+bi)(a-bi)&=a^2+abi-abi-b^2i^2\\&=a^2-b^2i^2\\&=a^2-b^2(-1)\\&=a^2+b^2\end{aligned}$$

Given the expression $\frac{1}{x+yi}$, rationalize the denominator.

(Most ACT questions won't ask you this question straight away. Instead, you'll have a "which of the following is equivalent to... same question, but multiple choice without the vocabulary). For the example above, our denominator is $x+yi$, so our complex conjugate is simply $x-yi$. Now let's multiply and FOIL:

$$\frac{1}{x+yi}\left(\frac{x-yi}{x-yi}\right)=\frac{x-yi}{x^2-xyi+xyi-y^2i^2}$$

Notice how the two middle terms in the denominator cancel out, leaving us with $\frac{x-yi}{x^2-y^2i^2}$.

But wait! There are still i's in the denominator! Remember that $i^2=-1$, so $(yi)^2=y^2i^2=-y^2$. We are left with two negative signs that become a positive:

$$\frac{x-yi}{x^2--y^2}=\frac{x-yi}{x^2+y^2}$$

Notice that we are left with only real terms in the denominator.

Using complex numbers, where $i^2 = -1$, $\frac{i+1}{i+5} \times \frac{i-5}{i-5} = ?$

MISTAKE ALERT: For this problem, we must use FOIL on both the top and bottom, as the pattern is not exactly the same as what we have in our equation box earlier (i.e. we can't square the i and the 5 and add together, our signs would then be wrong). Many students miss this problem as they blindly apply the "pattern" that does not exactly fit this circumstance.

Taking the numerator and denominator separately, we first expand the numerator by using FOIL.

$$\frac{(i+1)(i-5)}{(i+5)(i-5)} = \frac{i^2 + i - 5i - 5}{(i+5)(i-5)}$$

$$= \frac{-1 - 4i - 5}{(i+5)(i-5)}$$

$$= \frac{-6 - 4i}{(i+5)(i-5)}$$

Now moving onto the denominator, we get:

$$\frac{-6-4i}{(i+5)(i-5)} = \frac{-6-4i}{i^2 - 25}$$

$$= \frac{-6-4i}{-1-25}$$

$$= \frac{-6-4i}{-26}$$

$$= \frac{3+2i}{13}$$

Answer: $\frac{3+2i}{13}$.

USING THE COMPLEX PLANE

Problems will sometimes ask you to visualize complex numbers on a two-dimensional plane, where the x-axis is the real axis and the y-axis is the imaginary axis. It looks a lot like an ordinary two-dimensional plane, except points on this plane are denoted with the format $(a+bi)$ instead of (x,y).

Let's graph the point $(6-3i)$ for practice.

The 6 corresponds to the "x" or horizontal value, while the -3 corresponds to the "y" or vertical value. We graph as we would graph $(6,-3)$ in a regular coordinate plane. The y-axis is replaced with the "Imaginary Axis" and the x-axis with the "Real Axis."

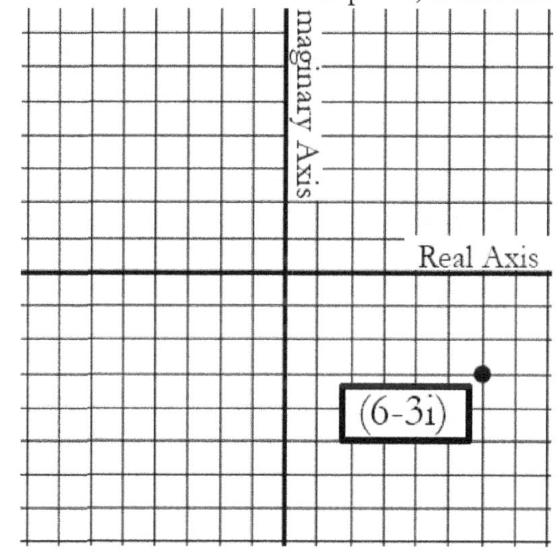

1. i is a complex number and n is an integer. Which of the following is not a possible value of i^{n+1}?

   A. 1
   B. i
   C. 0
   D. -1
   E. $-i$

2. 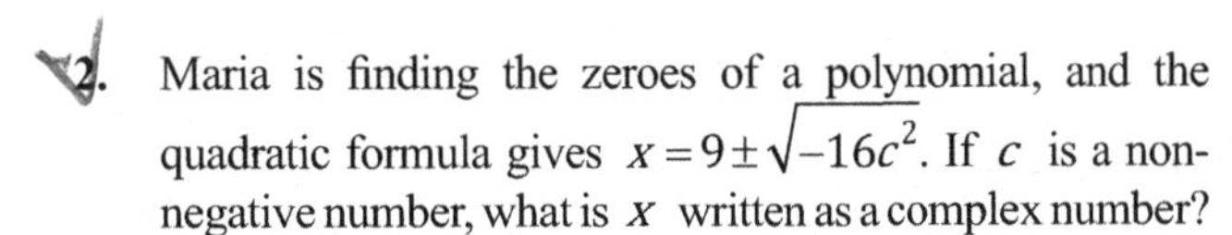
   Maria is finding the zeroes of a polynomial, and the quadratic formula gives $x = 9 \pm \sqrt{-16c^2}$. If c is a non-negative number, what is x written as a complex number?

   A. $9 \pm 4c$
   B. $9 \pm 4ci$
   C. $9 + 4c$
   D. $9 - 4ci$
   E. $9 + 4ci$

3. If $i = \sqrt{-1}$, then what does $\left(\frac{1-i}{1+i}\right)\left(\frac{-1}{1-i}\right) = ?$

   A. $\frac{1+i}{2}$
   B. $1 - i$
   C. $\frac{-1-i}{2}$
   D. $\frac{i-1}{2}$
   E. $-1 + i$

4. For all pairs of nonzero real numbers x and y, the product of the complex number $x - yi$ and which of the following complex numbers is a real number?

   A. $x + yi$
   B. $x - yi$
   C. xyi
   D. $y - xi$
   E. $x + i$

5. The product of two numbers is 41. One of the numbers is the complex number $5 + 4i$. What is the other number?

   A. $5 - 4i$
   B. $5 + 4i$
   C. $-5 - 4i$
   D. $-5 + 4i$
   E. $\frac{41}{5-4i}$

6. Which equation given in factored form has the roots $\frac{1}{4}, \frac{2}{3}, i$, and $-i$?

   A. $(4x-1)(3x-2)(x^2-1)$
   B. $(4x+1)(3x-2)(x^2-1)$
   C. $(4x-1)(3x-2)(x^2+1)$
   D. $(4x-1)(3x+2)(x^2+1)$
   E. $(4x-1)(3x+2)(x^2-1)$

7. For the complex number i such that $i^2 = -1$, what is the value of $i^8 - 2i^2 - 1$?

   A. -4
   B. -2
   C. 0
   D. 2
   E. 4

8. 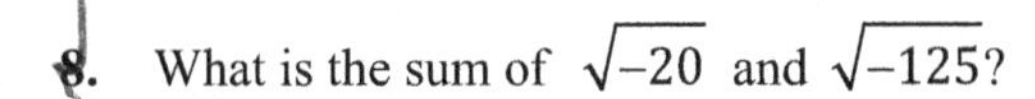
   What is the sum of $\sqrt{-20}$ and $\sqrt{-125}$?

   A. $-7i\sqrt{5}$
   B. $7i\sqrt{5}$
   C. $-21i\sqrt{5}$
   D. $21i\sqrt{5}$
   E. $i\sqrt{105}$

9. What is the square of the complex number $(2i - 4)$?

   A. $12 - 16i$
   B. $20 - 16i$
   C. -20
   D. 12
   E. 20

10. For $i^2 = -1$, $(3 - i)^2 = ?$

   A. 8
   B. 10
   C. $8 - 6i$
   D. $8 + 6i$
   E. $10 - 6i$

11. The solution set for the equation $3^{x^2+3}-1=0$ contains:

A. Only 1 imaginary number
B. 2 imaginary numbers
C. 1 imaginary and 1 real number
D. 1 negative real number and 1 imaginary number
E. 1 real number, which is 0.

12. For all $x<0$, which of the following expressions is equivalent to $\frac{\sqrt{x}}{\sqrt{x}-i}$?

A. $\frac{x-\sqrt{x}}{x-1}$

B. $\frac{x+\sqrt{x}}{x-1}$

C. $\frac{-x-\sqrt{x}}{-x+1}$

D. $\frac{x+\sqrt{-x}}{x+1}$

E. $\frac{\left(x-\sqrt{x}\right)}{x+1}$

13. Which of the following expressions is equivalent to $9x^2+169$?

A. $(3x+13)^2$
B. $(3x+13i)^2$
C. $(3x-13i)^2$
D. $(3x-13)(3x+13)$
E. $(3x-13i)(3x+13i)$

14. What complex number equals $(3-4i)(\pi+3i)$?

A. $(12+3\pi)i+(9-4\pi)$
B. $(12-3\pi)+(9-4\pi)i$
C. $(12+3\pi)+(9+4\pi)i$
D. $(12+3\pi)+(9-4\pi)i$
E. $(12-3\pi)i+(9-4\pi)i$

15. The figure below depicts a complex plane with the horizontal axis representing real values and the vertical axis representing imaginary values. The modulus of a complex number $a+bi$ is $\sqrt{a^2+b^2}$. By looking at the points below, which point has the smallest modulus?

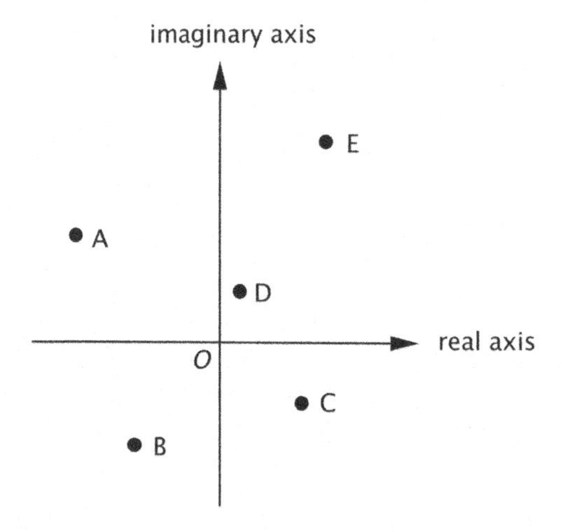

A. A
B. B
C. C
D. D
E. E

ANSWER KEY

1. C **2. B** **3. D** **4. A** **5. A** **6.C** **7. D** **8. B** **9. A** **10. C** **11. B** **12. D** **13. E** **14. D** **15. D**

ANSWER EXPLANATIONS

1. **C.** By definition, i is the complex number equal to $\sqrt{-1}$, and when taken to a power, equals one of four possible answers $(i,-1,-i,$ or $1)$. For A, $i^4=\sqrt{-1}\sqrt{-1}\sqrt{-1}\sqrt{-1}=(-1)(-1)=1$. B is wrong because $i^1=\sqrt{-1}=i$. D: $i^2=\sqrt{-1}\sqrt{-1}=-1$. E: $i^3=\sqrt{-1}\sqrt{-1}\sqrt{-1}=-1\sqrt{-1}=-i$. As such, the answer is C. Also, the only number that can equal 0 when taken to any power is zero. Therefore, 0 cannot be the answer.

2. **B.** $x=9\pm\sqrt{-16c^2}=9\pm4c\sqrt{-1}$. Since $i^2=-1$, we can rewrite $\sqrt{-1}=i$. This gives us $x=9\pm4ci$.

3. **D.** Canceling out $1-i$ from the numerator and denominator, we get $-\frac{1}{1+i}$. To get the i term out of the denominator, we multiply the fraction by the conjugate on the top and bottom to get $-\frac{1}{1+i}\left(\frac{1-i}{1-i}\right)=-\frac{1-i}{1+i-i+1}=-\frac{1-i}{2}=\frac{i-1}{2}$.

4. **A.** The product of a complex number and its conjugate is a real number because the $+$ and $-$ in front of the imaginary terms cancel out when applying FOIL. So, $(x-yi)(x+yi)$ will be equal to a real number. To verify this, we foil the factors and get $x^2-xyi+xyi-y^2i^2=x^2-(-1)y^2=x^2+y^2$. This is a real number because x and y are real numbers.

5. **A.** The product of a complex number and its conjugate is a real number because the $+$ and $-$ in front of the imaginary terms cancel out when expanded with FOIL. So, the only way $(5-4i)$ multiplied by something can yield a real number will be when it is multiplied by its conjugate. To verify, we multiply $(5-4i)(5+4i)$ and get $25-4i(5)+5(4i)-16i^2=25-(-1)16$ $\rightarrow25+16\rightarrow41$.

6. **C.** If a polynomial has roots equal to $\frac{1}{4},\frac{2}{3},i,$ and $-i$, it means that these terms make the polynomial equal to zero when plugged in. So, the following equations must be true: $x-\frac{1}{4}=0\rightarrow4x-1=0$, $x-\frac{2}{3}=0\rightarrow3x-2=0$, $x-i=0$, and $x+i=0$. So, the polynomial can be written as $(4x-1)(3x-2)(x-i)(x+i)$. Multiplying the last two factors, using FOIL, we get $(4x-1)(3x-2)(x^2-xi+xi-1)=(4x-1)(3x-2)(x^2+1)$.

7. **D.** The first term, $i^8=(i^2)^4=(-1)^4=1$. The second term, $-2i^2=-2(-1)=2$. By substituting these into the equation, we get $1+2-1$, which equals 2.

8. **B.** We can break down the square roots to $\sqrt{(-1)(4)(5)}$ and $\sqrt{(-1)(25)(5)}$. The $\sqrt{-1}$'s become i, and the perfect squares become their square roots. Thus, $\sqrt{-1}\sqrt{4}\sqrt{5}$ and $\sqrt{-1}\sqrt{25}\sqrt{5}$ simplify to $2i\sqrt{5}$ and $5i\sqrt{5}$. Their sum is $7i\sqrt{5}$.

9. **A.** Using FOIL, we get $(2i-4)^2=2i\times2i-8i-8i+16$. $2i\times2i$ is equal to $4\times-1=-4$, so simplifying, we get $-4-16i+16$. We combine the integers to get $12-16i$.

10. **C.** Using FOIL, we get $(3-i)^2=3\times3-3i-3i-i\times i$. Simplifying gets us $9-6i+i^2$. We are given that $i^2=-1$, so plugging that in: $9-6i-1=8-6i$.

11. **B.** In order to satisfy the equation, 3^{x^2+3} must equal 1. An exponential function only equals 1 when its exponent is equal to 0. This is called the zero power rule (see the first page of chapter 15). So if you have ANYTHING to a power and it equals one, you can just set the exponent equal to zero to solve. Solve for $x^2+3=0$. This becomes $x^2=-3$. x must then be equal to $i\sqrt{3}$ and $-i\sqrt{3}$. There are 2 imaginary numbers in the solution set.

12. D. In order to simplify the expression, we multiply the top and bottom of the fraction by the conjugate of the denominator. Since the denominator is $\sqrt{x}-i$, its conjugate is $\sqrt{x}+i$. Our expression now becomes $\frac{\sqrt{x}\left(\sqrt{x}+i\right)}{\left(\sqrt{x}-i\right)\left(\sqrt{x}+i\right)}$.

Distributing on the top and applying FOIL to the bottom gives us $\frac{x+i\sqrt{x}}{x+i\sqrt{x}-i\sqrt{x}-i^2}$, which we can simplify to $\frac{x+\sqrt{-x}}{x+1}$.

13. E. $9x^2+169$ can be expressed as the product of complex conjugates. The first term is the square of $3x$, and the second term is the product of $13i$ and $-13i$. Thus, we can set up our equation as $\left(3x+13i\right)\left(3x-13i\right)$. The 'O' and 'I' of FOIL cancel each other out, leaving us with $9x^2+169$.

14. D. Multiplying the expression out using FOIL, we get $(3-4i)(\pi+3i)=3\pi-4i\pi+9i-(-1)12$. Now, separating the real and imaginary terms, we get $3\pi+12-4\pi i+9i=(12+3\pi)+(9-4\pi)i$.

15. D. The modulus of the complex number is essentially the distance from the origin to the point. This can be seen since the value of the modulus, $\sqrt{a^2+b^2}$, is the Pythagorean theorem, which is used to find the distance of a point from the origin. Thus, we can tell what the smallest modulus is by seeing which point is the closest to the origin. In this case, it's D.

CHAPTER

19 RATIONAL EXPRESSIONS AND EQUATIONS

SKILLS TO KNOW

- Simplifying expressions by factoring/canceling
- Finding a common denominator
- Rationalizing the denominator in rational expressions

NOTE: This chapter builds on skills covered in chapters on **Fractions** and **LCM/GCF** in Book 2. For information on holes and asymptotes in rational expressions (or what happens when a denominator is zero) see **Chapter 22: Graph Behavior** in this book.

WHAT IS A RATIONAL EXPRESSION?

Rational expressions are fractions that have a polynomial either in the numerator, denominator, or both. To make identifying them easier, just look for **fractions that include variables**. Questions on rational expressions usually ask you to simplify or rewrite the expression or to add, subtract, multiply or divide fractions that contain variables.

Examples of rational expressions:

$$\frac{m^2-9}{3m^3+7m-2};\ (x+1)(x-3)^{-3};\ \frac{(x+2)(2x-5)}{(7x-1)(x+4)}$$

WAYS TO SIMPLIFY RATIONAL EXPRESSIONS

1. FACTORING AND CANCELING

We can cancel out factored elements in rational expressions. For example, when asked to simplify $\frac{x+2}{(x+2)(x-4)}$, we can simply cancel out the $x+2$ in the numerator and denominator to get $\frac{1}{x-4}$. Be careful to note, however, if you do cancel that term, x *cannot equal* -2. Remember that we can't divide by zero, so any factor in the bottom of a fraction must not be equal to zero. Because negative 2 would make the original expression undefined, it must be excluded from the possible x values in our answer. We thus must note this simplified expression has a domain such that $x \neq -2$.

The "solution" at negative 2 is called an **extraneous solution.** Extraneous solutions are solutions that we find when we take certain actions in problem solving that are necessary for solving but may introduce "answers" that are actually impossible. To avoid these here, again, take note of any values that could make the denominator equal to zero, even if they cancel, and exclude them from your final answer or your final answer's domain. We'll talk more on this topic in **Chapter 23, Graph Behavior.**

Simplifying Fractions with Factoring

When $\frac{x^2+6x}{x^3+x^2-30x}$ is defined, it is equivalent to which of the following expressions?

A. $\frac{6}{x-30}$ **B.** $\frac{1}{x^2-5}$ **C.** $\frac{1}{x-5}$ **D.** $\frac{1}{x+6}$ **E.** $-\frac{1}{5}$

First, I take note of the question's wording: **when the function "is defined."** This little phrase is very helpful: it means I don't have to worry about extraneous solutions or keeping track of elements I cancel. (Extraneous solutions occur when a function is NOT defined).

Now, let's completely factor both the numerator and denominator so we can see what cancels out:

$$\frac{x^2+6x}{x^3+x^2-30x}$$

$$\frac{\cancel{x}\cancel{(x+6)}\ 1}{\cancel{x}\cancel{(x+6)}(x-5)}$$

$$\frac{1}{x-5}$$

Answer: **C.**

TIP: **Peek ahead at the answer choices for ideas on what factors might work. Also, expect already factored pieces to cancel. Often these elements are factors.**

2. FORMING COMMON DENOMINATORS

Like regular fractions, rational expressions need to have common denominators before they are added or subtracted.

Which of the following expressions is equivalent to the expression $\frac{3}{n}+\frac{n}{8}-\frac{6}{r}$?

A. $\frac{n-3}{n+8-r}$ **B.** $\frac{24r+n^2r-48n}{8nr}$ **C.** $\frac{n-3}{8nr}$ **D.** $\frac{11n+n^2r-2n}{8nr}$ **E.** $-144n^2r$

To combine terms, find a common denominator. First, we find the least common multiple (LCM) of the denominators (n, r, and 8) (remember LCM and LCD are essentially the same). Because we have no common factors, we simply multiply all these terms together to find the LCM: $8nr$. Now, we turn each individual term into an equivalent fraction by multiplying by the missing "pieces" of this LCM on both the top and bottom. Remember, when we multiply a fraction by the same amount on the top and bottom, we are essentially multiplying by one. Here, the first term is missing is $8r$, so the top and bottom are multiplied by $8r$. The center term is missing nr, so it is multiplied on the top and bottom by nr, and the final term is missing $8n$, so it is multiplied on the top and bottom by $8n$.

$$\frac{3(8r)}{n(8r)}+\frac{n(nr)}{8(nr)}-\frac{6(8n)}{r(8n)}$$

After expanding we are left with:

$$\frac{24r}{8nr}+\frac{n^2r}{8nr}-\frac{48n}{8nr}$$

Which simplifies to:

$$\frac{24r+n^2r-48n}{8nr}$$

Answer: **B.**

These methods can also be applied to more complex rational expressions.

Simplify $\dfrac{2-\dfrac{3}{x+1}}{\dfrac{2}{x-1}+1}$.

First, we find a common denominator for all the terms in the numerator, and then the same for the denominator. In other words, we want to turn the numerator into a single fraction, and the denominator into a single fraction:

$$\frac{2\times\frac{(x+1)}{(x+1)}-\frac{3}{x+1}}{\frac{2}{x-1}+1\times\frac{(x-1)}{(x-1)}}$$

$$\frac{\frac{2(x+1)-3}{x+1}}{\frac{2+(x-1)}{x-1}}$$

$$\frac{\frac{2x+2-3}{x+1}}{\frac{2+x-1}{x-1}}$$

$$\frac{\frac{2x-1}{x+1}}{\frac{x+1}{x-1}}$$

Now we can use the principle of dividing fractions to further simplify. The "fraction" in the numerator is being divided by the "fraction" in the denominator.

$$\frac{2x-1}{x+1}\div\frac{x+1}{x-1}$$

Therefore, we can multiply by the reciprocal of $\dfrac{x+1}{x-1}$ to calculate this division:

$$\frac{2x-1}{x+1}\times\frac{x-1}{x+1}=\frac{(2x-1)(x-1)}{(x+1)^2}$$

3. RATIONALIZING THE DENOMINATOR IN RATIONAL EXPRESSIONS

To simplify fractions containing square roots, we must eliminate radicals in the denominator. This is usually achieved by multiplying the original expression by the **conjugate** divided by itself, i.e. we use the idea of equivalent fractions to multiply by an amount equal to 1 that will eliminate the square root in the denominator.

DEFINITION OF A CONJUGATE

If a number can be expressed as $a+b\sqrt{c}$ where a, b, and c are rational numbers, then its conjugate is:

$$a-b\sqrt{c}$$

Simplify $\frac{1}{5+9\sqrt{3}}$

To simplify $\frac{1}{5+9\sqrt{3}}$ we first need to figure out the conjugate of its denominator. Here, the conjugate of $5+9\sqrt{3}$ is $5-9\sqrt{3}$. Now we multiply the original expression by the conjugate over the conjugate. In this case, $\frac{5-9\sqrt{3}}{5-9\sqrt{3}}$.

Here we proceed by multiplying the original expression by the **conjugate over the conjugate**:

$$\frac{1}{5+9\sqrt{3}}\times\frac{5-9\sqrt{3}}{5-9\sqrt{3}}$$

But where does this "conjugate" come from? The idea of conjugates is based on the special product, **The Difference of Squares** (or the **Product of a Sum and a Difference** as some may call it). (For more, see **FOIL and Factoring** Chapter of this book).

The Difference of Squares

$$a^2-b^2=(a-b)(a+b)$$

Knowing and applying this pattern, letting letting $a=5$ and $b=9\sqrt{3}$, we can see how the square roots in the denominator disappear as we continue to simplify:

$$\frac{5-9\sqrt{3}}{5^2-\left(9\sqrt{3}\right)^2}$$

$$\frac{5-9\sqrt{3}}{(25-243)}$$

$$\frac{5-9\sqrt{3}}{(-218)}$$

If you find this confusing, my best advice is to use your calculator in squaring the expression with the root, to study the examples below, and to practice!

Answer: $\frac{5-9\sqrt{3}}{(-218)}$

The expression $\frac{11}{10-\sqrt{7}}$ is equivalent to:

A. $\frac{110+11\sqrt{7}}{93}$ B. $\frac{110+\sqrt{7}}{107}$ C. $\frac{100+\sqrt{7}}{100}$ D. $\frac{100+\sqrt{7}}{93+20\sqrt{7}}$ E. $\frac{110+11\sqrt{7}}{10\sqrt{7}}$

$\frac{11}{10-\sqrt{7}}$ can be simplified by multiplying the fraction by the denominator's conjugate, $\frac{10+\sqrt{7}}{10+\sqrt{7}}$:

$$\frac{11}{10-\sqrt{7}} \times \frac{10+\sqrt{7}}{10+\sqrt{7}}.$$

This simplifies to:

$$\frac{11(10+\sqrt{7})}{(10-\sqrt{7})(10+\sqrt{7})}$$

Now, let's use FOIL to expand the bottom (you can also use knowledge of the Difference of Squares pattern if you prefer).

$$\frac{110+11\sqrt{7}}{100-10\sqrt{7}+10\sqrt{7}-7}$$

$$\frac{110+11\sqrt{7}}{100-7}$$

$$\frac{110+11\sqrt{7}}{93}$$

Answer: **A.**

1. The expression $\frac{6x+12}{24x^3}$ is equivalent to which of the following?

A. $\frac{1}{x^2}+\frac{1}{2x^3}$

B. $\frac{1}{2x^2}+\frac{1}{4x^3}$

C. $\frac{1}{4x^2}+\frac{1}{2x^3}$

D. $\frac{1}{2x^2}+\frac{1}{2x^3}$

E. $\frac{1}{x^2}-\frac{1}{2x^3}$

2. For $x \neq 15$, $\frac{x^2-225}{(x-15)^3}=?$

A. $\frac{x+15}{(x-15)^2}$

B. $\frac{x-15}{(x+15)^2}$

C. $\frac{x+15}{(x+15)^2}$

D. $\frac{x-15}{(x-15)^2}$

E. $\frac{x+15}{(x-15)^3}$

3. When $x \neq 3$ and $x \neq -3$, $\frac{5x}{x^2-6x+9}+\frac{5x}{9-x^2}$ is equivalent to:

A. 0

B. $\frac{30x}{(x-3)^2(x+3)}$

C. $\frac{-30x^2}{(x-3)^2(x+3)}$

D. $\frac{10x^2+30x}{(x-3)^2(x+3)}$

E. $\frac{15x}{(x-3)^2(x+3)}$

4. For all x in the domain of the function $\frac{(x-3)^2}{x^4-9x^2}$, this function is equivalent to:

A. $\frac{2}{x^2}$

B. $\frac{1}{x^2}$

C. $\frac{(x-3)}{x^3+3x^2}$

D. $\frac{(x-3)}{(x+3)}$

E. $\frac{1}{x^4}-\frac{1}{9x^2}$

5. For all $x>6$, $\frac{(x^2+7x+10)(x-2)}{(x^2+x-6)(x+5)}=?$

A. $\frac{x+2}{x+3}$

B. $\frac{x+3}{x+2}$

C. $\frac{x-2}{x+5}$

D. $\frac{x+2}{x+5}$

E. $\frac{x+2}{x-5}$

6. If $\frac{A}{15}+\frac{B}{100}=\frac{20A+3B}{5x}$ and $A, B,$ and x are integers greater than 1, then what must x be equal to?

A. 23

B. 60

C. 115

D. 300

E. 1500

7. When $\frac{x+2}{-x^3-5x^2-6x}$ is defined, it is equivalent to which of the following expressions?

A. $\frac{-1}{x+3}$

B. $\frac{1}{x+3}$

C. $\frac{-x}{x+3}$

D. $\frac{-1}{x^2+3x}$

E. $\frac{1}{x^2+3x}$

8. If $m>0$, $\frac{4}{m}-\frac{m}{7}=?$

A. $\frac{28-4m}{7m}$

B. $\frac{28-4m}{m^2}$

C. $\frac{4-m}{7m}$

D. $\frac{28-m^2}{7m}$

E. $\frac{4-m}{m-7}$

9. When $x \neq 4$ and $x \neq -4$, $\frac{3x}{x^2-16}+\frac{3x}{4-x}$ is equivalent to:

A. $\frac{3x^2}{x^2-16}$

B. $\frac{-3x^2-12x}{x^2-16}$

C. $\frac{-3x^2-9x}{x^2-16}$

D. $\frac{3x^2-15x}{x^2-16}$

E. $\frac{-9x}{x^2-16}$

10. Which of the following expressions is equal to $\frac{5}{7-\sqrt{8}}$?

A. $\frac{35-\sqrt{40}}{41}$

B. $\frac{35+5\sqrt{8}}{41}$

C. $\frac{5}{33}$

D. $\frac{35-5\sqrt{8}}{33}$

E. $\frac{5}{41}$

ANSWER KEY

1. C **2. A** **3. B** **4. C** **5. A** **6. B** **7. D** **8. D** **9. C** **10. B**

ANSWER EXPLANATIONS

1. **C.** We can split the fraction up into two separate fractions to get $\frac{6x}{24x^3}+\frac{12}{24x^3}$. Simplifying each fraction, we get $\frac{1}{4x^2}+\frac{1}{2x^3}$.

2. **A.** We must first recognize that the numerator is a difference of two squares. We can then factor it to be $x^2-225=(x+15)(x-15)$. Rewriting the fraction, we get $\frac{(x+15)(x-15)}{(x-15)^3}=\frac{x+15}{(x-15)^2}$.

3. **B.** In order to find a common denominator for both fractions, we first factor both denominators. We get $x^2-6x+9=(x-3)(x-3)$ for the first denominator, and $9-x^2=(3-x)(x+3)$ for the second denominator. We can write the second denominator as $9-x^2=-(x-3)(x+3)$. So, the equation can be rewritten as $\frac{5x}{(x-3)(x-3)}-\frac{5x}{(x-3)(x+3)}$. The least common multiple of both denominators is $(x-3)(x-3)(x+3)$. Writing both fractions with the new denominator, we must multiply the fraction by $\frac{x+3}{x+3}$ and the second fraction by $\frac{x-3}{x-3}$ which gives us $\frac{5x}{(x-3)(x-3)}\times\frac{x+3}{x+3}-\frac{5x}{(x-3)(x+3)}\times\frac{x-3}{x-3}=\frac{5x(x+3)}{(x-3)(x-3)(x+3)}-\frac{5x(x-3)}{(x-3)(x-3)(x+3)}$.

   Now, we can add the numerators to get $\frac{5x(x+3)-5x(x-3)}{(x-3)(x-3)(x+3)}$. Distributing $5x$, we get $\frac{5x^2+15x-(5x^2-15x)}{(x-3)(x-3)(x+3)}=$ $\frac{5x^2+15x-5x^2+15x}{(x-3)(x-3)(x+3)}=\frac{30x}{(x-3)(x-3)(x+3)}=\frac{30x}{(x-3)^2(x+3)}$.

4. **C.** Factor out an x^2 from the denominator $\frac{(x-3)^2}{x^4-9x^2}=\frac{(x-3)^2}{x^2(x^2-9)}$. x^2-9 is the difference of squares, so your expression becomes $\frac{(x-3)^2}{x^2(x+3)(x-3)}$. Simplify to get $\frac{(x-3)}{x^3+3x^2}$, answer (C). Answer (A) incorrectly tries to square the two terms in the numerator instead of expanding, and then tries to split the sums and differences into two different fractions. Answer (B) again tries to square the numerator incorrectly, although it simplifies the resultant expression correctly. Answer (D) forgets the x^2 term that was factored out in the beginning, and answer (E) disregards the numerator and incorrectly assumes that the denominator sum is the same as the sum of fractions.

5. **A.** We want to factor $(x^2+7x+10)$ and (x^2+x-6) into something we can cancel out. To factor $(x^2+7x+10)$, we need two numbers that add up to equal 7 and multiply to equal 10. The numbers 5 and 2 work, so we can factor it as $(x+2)(x+5)$. To factor (x^2+x-6), we need two numbers that add up to equal 1 and multiply to equal -6. The numbers -2 and 3 work, so we can factor it as $(x+3)(x-2)$. Substituting these factored forms in, we get $\frac{(x+2)(x+5)(x-2)}{(x+3)(x-2)(x+5)}$. Canceling out terms, we get $\frac{x+2}{x+3}$.

6. **B.** Instead of trying to find a common denominator to add the fractions, let's find a way to set the numerators equal. Multiply the A term top and bottom by 20, and the B term top and bottom by 3: $\frac{A}{15}\left(\frac{20}{20}\right)+\frac{B}{100}\left(\frac{3}{3}\right)=\frac{20A+3B}{5x}\to\frac{20A}{300}+\frac{3B}{300}=\frac{20A+3B}{5x}$. Without even trying, we have a common denominator and can thus combine these two fractions! $\frac{20A}{300}+\frac{3B}{300}=\frac{20A+3B}{5x}\to\frac{20A+3B}{300}=\frac{20A+3B}{5x}$. Since the numerators are equal, the denominators must be equal as well, giving is the equation: $300=5x\to x=60$. Remember to put what the question asks for, not the value of the denominator!

7. **D.** Because the expression is defined, we don't have to worry about any holes or asymptotes, we can just simplify the expression as far as we can go. Factor the denominator and simplify: $\frac{x+2}{-x^3-5x^2-6x}\to\frac{x+2}{-x(x+3)(x+2)}\to\frac{x+2}{-x(x+3)(x+2)}\to\frac{-1}{x^2+3x}$.

8. **D.** Multiplying the first fraction by $\frac{7}{7}$ and the second fraction by $\frac{m}{m}$, we get $\frac{4}{m}\left(\frac{7}{7}\right)-\frac{m}{7}\left(\frac{m}{m}\right)=\frac{28-m^2}{7m}$.

9. **C.** To make things easier, we can rewrite $\frac{3x}{4-x}$ as $\frac{3x}{-x+4}$. We want the two fractions have a common denominator, so we multiply $\frac{3x}{-x+4}$ by $\frac{-x-4}{-x-4}$ to have a denominator of x^2-16. This equals $\frac{-3x^2-12x}{x^2-16}$. Now with a common denominator, we can add the two fractions together: $\frac{3x}{x^2-16}+\frac{-3x^2-12x}{x^2-16}=\frac{-3x^2-9x}{x^2-16}$.

10. **B.** When a square root added to another term is in the denominator, we multiply the top and bottom by the conjugate (defined earlier in the chapter or same expression with the sign for the square root flipped). In this problem, we would multiply $\frac{5}{7-\sqrt{8}}$ by $\frac{5}{7+\sqrt{8}}$, which equals $\frac{5(7+\sqrt{8})}{(7-\sqrt{8})(7+\sqrt{8})}=\frac{35+5\sqrt{8}}{49-8}\to\frac{35+5\sqrt{8}}{41}$.

CHAPTER

20 LOGARITHMS

SKILLS TO KNOW

- Definition of a logarithm
- Logarithm rules & their application
- Solving problems with logarithms
- Definition of a natural log

WARNING: Before tackling this chapter, it is important that you have a strong knowledge of exponents (**Chapter 15**).

LOGARITHM BASICS

Whenever you must solve for an exponent, using logarithms or "logs" is one path to doing so. A logarithm is essentially an exponential expression but expressed in a different way to facilitate finding the exponent when you know the other parts of the expression.

For example, a logarithm could be used to answer the question: "What power do we have to raise 2 to in order to get 128?" Using exponential expression I could turn that sentence into this equation:

$$2^x = 128$$

As you can see, however, using our normal old math skills or even the rules of exponents will not give us an easy answer unless we have this relationship memorized (or we start dividing like mad by two, but that only works with integers…). We can rewrite this expression using logs as follows:

$$log_2 128 = x$$

Doing so makes it easier to solve for x. (We'll come back to this problem later in the chapter.)

Re-writing exponents into log form is something like rearranging the numbers according to a secret code. You need to have this basic structure memorized. Here is a simple exponential expression most people know the answer to, and how it translates to log form:

$$2^3 = 8$$

$$\log_2 8 = 3$$

The lower expression reads: "log base 2 of eight equals 3." Let's look at each piece of these two **mathematically equivalent** equations:

2 is our **base**. I think the base is "low" and it's the part that is raised to a power, and also sits "low" next to the word log.

3 is our **exponent**. The phrase I remember to write in log form is "the exponent is the answer." Remember logs solve for exponents.

8 is the number we are taking our log **"of."** We say "log base 2 of 8 is…" It is the "answer" of the original exponentially expressed equation.

THE DEFINITION OF A LOGARITHM

$\log_c a = b$ means that $c^b = a$

Here, c is our **base**, a is our **original "answer"** we are taking the log **"of"** and **"b"** our exponent is the answer to the logarithmic expression.

Again, placing numbers in log form makes it easier to solve for the exponent as a calculator has the ability to solve a logarithmic expression. However, the ACT will never require a calculator to solve any log problems. But that doesn't mean using your calculator isn't a good way to sometimes solve these problems.

What is the value of x in the equation $100 = 10^x$?

If we didn't know the answer (2), we could solve it using a log and our calculator: $\log_{10}(100) = x$. But how? Let's learn about Common Logarithms and the built in function LOG on your calculator to find out.

Common Logarithms

$$\log(x) = \log_{10}(x)$$

The word **"log"** without any base next to it **is short for base 10**. In other words, $\log_{10}(100) = x$. can be expressed as $\log(100) = x$. So the word "log" on your calculator will automatically solve for anything in base 10, such as the expression above. We call this a **common logarithm.**

Plugging $\log(100)$ into the calculator (hit LOG then 100 then ENTER), we get $x = 2$. This makes sense because 10 raised to the power of 2 is 100. Thus the answer to the question above is two.

Answer: 2.

If $x = 50^{(b-c)}$, what is $\log_{50}(x) + c$?

A. $-b$ **B.** b **C.** $\log_{50} b$ **D.** $\log_b 50$ **E.** $\frac{\ln b}{\ln 50}$

Using the definition of logarithms: $\log_c a = b$ means that $c^b = a$

$x = 50^{(b-c)}$ can be rewritten as $\log_{50} x = b - c$

Since $\log_{50} x = (b - c)$ and we're looking for $\log_{50}(x) + c$, we can just add c to both sides of the equation:

$$\log_{50} x + c = b - c + c$$

Isolating the second half of the equation we can simplify to get what we need:

$$(b - c) + c = b$$

If you can't see why this works, you can also think of this as substituting into the expression. Remember the pattern of equation-expression (we cover this in **Basic Algebra** in Part One). Whenever you have an equation and are asked for an expression: substitute! That pattern holds here. "If $\log_{50} x = (b-c)$ what is $\log_{50}(x)+c$? See the big ugly clump that is present in both the equation and the expression? $\log_{50}(x)$? We are essentially group substituting in $b-c$ for that big ugly clump to get $b-c+c$ or b as our answer.

Answer: **B**.

Natural Logarithms

A special kind of logarithm is the **natural log**, which is simply a logarithm with a **base of** e.

$$\ln(x) = \log_e(x)$$

A natural log is denoted as $\ln(\ )$. This is the same as writing $\log_e(\ )$. The natural log (as well as e in general) is useful in continuous growth problems (think compounding interest), and e is a defined number around 2.7 in value, but all that's not terribly important. What's more important is that you can you solve problems that involve e and $\ln$ should they appear on the ACT®.

CHANGE OF BASE FORMULA

Your calculator knows every value of log (log base 10) and ln (natural log base e), but in reality, many log problems are not base 10 or base e. So what do you do if you don't have a log conveniently in base 10 or base e? True, some advanced calculators may perform logs of any base (though often this function is hidden in a menu somewhere). For all other cases, though, you'll need to use the **Change of Base Formula**. Using this formula, you can calculate logs of any base by converting them to a different base.

CHANGE OF BASE FORMULA

For all positive numbers a, b, and c, where $b \neq 1$, and $c \neq 1$:

$$\log_b a = \frac{\log_c a}{\log_c b}$$

Example: $\log_3 9 = \dfrac{\log_{10} 9}{\log_{10} 3}$

This formula is pretty easy to remember, if you look at the taller number in the original expression first (that goes on top) and the lower number in the original expression (the base) which goes on the bottom. Now let's get back to that problem from the beginning:

What power do we have to raise 2 to in order to get 128?

Again I could start by setting up:

$$2^x = 128$$

Then I rearrange in log form:

$$\log_2 128 = x$$

Now I apply the Change of Base Formula:

$$\log_2 128 = \frac{\log 128}{\log 2}$$

And use my calculator: (LOG 128)/(LOG 2) press ENTER and I get 7 .

Answer: **7.**

If the number of bacteria in a colony, n, is a function f such that $n = 2^t$, after how many hours, to two decimal places, will there be 100,000 bacteria? NOTE: $LOG(2) \approx .3010$

A. 16.27 hours **B.** 16.61 hours **C.** 17.11 hours **D.** 23.46 hours **E.** 50.00 hours

$100{,}000 = 2^t$, which implies $\log_2 100{,}000 = t$.

If you plug this equation into your calculator using the **Change of Base Formula** (Enter (LOG 100000)/(LOG 2) and hit ENTER), you'll find that it equals 16.61. Thus, after 16.61 hours, there will be 100,000 bacteria. (You can use the LOG values in your calculator, or the ones ACT gives you if you prefer; ACT will never require a calculator, hence the value of LOG(2) given).

This can be checked by plugging in $t = 16.61$ to see that $2^{16.61} = 100{,}000$.

Answer: **B.**

MORE PROPERTIES OF LOGARITHMS

POWER PROPERTY

$$a \log_b x = \log_b x^a$$

PRODUCT PROPERTY

$$\log_a x + \log_a y = \log_a xy$$

QUOTIENT PROPERTY

$$\log_a x - \log_a y = \log_a \frac{x}{y}$$

These rules, in addition to the **definition of a log** and the **Change of Base Formula** discussed earlier, allow you to manipulate and solve many problems with logarithms, so memorize them for the ACT®!

Each of these rules is essentially related to the exponent rules, but each rule is written differently in the same way that a log rewrites the mathematical information contained in an exponential expression.

Additionally, it's good to understand rules that derive from the definition of logs:

RULES BASED ON THE DEFINITION OF LOGARITHMS

$$n^{\log_n a} = a$$

$$\log_x x^n = n$$

Logarithm of the Base:

$$\log_x x = 1$$

We assume you've learned these in class (typically Algebra II) and just need a refresher. If this is all new to you, you may need to look up some more videos or information on each of these properties and go back to the basics and then complete at least a worksheet's worth of practice on a**t least the first three rules (Google "properties of logarithms worksheet or practice")** in addition to working through this chapter.

Which of the following expressions is equivalent to $3\log\sqrt[9]{x}$?

A. $\log\sqrt[9]{3x}$ **B.** $\log\sqrt[9]{9x}$ **C.** $\log x^{\frac{9}{3}}$ **D.** $\log x^{\frac{1}{3}}$ **E.** $3\log x$

We can rewrite $3\log\sqrt[9]{x}$ as $3\log x^{\frac{1}{9}}$ since we know roots and fractional exponents are equivalent.

Using the **Power Property of Logarithms**, we can then simplify where $a = 3$, $b = \text{base } 10$ (nothing needed, thus ignore b), and $x = x^{\frac{1}{9}}$:

$$a\log_b x = \log_b x^a$$

$$3\log x^{\frac{1}{9}} = \log\left(x^{\frac{1}{9}}\right)^3$$

Distributing the exponent 3, we get the answer:

$$\log x^{\frac{3}{9}} = \log x^{\frac{1}{3}}$$

Answer: **D.**

What is $\log\sqrt{x}+\log x^3$?

A. $\log\left(x^3+\sqrt{x}\right)$ **B.** $\log\left(x^3-\sqrt{x}\right)$ **C.** $\log x^{\frac{7}{2}}$ **D.** $\log x^{\frac{5}{2}}$ **E.** $\log x^{\frac{3}{2}}$

This question is testing our knowledge of general logarithm rules. In this case, we need to know the Product Property of Logarithms:

$$\log_a x+\log_a y=\log_a xy$$

Here we have:

$$\log\sqrt{x}+\log x^3$$

Which we can rewrite as:

$$\log x^{\frac{1}{2}}+\log x^3$$

to make things a bit easier.

Now we can see the pattern matches the property rule above. Our base is 10 (because no base is written, we assume 10), so we can effectively ignore the "a" value, and simply multiply the elements:

$$\log x^{\frac{1}{2}}+\log x^3=\log\left(x^{\frac{1}{2}}x^3\right)$$

Now we apply an exponent rule, the Product of Powers ($a^b a^c=a^{b+c}$) (see Exponents chapter):

$$\log\left(x^{\frac{1}{2}}x^3\right)=\log x^{3+\frac{1}{2}}$$
$$=\log x^{\frac{7}{2}}$$

Answer: **C.**

What is $\log_3 53-\log_3 159$?

A. -1 **B.** $\frac{1}{3}$ **C.** $\log(-106)$ **D.** $\log_3 8427$ **E.** 3

We can simplify $\log_3 53-\log_3 159$ using the Quotient Property of Logarithms:

$$\log_a x-\log_a y=\log_a \frac{x}{y}$$

$$\log_3 53-\log_3 159=\log_3 \frac{53}{159}$$

Simplifying the fraction $\frac{53}{159}$, we get $\log_3 \frac{1}{3}$.

We can then use our calculator to solve ((LOG 1/3) / (LOG 3) ENTER) to get our answer: -1.

Alternatively, we could rewrite this, pulling the negative one outside using the Power Property:

$$\log_b x^a = a\log_b x$$

$$\log_3 3^{-1} = -1(\log_3 3)$$

Then we apply Logarithm of the Base:

$$\log_x x = 1$$

$$-1(\log_3 3) = -1(1)$$

And our answer again is -1.

Answer: **A.**

What is $10^{\frac{3}{2}\log 16}$?

A. $\frac{32}{3}$ **B.** $16^{\frac{2}{3}}$ **C.** 24 **D.** 64 **E.** 10^{24}

One way to solve this is to let your calculator do the work. Solve for $\log 16$, multiply by 1.5, and then store that in your calculator. Then take 10 to that power. **The answer is 64.**

But we can also do this without a calculator. I'll make up a variable "x" to help:

$$10^{\frac{3}{2}\log 16} = x$$

Now let's put this in log form. Remember, the exponent is the answer, and 10 is the base so we can use the common logarithm (i.e. we don't have to write "10" next to log).

$$\log x = \frac{3}{2}\log 16$$

Now we can move the $\frac{3}{2}$ into the "$\log 16$" portion by applying the Power Property of Logarithms.

$$a\log_b x = \log_b x^a$$

$$\frac{3}{2}\log 16 = \log 16^{\frac{3}{2}}$$

Back to our equation, substituting this in we get:

$$\log x = \log 16^{\frac{3}{2}}$$

Now we can drop the word "log":

$$x = 16^{\frac{3}{2}}$$

Now I split the root off from the fractional exponent, factoring out the $\frac{1}{2}$ power:

$$x = \left(16^{\frac{1}{2}}\right)^3$$
$$x = \left(\sqrt{16}\right)^3$$
$$x = 4^3$$
$$x = 64$$

A third way to approach this would be to rewrite $10^{\frac{3}{2}\log 16}$ as $10^{\log 16^{\frac{3}{2}}}$ and then use the rule $n^{\log_n a} = a$. Because a log's base is 10 (unless otherwise indicated), we know that $10^{\log 16^{\frac{3}{2}}}$ can be simplified to $16^{\frac{3}{2}}$, which then equals 64.

Answer: **D**.

Which of the following is equal to $\log_{100} 548$?

A. $\log 5.48$ **B.** $\log 548 - \log 100$ **C.** $\log 548^2$ **D.** $\frac{\log 548}{2}$ **E.** $\frac{\log 548}{\log 10}$

For this problem, we could plug this into our calculator using the **Change of Base Formula** (LOG 548 divided by LOG 100), and then back solve each answer choice by plugging each one into our calculator to find the matching answer.

Alternatively, we probably save some time by working through some algebra. We can use the **Change of Base Formula** $\log_b a = \frac{\log_c a}{\log_c b}$ to convert $\log_{100} 548$ to $\frac{\log 548}{\log 100}$, and because the we know that a log without an explicit base is in base 10, we can solve for the bottom knowing that 10 squared is 100, so the exponent 2 is equivalent to $\log 100$. Thus, we can simplify to $\frac{\log 548}{2}$.

Answer: **D**.

1. If x, y, z are positive real numbers, which of the following expressions is equal to:

$$3\log_2 x - \log_4 y + \frac{1}{2}\log_2 z$$

A. $\log_2 \frac{x^2\sqrt{z}}{2y}$
B. $\log_2 \frac{x^3 z}{2} - \log_4 y$
C. $\frac{3}{2}\log_2(x+z) - \log_4 y$
D. $\log_2 x^3\sqrt{z} - \log_4 y$
E. $\log_2\left(x^3 + \sqrt{z}\right) - \log_4 y$

2. If $\log_4 3 = a$ and $\log_4 5 = b$, which of the following is equal to 8?

A. 4^{a+b}
B. $4^a + 4^b$
C. 16^{a+b}
D. ab
E. $a+b$

3. If $\log_a x = n$ and $\log_a y = m$ then $\log_a\left(\frac{x}{y}\right)^3 = ?$

A. $3(n-m)$
B. $3(n+m)$
C. $n-m$
D. $3(m-n)$
E. $\frac{n}{m}$

4. If $3^{x-1} = 3y$, what is 3^{x+1} in terms of y?

A. $27y$
B. $3y$
C. $3y+2$
D. $(3y)^2$
E. $9y$

5. If $2^{a+2} = 4b$, which of the following is an expression for b^2 in terms of a?

A. $\frac{1}{2^{2a}}$
B. 4^{2a}
C. 2^{a+1}
D. 2^{a+2}
E. 2^{2a}

6. If $2^n = 53$, then which of the following must be true?

A. $2 < n < 3$
B. $3 < n < 4$
C. $4 < n < 5$
D. $5 < n < 6$
E. $6 < n$

7. Which of the following is a value of x that satisfies $\log_x 27 = 3$?

A. 3
B. 6
C. 9
D. 24
E. 27

8. If $\log_x 625 = 4$, then $x = ?$

A. 5
B. 25
C. $\frac{625}{4}$
D. $\frac{625}{\log 4}$
E. 625^4

9. What is x if $\log_6 x = 2$?

A. 3
B. $\sqrt{6}$
C. $\sqrt[6]{2}$
D. 36
E. 12

10. For all $x>0$, which of the following expressions is equivalent to $\log\left[\left(\frac{3}{x}\right)^{\frac{1}{3}}\right]$?

A. $\log\frac{1}{x}$
B. $\log 1-\log\frac{x}{3}$
C. $\frac{1}{3}\left[(\log 3)+(\log x)\right]$
D. $\frac{1}{3}(\log 3-\log x)$
E. $\log 3-\frac{1}{3}\log x$

11. What is the value of $\log_4 64$?

A. 2
B. 3
C. 60
D. 4
E. 16

12. What value of x satisfies the following equation $\log_{16} x=\frac{-3}{4}$?

A. $\frac{-16}{3}$
B. -4
C. $\frac{1}{8}$
D. $\frac{1}{4}$
E. 4

13. If a is a positive number such that $\log_a\left(\frac{1}{125}\right)=-3$, then $a=$?

A. 5
B. 25
C. 128
D. $\frac{1}{5}$
E. $\frac{1}{25}$

14. What is the set of all values of a that satisfy the equation $\left(y^2\right)^{a^2+10a+25}=1$ if $y\neq 1$?

A. $\{0\}$
B. $\{5\}$
C. $\{-10\}$
D. $\{-5\}$
E. $\{-5,5\}$

15. What is the real value of a in the equation $\log_3 54-\log_3 6=\log_6 a$?

A. 3
B. 12
C. $\frac{1}{3}$
D. 36
E. $\frac{8}{3}$

ANSWER KEY

1. D **2. B** **3. A** **4. A** **5. E** **6. D** **7. A** **8. A** **9. D** **10. D** **11. B** **12. C** **13. A** **14. D** **15. D**

ANSWER EXPLANATIONS

1. **D.** Since $a\log_b x = \log_b x^a$, we can rewrite $3\log_2 x$ as $\log_2 x^3$ and $\frac{1}{2}\log_2 z$ as $\log_2 \sqrt{z}$. Our equation can now be written as $\log_2 x^3 - \log_4 y + \log_2 \sqrt{z}$. Combining the two terms with log base 2, we use the property $\log_a x + \log_a y = \log_a xy$ to rewrite the expression: $\log_2 x^3 - \log_4 y + \log_2 \sqrt{z} \rightarrow \log_2 x^3 + \log_2 \sqrt{z} - \log_4 y \rightarrow \log_2 x^3\sqrt{z} - \log_4 y$.

2. **B.** By the definition of a logarithm, $y = b^x$ is equivalent to $log_b(y) = x$. Thus, $log_4(3) = a$ is equivalent to $4^a = 3$, and $log_4(5) = x$ is equivalent to $4^b = 5$. We can then add the two equations.

$$\begin{array}{r} 4^a = 3 \\ + \; 4^b = 5 \\ \hline 4^a + 4^b = 3 + 5 \end{array}$$

Thus, $8 = 4^a + 4^b$.

3. **A.** Since $a\log_b x = \log_b x^a$, we can write $\log_a\left(\frac{x}{y}\right)^3$ as $3\log_a\left(\frac{x}{y}\right)$. Since $\log_a\left(\frac{x}{y}\right) = \log_a x - \log_a y$, we can write $3\log_a\left(\frac{x}{y}\right)$ as $3(\log_a x - \log_a y)$. Now, substituting in $\log_a x = n$ and $\log_a y = m$, we get $3(\log_a x - \log_a y) = 3(n - m)$.

4. **A.** $3^{x+1} = 3^{x-1}(3^2)$ so substituting $3y$ for 3^{x-1}, we get $3^{x+1} = 3y(3^2) = 27y$.

5. **E.** We can write 2^{a+2} as $2^a 2^2$ which simplified becomes $2^a(4)$. Dividing both sides of the equation by 4, we get $2^a = b$. Now, squaring both sides, we get $2^{2a} = b^2$.

6. **D.** Looking at the powers of 2, we know that $2^5 = 32$ and $2^6 = 64$. Since $2^5 = 32 < 2^n = 53 < 2^6 = 64$, $5 < n < 6$.

7. **A.** Raising x to the values on both sides of the equation, we get $x^{(\log_x 27)} = x^3 \rightarrow 27 = x^3$. Taking the cube root of both sides, we get $3 = x$.

8. **A.** By the definition of a logarithm, $\log_x 625 = 4$ is equivalent to $x^4 = 625$. Taking the 4th root of both sides, we get $x = \sqrt[4]{625} \rightarrow x = 5$.

9. **D.** By the definition of a logarithm, $\log_6 x = 2$ is equivalent to $6^2 = x$, so $x = 36$.

10. **D.** Since $a\log_b x = \log_b x^a$, we can write $\log\left[\left(\frac{3}{x}\right)^{\frac{1}{3}}\right] = \frac{1}{3}\log\left(\frac{3}{x}\right)$. Since $\log_a\left(\frac{x}{y}\right) = \log_a x - \log_a y$, we can write $\frac{1}{3}\log\left(\frac{3}{x}\right) = \frac{1}{3}(\log 3 - \log x)$.

11. **B.** We want to find the value that $\log_4 64$ is equal to, which we will call x. By the definition of a logarithm, $\log_4 64 = x$ is equivalent to $4^x = 64$. Since we know that $4^3 = 64$, we know $x = 3$.

12. **C.** Because we understand what a logarithm represents, we know that $\log_{16} x = \frac{-3}{4}$ is equivalent to:

$$x = 16^{-\frac{3}{4}} = \frac{1}{16^{\frac{3}{4}}} = \frac{1}{2^3} = \frac{1}{8}$$

13. A. Again, because we know the definition of a logarithm, we know that $\log_a\left(\frac{1}{125}\right)=-3$ is equivalent to $a^{-3}=\frac{1}{125}$. This implies that $\frac{1}{a^3}=\frac{1}{125}\rightarrow a^3=125\rightarrow a=5$.

14. D. If $y\neq 1$, then the only way the equation is true is if the exponent equals 0, because $y^0=1$. Thus we know that $2\left(a^2+10a+25\right)=0$. Factoring, we get $2(a+5)(a+5)=0$, which means $y=-5$ only. This question is more of an Exponents question, so see that chapter for more similar problems. You could also take the log of both sides, knowing that the log of 1 is zero and proceed in a likewise fashion, but the exponents method is likely more simple and efficient.

15. D. First I'll simplify the left side of the equation. Since $\log_a\left(\frac{x}{y}\right)=\log_a x-\log_a y$, we can write $\log_3 54-\log_3 6$ as $\log_3\left(\frac{54}{6}\right)\rightarrow\log_3 9$. Using the definition of a logarithm, we can see that this expression is equal to the exponent we must raise 3 to in order to get 9. Thus because $3^2=9$ then $\log_3 9=2$. Now that we know the value of the left side of the equation, $\log_3 54-\log_3 6=2$, let's plug this value into our original equation to get $2=\log_6 a$. Rearranging this given the definition of a log, $6^2=a$. So, $a=36$.

CHAPTER

21 CONICS: CIRCLES AND ELLIPSES

SKILLS TO KNOW

- The Circle Equation and its application
- The Ellipse Equation and its application
- How to approach miscellaneous conic section problems
- Identifying what type of graph (circle, ellipse, hyperbola, polynomial, etc.) an equation corresponds to

NOTE: Conics problems lie at the intersection of coordinate geometry and geometry itself. If you're looking for more work on **circles,** we have an entire chapter on that in our 2nd book in this series. In this chapter, however, we will focus on circle equations, and briefly on ellipses and other graph identification problems. Some of these questions require the distance formula (see **chapter 6**) or knowledge of intercepts (see **chapter 6**). Circle equations appear on the ACT® far more often than the other types of problems covered in this chapter.

If you are short on prep time, or scoring below a 32, focus on circles!

THE CIRCLE EQUATION

The formula for a circle is $(x-h)^2+(y-k)^2=r^2$ with the center point at (h,k).
The radius, r, is the distance from the center point to the edge of the circle.
This graphic shows a visual representation of the variables:

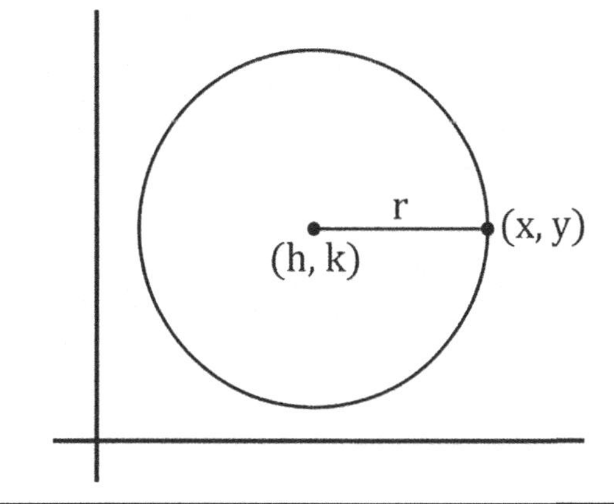

A circle in the standard (x,y) coordinate plane has an equation of $(x-3)^2+(y+7)^2=32$. What are the radius of the circle and the coordinate of the center of the circle?

	Radius	Center
A.	32	$(-3,7)$
B.	32	$(3,-7)$
C.	6	$(3,-7)$
D.	$\sqrt{32}$	$(3,-7)$
E.	$\sqrt{32}$	$(-3,-7)$

We know that the formula for a circle is $(x-h)^2+(y-k)^2=r^2$.

We can match this with the equation given in the problem:

$$(x-h)^2+(y-k)^2=r^2$$
$$\downarrow \qquad \downarrow \quad \downarrow$$
$$(x-3)^2+(y+7)^2=32$$

Matching the terms, we know that $h=3$, $k=-7$ and $r^2=32$. Now we know the points of the circle's center $(3,-7)$ and that the radius $=\sqrt{32}$. Sometimes a problem like this will simplify the radical, but sometimes it won't. Always check your answers before completing that step.

$\sqrt{32}$ can be simplified to $4\sqrt{2}$—but as you can see that is not an option.

Answer: **D.**

A circle in the standard (x,y) coordinate plane has center $(9,-5)$ and passes through the point $(1,1)$. Which of the following equations represents this circle?

A. $(x-9)^2-(y+5)^2=100$ **B.** $(x-9)^2+(y-5)^2=100$ **C.** $(x-9)^2-(y+5)^2=10$

D. $(x-9)^2+(y+5)^2=100$ **E.** $(x+9)^2+(y-5)^2=10$

We know that the equation of a circle is:

$$(x-h)^2+(y-k)^2=r^2$$

The problem tells us the center of the circle, so:

$$h=9;\ k=-5$$

First, we can plug this into the x and y in the equation to get:

$$(x-9)^2+(y--5)^2=r^2$$
$$(x-9)^2+(y+5)^2=r^2$$

Now all we need is the radius. We find this by plugging into this equation the given point the graph passes through, (1, 1) for x and y:

$$(1-9)^2+(1+5)^2=r^2$$

Now we solve for the radius, r.

$$(-8)^2+(6)^2=r^2$$

$$64+36=r^2$$

$$100=r^2$$

We can actually just leave the r squared and plug back in to the equation we have from earlier.

Thus, our answer is $(x-9)^2+(y+5)^2=100$.

(Note: You could also calculate the radius by using the distance formula for the center point and the given point on the circle.)

Answer: **D.**

In the standard (x,y) coordinate plane below, the vertices of a square have coordinates $(0,0)$, $(0,8)$, $(8,8)$ and $(8,0)$. Which of the following is the equation of the circle that circumscribes the square?

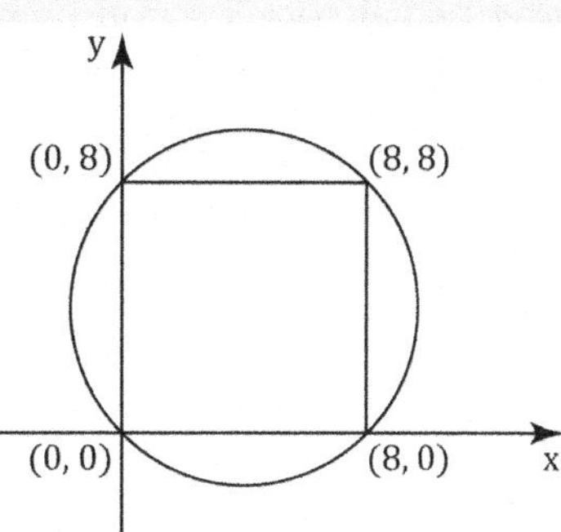

A. $(x-4)^2+(y-4)^2=32$ **B.** $(x-4)^2+(y-4)^2=\sqrt{32}$ **C.** $(x+4)^2+(y+4)^2=32$

D. $(x-4)^2+(y-4)^2-16$ **E.** $(x+4)^2+(y+4)^2=4$

Although the center point and radius aren't given explicitly, we know that the center is the midpoint of the two corners of the square (either diagonal works).

The midpoint formula is $\left(\frac{x_1+x_2}{2},\frac{y_1+y_2}{2}\right)=(a,b)$

Let's use points $(0,0)$ and $(8,8)$:

$$\left(\frac{8-0}{2},\frac{8-0}{2}\right)=(4,4)$$

The center of the circle is $(4,4)$. Now, our circle's formula is $(x-4)^2+(y-4)^2=r^2$.

We can plug in any of the 4 points to solve for the radius. We'll plug in $(8,8)$:

$$(8-4)^2+(8-4)^2=r^2$$
$$4^2+4^2=r^2$$
$$16+16=r^2$$
$$32=r^2$$

Therefore, $(x-4)^2+(y-4)^2=32$.

We could also calculate the radius by finding the length of the diagonal of the square. Using the 45-45-90 triangle ratio, a square with side 8 would have a diagonal of $8\sqrt{2}$. The length of the diagonal is also the diameter of the circle. To get the radius, we divide by 2 to get $4\sqrt{2}$. We then square this to get the radius squared for our equation, 32. The techniques for this method are covered in Book 2.

Answer: **A.**

ELLIPSES

An ellipse is defined as a regular oval shape or an oblong circle. Each ellipse has a horizontal range ("width") and a vertical range ("height") and two focal points, or foci. What is most important is that you know the equation for an ellipse. I usually don't recommend students focus on learning about these until they are comfortably scoring above about a 32 on the section; these questions are rare.

THE ELLIPSE EQUATION

The equation for an ellipse is $\frac{(x-h)^2}{a^2}+\frac{(y-k)^2}{b^2}=1$, with the center labeled as (h,k) and the distance from that center to the end points of the major and minor axes denoted by a and b.

A less confusing way to think of a and b is to imagine that each is a sort of radius, and each corresponds to the axis that its numerator refers to. So in the above equation, if you move a units to the right or left of the center point (h), you'll reach the end points of the x-axis of the ellipse. If you move b units up or down from the center point (k), you'll reach the end points of the y-axis of the ellipse. Whatever number is squared under the numerator with an x in it will always be a horizontal, x-axis move. Whatever number is squared beneath the numerator that includes a y element will always be a vertical, y move.

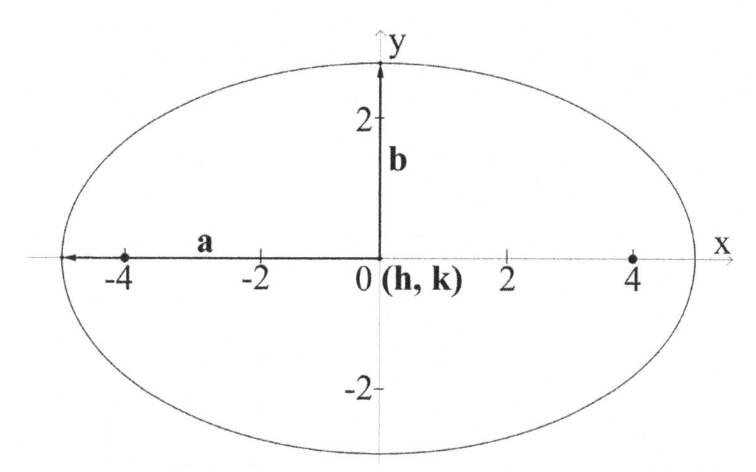

Whichever axis is longer we call the "major axis"—here that is the x-axis. Whichever axis is shorter we call the "minor axis." These terms probably won't be on the ACT®, but I'm mentioning them here to better explain the ideas that might be: how to figure out the equation that matches a particular ellipse graph.

If you think about it, the circle equation is actually a special case of an ellipse, one in which a and b are identical. If we now multiply by r^2 on both sides and clear the fraction, we get $(x-h)^2+(y-k)^2=r^2$, the circle equation.

Though it's unlikely the ACT® will test you on it, the foci of an ellipse are defined such that the distance from any point on the circumference to the two foci added together is equal. For this ellipse, the two foci are at $(-4,0)$ and $(4,0)$. As mentioned above, the summed distance from the two foci to any point on the circumference (denoted by (x,y)), is the same.

Which of the following is the equation of the ellipse that is graphed in the standard (x,y) coordinate plane below?

A. $\frac{(x-2)^2}{81}+\frac{(y-2)^2}{36}=1$ B. $\frac{(x+2)^2}{81}+\frac{(y+2)^2}{36}=1$ C. $\frac{(x+2)^2}{9}+\frac{(y+2)^2}{6}=1$

D. $\frac{(x-2)^2}{9}+\frac{(y-2)^2}{6}=1$ E. $\frac{(x+2)^2}{36}+\frac{(y+2)^2}{81}=1$

First you can calculate out or sketch the center. One way to do this quickly would be to sketch a midline both vertically and horizontally on your test booklet and find their point of intersection. This problem has drawn these lines for us already, but if they're not there, you can draw them yourself.

Here it's clear they intersect at the center point: $(-2,-2)$

TIP: most all pictures on the ACT® are pretty close to scale unless the picture specifically mentions it's not, so you can usually rely on the picture.

Another way to find the center would be to pluck the points on the maximum and minimum points on the ellipse (the end points of the major and minor axes), and then figure out the midpoint of these extremes—that will be your center.

$$\text{Horizontal center: } \frac{-11+7}{2}=-2$$

$$\text{Vertical center: } \frac{-8+4}{2}=-2$$

$$\text{Center: } (-2,-2)$$

Now we find our vertical and horizontal range, and divide by two to find our axes' half lengths (pseudo radii if you will), a and b.

Vertical range: -8 to $4=12$—we divide this by 2 to get our y-axis half length, $b=6$.

$$b=\frac{12}{2}=6$$

Horizontal range: -11 to $7=18$—we divide this by 2 to get our x-axis half length, $a=9$.

$$a=\frac{18}{2}=9$$

Now we plug in our center point and square a and b to get our general form:

$$\frac{(x-h)^2}{a^2}+\frac{(y-k)^2}{b^2}=1$$

$$\frac{(x+2)^2}{81}+\frac{(y+2)^2}{36}=1$$

Answer: **B.**

MISCELLANEOUS QUESTIONS ON CONIC SECTIONS

Occasionally, you might see a problem on the ACT® that involves other conic sections, such as the hyperbola. I won't get into to much detail on this graph, but you should know that **it looks like an ellipse equation or circle equation but has a negative sign between the two fractions with x and y squared terms.**

EQUATION OF A HYPERBOLA

$$\frac{(x-h)^2}{a^2}-\frac{(y-k)^2}{b^2}=1 \text{ or } \frac{(y-k)^2}{b^2}-\frac{(x-h)^2}{a^2}=1$$

I'm not going to get into the details of hyperbolas as they are insanely rare on this test, but you should vaguely know what a graph of one looks like:

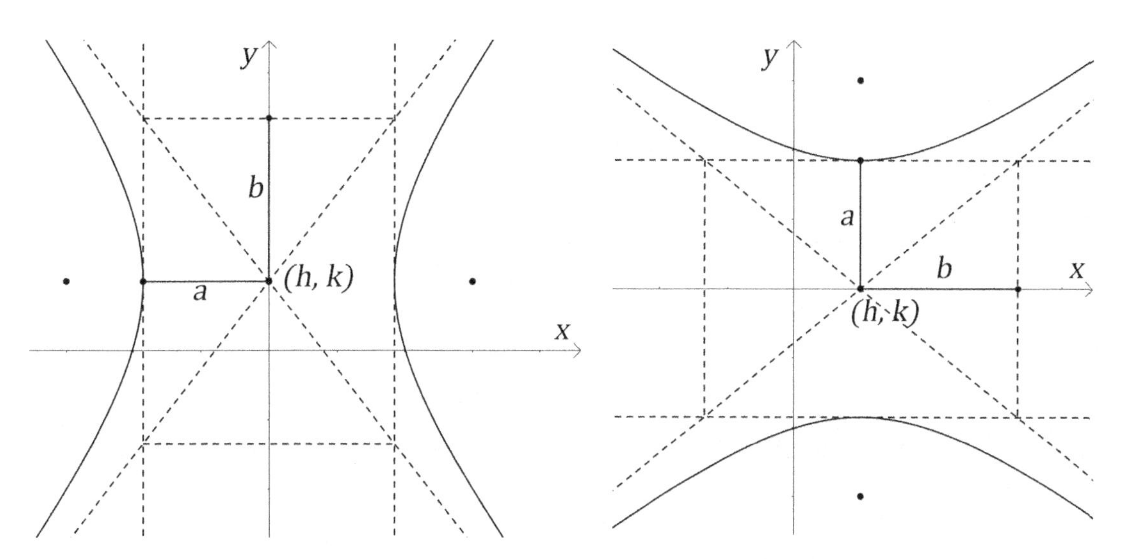

Which of the following points is a possible y-intercept of the graph of $\frac{y^2}{17}-\frac{x^2}{9}=1$?

A. $(0,9)$ **B.** $(0,3)$ **C.** $(0,\sqrt{17})$ **D.** $(0,17)$ **E.** $(0,-17)$

For this question, you don't actually have to know much at all about hyperboles to solve. But you do need to know what a y-intercept is. Remember y "i"ntercepts occur when y "is" the number (I think of the "i" in intercept matching the "i" in is) and x equals zero. x-intercepts occur when x is the number and y equals zero. Don't confuse the two, and you'll be in fine shape. Here I need to simply plug in zero for x and solve for y.

Because all the answers have 0 for x, we can go ahead and plug it in:

$$\frac{y^2}{17}-\frac{0^2}{9}=1$$

$$\frac{y^2}{17}=1$$

$$y^2=17$$

$$y=\pm\sqrt{17}$$

So our answer is either $(0,-\sqrt{17})$ or $(0,\sqrt{17})$.

Answer: **C.**

In the standard (x,y) coordinate plane, the graph of the equation $x^2+y^2+18x-2y=-73$ is a/an:

A. Ellipse **B.** Straight line **C.** Hyperbola **D.** Parabola **E.** Circle

Sometimes, the ACT® will ask you to determine what kind of conic an equation is.

When you must do so, the standard way to solve is to complete the square. No, I don't like completing the square (who does?), but here it may be your best option.

We should already know this is not a straight line because of the x^2 term, and that it is not a parabola because of the y^2 term in addition to the x^2 term. A parabola will only have one squared variable. Therefore, we know the graph must be a conic section—and ellipse, parabola, or circle.

For this problem, we'll simplify to a recognizable conic section form by completing the square. Once we complete the square, we'll analyze the form of the equation and knowing what the standard form of all these equation types are (look earlier in this chapter if you need to review those standard forms), will be able to deduce which answer is correct.

First, we combine like variables: $\left(x^2+18x\right)+\left(y^2-2y\right)=-73$.
Then we complete the square by taking the x terms first.

Completing the Square

STEP 1: Check your coefficient of the squared term, a, and if necessary, divide all elements of the expression or equation by a to make the coefficient of the squared term 1 .

$$ax^2+bx$$
$$x^2+18x$$
$$a=1 \quad b=18$$

Because the coefficient of x is 1, we do not need to divide all terms by "a." If this were not the case, this problem would be much more difficult, and we would need to divide every term in the entire expression by this number and then factor that number out to get:

$$a\left(x^2+\left(\frac{b}{a}\right)x\right)$$

We would then complete the square off of the piece in parenthesis but that can get confusing. Luckily, we don't have to do that here.

STEP 2: Find the number to complete the square.

We now take half of the coefficient of the second term, which in our case is simply $b=18$, and square it. The goal is to create a perfect square, so we're working off the idea of the square of a sum special pattern.

$$\left(a^2+2ab+b^2\right)=\left(a+b\right)^2$$
$$\downarrow \quad \downarrow \quad \downarrow$$
$$x^2+18x+?$$

What we're trying to do is engineer a b squared term that "completes" the square.

Matching the middle terms, because $a=x$, we see $2b=18$. 18 divided by 2 is 9. $b=9$; so $b^2=81$.

STEP 3: Then you add this amount in, and subtract it at the same time, as so, to create an equivalent expression:

$$\left(x^2+18x\right)=\left(x^2+18x+81\right)-81$$

STEP 4: Factor
This gives us a perfect square—remember the special pattern called the "Square of a Sum":

$$\left(x^2+2xy+y^2\right)=\left(x+y\right)^2$$

Now we can factor that perfect square piece according to the pattern:

$$\left(x^2+18x+81\right)=\left(x+9\right)^2$$

And now we substitute this in to our expression:

$$\left(x^2+18x\right)=\left(x+9\right)^2-81$$

We now apply the same process to the y-terms:

$$\left(y^2-2y\right)$$

Take -2, divide it by 2 to get -1, and square it to get 1.
Now add one, and subtract one as so:

$$\left(y^2-2y\right)=\left(y^2-2y+1\right)-1$$

Now we use the pattern from the "Square of a Sum" to make this:

$$\left(y^2-2y\right)=\left(y-1\right)^2-1$$

At this point, we can plug both the x portion and y portion back into our original equation, and simplify:

$$\left(x^2+18x\right)+\left(y^2-2y\right)=-73$$

$$\left(x+9\right)^2-81+\left(y-1\right)^2-1=-73$$

$$\left(x+9\right)^2+\left(y-1\right)^2=-73+81+1$$

$$(x+9)^2+(y-1)^2=9$$

Because this follows the standard equation form of a circle, we know it is a circle:

$$(x-h)^2+(y-k)^2=r^2$$

Answer: **E.**

There is also an alternative, incredibly quick method to solve this problem, but it is hyper-specific to this type of problem and requires memorization. I have it here if you're in the mood for such memory tricks, but completing the square is a more useful skill that will apply to more types of questions and thus I focus on that method above.

CONIC EQUATION TYPES

A conic section of the form $Ax^2+By^2+Cx+Dy+E=0$, in which A and B are both not zero is: A **circle** if $A=B$. A **parabola** if $AB=0$. An **ellipse** if $A\neq B$ and $AB>0$. A **hyperbola** if $AB<0$.

Because A and B in our equation are 1 and 1, we can see that the equation is a circle.

1. In the standard (x,y) coordinate plane, what is the radius of the circle $(x-7)^2+(y+9)^2=169$?

A. 169
B. 3
C. $\sqrt{7}$
D. 7
E. 13

2. A certain circle has equation of $\left(x-\sqrt{5}\right)^2+\left(y+2\sqrt{3}\right)^2=43$ in the standard (x,y) coordinate plane. What are the coordinates of the center of the circle, and the radius of the circle, in coordinate units?

	Center	Radius
A.	$\left(\sqrt{5},-2\sqrt{3}\right)$	$\sqrt{43}$
B.	$\left(-\sqrt{5},2\sqrt{3}\right)$	$\sqrt{43}$
C.	$\left(\sqrt{5},-2\sqrt{3}\right)$	43
D.	$\left(\sqrt{5},-2\sqrt{3}\right)$	21.5
E.	$\left(-\sqrt{5},2\sqrt{3}\right)$	21.5

3. A circle in the standard (x,y) coordinate plane has center $(-4,7)$ and radius 3 units. Which of the following equations represents this circle?

A. $(x+4)^2+(y-7)^2=9$
B. $(x-4)^2+(y+7)^2=9$
C. $(x-4)^2+(y+7)^2=3$
D. $(x+4)^2+(y-7)^2=3$
E. $(x+4)^2-(y-7)^2=9$

4. A circle in the standard (x,y) coordinate plane has center $(-12,8)$ and radius 11 units. Which of the following equations represents this circle?

A. $(x-12)^2-(y+8)^2=121$
B. $(x-12)^2-(y+8)^2=11$
C. $(x-12)^2+(y+8)^2=11$
D. $(x+12)^2+(y-8)^2=121$
E. $(x+12)^2+(y-8)^2=11$

5. What is the equation of a circle in the standard (x,y) coordinate plane that has a radius of 8 and the same center as the circle described as $x^2-4x+y^2+8y+10=0$?

A. $x^2+y^2=64$
B. $(x+4)^2+(y-2)^2=64$
C. $(x-4)^2+(y+2)^2=64$
D. $(x-2)^2+(y+4)^2=64$
E. $(x+2)^2+(y-4)^2=64$

6. In the standard (x,y) coordinate plane, what are the coordinates of the center of the circle whose equation is $x^2+12x+y^2-2y-15=0$?

A. $(6,-1)$
B. $(-1,-6)$
C. $(1,-6)$
D. $(-6,1)$
E. $(6,1)$

7. A circle with an area of 9π square in the standard (x,y) coordinate plane has its center at $(8,7)$. Which of the following is an equation for this circle?

A. $(x-7)^2+(x-8)^2=81$

B. $(x-7)^2+(y-8)^2=9$

C. $(x-7)^2-(y-8)^2=81$

D. $(x-8)^2+(y-7)^2=9$

E. $(x-7)^2+(y-8)^2=81$

8. In the (x,y) coordinate plane, what is the radius of the circle having the points $(0,-8)$ and $(-24,24)$ as endpoints of a diameter?

A. 10
B. 20
C. 40
D. 49
E. 60

9. A circle in the standard (x,y) coordinate plane is tangent to the x-axis at -7 and tangent to the y-axis at -7. Which of the following is an equation of the circle?

A. $x^2+y^2=7$

B. $x^2+y^2=49$

C. $(x+7)^2+(y+7)^2=7$

D. $(x+7)^2+(y+7)^2=49$

E. $(x-7)^2+(y-7)^2=49$

10. What is the center of the circle with the equation $(x+3)^2+(y-3)^2=27$ in the standard (x,y) coordinate plane?

A. $(-\sqrt{3},\sqrt{3})$

B. $(-3,3)$

C. $(\sqrt{3},-\sqrt{3})$

D. $(3,-3)$

E. $(3,3)$

11. Which of the following is an equation of the largest circle that can be inscribed in an ellipse with the equation $\frac{(x+5)^2}{36}+\frac{(y-7)^2}{25}=1$?

A. $(x+5)^2+(y-7)^2=900$

B. $(x+5)^2+(y-7)^2=36$

C. $(x+5)^2+(y-7)^2=25$

D. $x^2+y^2=36$

E. $x^2+y^2=25$

12. Which of the following points is a x-intercept of the graph of $\frac{x^2}{81}+\frac{y^2}{49}=1$?

A. $(81,0)$

B. $(49,0)$

C. $(15,0)$

D. $(9,0)$

E. $(7,0)$

13. Which of the following points is a y-intercept of the graph of $\frac{x^2}{13}+\frac{y^2}{21}=1$?

A. $(0,21)$
B. $(0,13)$
C. $(0,-\sqrt{13})$
D. $(0,\sqrt{21})$
E. $(0,\sqrt{13})$

14. Tyler's Tattoo Parlor's logo consists of three concentric circles. The radius of the innermost circle of the logo is 1 foot. The distance between the innermost circle and the outermost circle is 5.5 feet and the distance between the outermost circle and middle circle is 2.5 feet. Which of the following is an expression for the area, in square feet, of the middle circle of the logo of Tyler's Tattoo Parlor?

A. $(4)^2\pi$
B. $(1+4)\pi$
C. $(1^2+4^2)\pi$
D. $(1+5.5)^2\pi$
E. $(1+5.5)\pi$

15. In the standard (x,y) coordinate plane, the graph of the equation $x^2+8x-5y^2-10y+12=0$ is a(n):

A. Linear line
B. Parabola
C. Circle
D. Ellipse
E. Hyperbola

16. A circle in the standard (x,y) coordinate plane has an equation of $x^2+(y+2)^2=24$. What are the radius of the circle and the coordinate of the center of the circle?

	Radius	Center
A.	24	$(0,2)$
B.	24	$(0,-2)$
C.	$\sqrt{24}$	$(0,-2)$
D.	$\sqrt{24}$	$(0,2)$
E.	12	$(0,2)$

17. A circle in the standard (x,y) coordinate plane has center $(-4,3)$ and is tangent to the y–axis. The point (x,y) is on the circle if and only if x and y satisfy which of the following equations?

A. $(x+4)^2+(y+3)^2=16$
B. $(x-4)^2+(y+3)^2=16$
C. $(x+4)^2+(y-3)^2=16$
D. $(x-4)^2+(y+3)^2=9$
E. $(x+4)^2+(y-3)^2=9$

18. In the standard (x,y) coordinate plane there is only one circle centered at the point $(5,2)$ that also passes through the point $(9,-1)$. Which of the following is the equation for that circle?

A. $(x+5)^2+(y+2)^2=25$
B. $(x-5)^2+(y+2)^2=5$
C. $(x-5)^2+(y+2)^2=25$
D. $(x-5)^2+(y-2)^2=25$
E. $(x-5)^2+(y-2)^2=5$

19. In the standard (x,y) coordinate plane, what is another way of writing the equation for a circle whose equation is $x^2+y^2+4x-2y=4$?

A. $(x-1)^2+(y+2)^2=9$
B. $(x+2)^2+(y+1)^2=3$
C. $(x-1)^2+(y-2)^2=3$
D. $(x+2)^2-(y-1)^2=9$
E. $(x+2)^2+(y-1)^2=9$

20. What is the largest value of x for which there exists a real value of y such that $x^2+y^2=100$?

A. 6
B. 8
C. 10
D. 99
E. 100

21. Which of the following is the equation of the ellipse that is graphed in the standard (x,y) coordinate plane below?

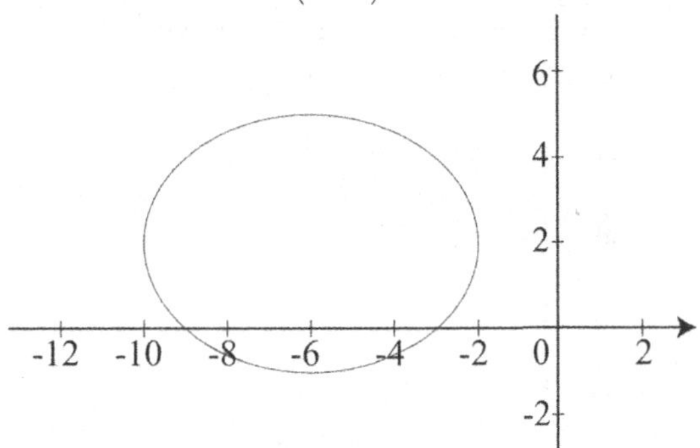

A. $\frac{(x-6)^2}{16}+\frac{(y-2)^2}{9}=1$

B. $\frac{(x-6)^2}{4}+\frac{(y+2)^2}{3}=1$

C. $\frac{(x+6)^2}{16}+\frac{(y+2)^2}{9}=1$

D. $\frac{(x+6)^2}{16}+\frac{(y-2)^2}{9}=1$

E. $\frac{(x+6)^2}{4}+\frac{(y-2)^2}{3}=1$

22. Jenny decides to frame the edge of her elliptical mirror. The perimeter of an ellipse is given by the formula $p=\frac{\pi}{2}\sqrt{2(h^2+w^2)}$, where h is the height and w is the width, as shown in the diagram below. If the mirror has an outside height equal to 5 feet and an outside width equal to 2 feet, what is the outside perimeter, in feet?

A. $\frac{\pi}{2}\sqrt{29}$

B. $\pi\sqrt{29}$

C. $\pi\sqrt{58}$

D. $\frac{\pi}{2}\sqrt{58}$

E. $\frac{\pi}{2}\sqrt{98}$

23. How many points lie on an ellipse?

A. 4
B. 2
C. 1
D. 0
E. Infinitely many

24. One of the following equations determines the graph in the standard (x,y) coordinate plane below. Which one?

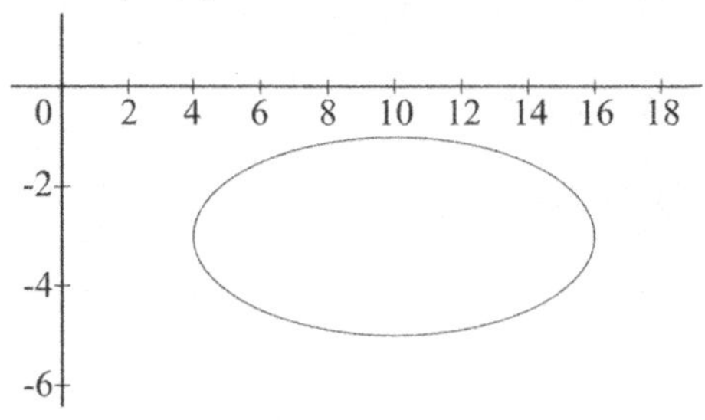

A. $4(x-10)^2+36(y+3)^2=144$

B. $4(x+10)^2+36(y-3)^2=144$

C. $\frac{(x+10)^2}{4}+\frac{(y-3)^2}{36}=1$

D. $\frac{(x-10)^2}{4}+\frac{(y+3)^2}{36}=1$

E. $\frac{(x+10)^2}{36}+\frac{(y-3)^2}{4}=1$

25. Which of the following is the equation of the ellipse that is graphed below?

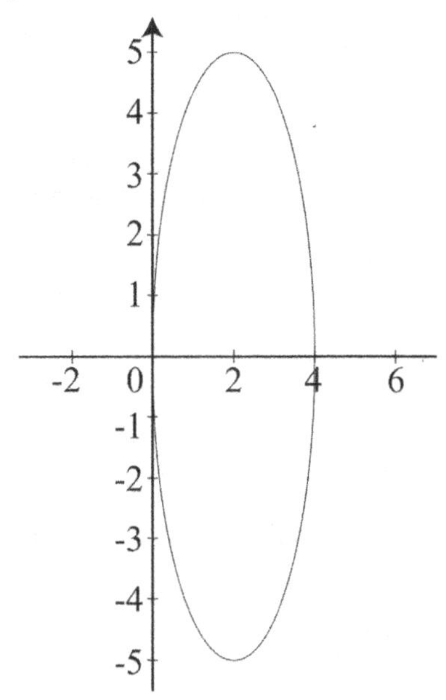

A. $25x^2-100x+y^2=50$

B. $4(x-2)^2+25y^2=100$

C. $25(x-2)^2+4y^2=100$

D. $25x^2-100x+4y^2=100$

E. $\frac{(x+2)^2}{4}+\frac{y^2}{25}=1$

26. In the standard (x,y) coordinate plane, the center of the circle shown below lies on the y -axis at $y=7$. If the circle is tangent to the x -axis, which of the following is an equation of that circle?

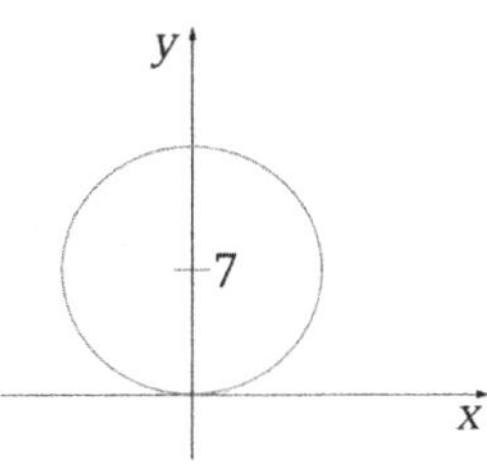

A. $x^2+(y-7)^2=49$
B. $x^2+(y+7)^2=49$
C. $x^2+(y-7)^2=7$
D. $x^2+(y+7)^2=7$
E. $(x-7)^2+y^2=49$

27. Which of the following equations represents the circle with center $(-3,4)$ shown in the standard (x,y) coordinate plane below?

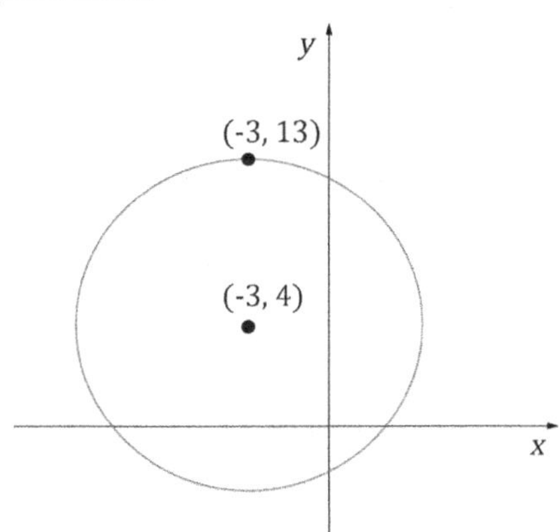

A. $(x-3)^2+(y+4)^2=81$
B. $(x+3)^2+(y-4)^2=81$
C. $(x-3)^2+(y+4)^2=9$
D. $(x+3)^2+(y-4)^2=9$
E. $(x+3)^2-(y-4)^2=81$

28. A diameter of a circle in the standard (x,y) coordinate plane has end points at $(3,8)$ and $(-3,4)$. Which of the following points must also be on the circle?

A. $(0,6)$
B. $(-3,6)$
C. $(-3,8)$
D. $(3,6)$
E. $(0,12)$

29. A circle in the standard (x,y) coordinate plane crosses the points $(-1,-1)$ and $(5,5)$. If the radius of the circle is 6 units, which of the following could be the center of the circle?

I. $(-1,2)$
II. $(-1,5)$
III. $(5,-1)$

A. I only
B. I and II only
C. I and III only
D. II and III only
E. I, II, and III

30. What is the largest value of x for which there exists a real value of y such that $x^2+y^2=100$?

A. 6
B. 8
C. 10
D. 99
E. 100

ANSWER KEY

1. E	**2. A**	**3. A**	**4. D**	**5. D**	**6. D**	**7. D**	**8. B**	**9. D**	**10. B**	**11. C**	**12. D**	**13. D**	**14. A**
15. E	**16. C**	**17. C**	**18. D**	**19. E**	**20. C**	**21. D**	**22. D**	**23. E**	**24. A**	**25. C**	**26. A**	**27. B**	**28. C**
29. D	**30. C**												

ANSWER EXPLANATIONS

1. **E.** The equation of a circle is $(x-a)^2+(y-b)^2=r^2$ where a and b represent the displacement of the center of the circle from the origin and r is the circle's radius. From the equation they give us, we see that $r^2=169$ so $r=13$.

2. **A.** The equation of a circle is in the form $(x-a)^2+(y-b)^2=r^2$ where a and b represent the displacement of the center of the circle from the origin and r is the circle's radius. From the equation they give us, we see that the center is at $(\sqrt{5},-2\sqrt{3})$, and $r^2=43$ so $r=\sqrt{43}$.

3. **A.** The equation of a circle is in the form $(x-a)^2+(y-b)^2=r^2$ where a and b represent the displacement of the center of the circle from the origin and r is the circle's radius. So, the equation of a circle with center at $(-4,7)$ and radius 3 is $(x-(-4))^2+(y-7)^2=3^2$. This can be simplified to be $(x+4)^2+(y-7)^2=9$.

4. **D.** The equation of a circle is in the form $(x-a)^2+(y-b)^2=r^2$ where a and b represent the displacement of the center of the circle from the origin and r is the circle's radius. So, the equation of a circle with center at $(-12,8)$ and radius 11 is $(x-(-12))^2+(y-8)^2=11^2$. This can be simplified to $(x+12)^2+(y-8)^2=121$.

5. **D.** The equation of a circle is in the form $(x-a)^2+(y-b)^2=r^2$ where a and b represent the displacement of the center of the circle from the origin and r is the circle's radius. So, we wish to write the circle described as $x^2-4x+y^2+8y+10=0$ in that form. First, we must complete the square in both the x and y terms. We do this by adding constant terms to make the x and y polynomials perfect squares. These constant terms can be calculated by taking the coefficient of the x term and y term, dividing these by 2, and then squaring them. So, in the case of the x term we have $\left(-\frac{4}{2}\right)^2=4$ and in the case of the y term, we have $\left(\frac{8}{2}\right)^2=16$. Now we add these amounts in, and then subtract them on the back end of the equation $(x^2-4x+4)+(y^2+8y+16)+10-4-16=0$. This factors to $(x-2)^2+(y+4)^2=10$. So, the center of this circle is at $(2,-4)$. We now want to construct the equation for a circle with the same center but with radius 8. This would be $(x-2)^2+(x+4)^2=8^2$ or $(x-2)^2+(x+4)^2=64$.

6. **D.** The equation of a circle is in the form $(x-a)^2+(y-b)^2=r^2$ where a and b represent the displacement of the center of the circle from the origin and r is the circle's radius. So, we wish to write the circle described as $x^2+12x+y^2-2y-15=0$ in that form. First, we must complete the square in both the x and y terms. We do this by adding constant terms to make the x and y polynomials perfect squares. These constant terms can be calculated by taking the coefficient of the x term and y term, dividing them by 2, and then squaring them. So, in the case of the x term we have $\left(-\frac{12}{2}\right)^2=36$ and in the case of the y term, we have $\left(-\frac{2}{2}\right)^2=1$. Now we add in these values to complete the square and simultaneously subtract them from the end of the equation so that the value of the equation is not overall affected: $(x^2+12x+36)+(y^2-2y+1)-15-36-1=0$. This factors to $(x+6)^2+(y-1)^2=52$. So, knowing the center is at (a, b) we can tell that the center of this circle is $(-6,1)$.

7. **D.** The equation of a circle is in the form $(x-a)^2+(y-b)^2=r^2$ where a and b represent the displacement of the center of the circle from the origin and r is the circle's radius. We find the radius of the circle first using the formula $A=\pi r^2$. Plugging in $A=9\pi$, we get $9\pi=\pi r^2 \rightarrow 9=r^2 \rightarrow r=3$. So, the circle with radius 3 and center at $(8,7)$ has the equation $(x-8)^2+(y-7)^2=3^2 \rightarrow (x-8)^2+(y-7)^2=9$.

8. **B.** To find the radius of the circle, we first find the diameter of the circle and divide it by 2 to get the radius. The end points of the diameter are given, so we can find the length of the diameter by finding the distance between the two points $(0,-8)$ and $(-24,24)$. The distance formula is $d=\sqrt{(x_1-x_2)^2+(y_1-y_2)^2}$. Following this formula, we get: $\sqrt{(0-(-24))^2+(-8-24)^2} \to \sqrt{24^2+32^2} \to \sqrt{576+1024} \to \sqrt{1600}=40$. So, $r=\frac{d}{2}=\frac{40}{2}=20$.

9. **D.** If the circle is tangent to the x-axis at -7 and tangent to the y-axis at -7, then the circle is in the third quadrant with a center at $(-7,-7)$ and radius 7. The equation of a circle is in the form $(x-a)^2+(y-b)^2=r^2$ where a and b represent the displacement of the center of the circle from the origin and r is the circle's radius. So, the equation of this circle is $(x-(-7))^2+(y-(-7))^2=7^2 \to (x+7)^2+(x+7)^2=49$.

10. **B.** The equation of a circle is $(x-a)^2+(y-b)^2=r^2$ where a and b indicate the displacement of the center of the circle from the origin and r is the circle's radius. So, the center of the circle of $(x+3)^2+(y-3)^2=27$ is at $(-3,3)$.

11. **C.** The equation of an ellipse is in the form $\frac{(x-a)^2}{c^2}+\frac{(y-b)^2}{d^2}=1$ where a and b represent the displacement of the center of the circle from the origin and c and d represent the horizontal and vertical radii of the ellipse respectively. So, the ellipse represented by $\frac{(x+5)^2}{36}+\frac{(y-7)^2}{25}=1$ has a horizontal "radius" of 6, a vertical "radius" of 5, and a center at $(-5,7)$. Since the ellipse has a shorter vertical "radius," the biggest circle that can be inscribed in the ellipse is limited to the vertical "radius." So, this circle will have radius 5 and center at $(-5,7)$. The equation for this circle is $(x+5)^2+(y-7)^2=25$.

12. **D.** The equation of an ellipse is in the form $\frac{(x-a)^2}{c^2}+\frac{(y-b)^2}{d^2}=1$ where a and b represent the displacement of the center of the ellipse from the origin and c and d represent the horizontal and vertical "radii" of the ellipse respectively. So, the ellipse represented by $\frac{x^2}{81}+\frac{y^2}{49}=1$ has center at $(0,0)$ with a horizontal "radius" of $\sqrt{81}=9$. So, the ellipse intersects the x-axis at $x=9$. This is at the point $(9,0)$. Alternatively, know the x intercept means x is a number and y is zero, so plug in zero for y and solve for x. (See more in Ch 5 for this method).

13. **D.** The equation of an ellipse is in the form $\frac{(x-a)^2}{c^2}+\frac{(y-b)^2}{d^2}=1$ where a and b represent the displacement of the center of the circle from the origin and c and d represent the horizontal and vertical "radii" of the ellipse respectively. So, the circle represented by $\frac{x^2}{13}+\frac{y^2}{21}=1$ has center at $(0,0)$ with a vertical "radius" of $\sqrt{21}$. So, the ellipse intersects the x-axis at $y=\sqrt{21}$. This is at the point $(0,\sqrt{21})$. Alternatively, know the y intercept means y is a number and x is zero, so plug in zero for x and solve for y. (See more in Ch 5 for this method).

14. **A.** The radius of the innermost circle is 1 and the distance between the innermost circle and the outermost circle is 5.5, and the distance between the middle circle and the outermost circle is 2.5. So the distance between the middle circle and the innermost circle is $5.5-2.5=3$. The radius of the middle circle is then $3+1=4$. The area of the middle circle can then be calculated as $A=\pi r^2=(4)^2\pi$. The diagram below illustrates the logo and its dimensions.

1 3 2.5

15. E. Completing the square for the equation, we get $x^2+8x+16-5\left(y^2+2y+1\right)+12-16+5=0$. This is simplified to be $\left(x+4\right)^2-5\left(y+1\right)^2=1$. The equation of a hyperbola is in the form $\frac{\left(x-a\right)^2}{c^2}-\frac{\left(y-b\right)^2}{d^2}=1$, so our equation matches this form. We can also use the shortcut at the end of the chapter, or see that the signs on the x squared and y squared terms are opposites, which indicates a hyperbola.

16. C. The equation provided follows the format of the basic circle formula, where $\left(x-a\right)^2+\left(y-b\right)^2=r^2$, and the circle's center is at $\left(a,b\right)$. Thus, in our equation, $r^2=24$, so $r=\sqrt{24}$. $a=0$ and $b=-2$, since the addition implies double negation (subtracting a negative number), placing the circle's center at $\left(0,-2\right)$.

17. C. When given the description of a conic without a picture, draw a little diagram. This circle is tangent to the y–axis, which means it must have a radius of 4 (because the center is 4 units away from the y axis). Looking at the same equation from the question above and plugging in -4 for a, 3 for b, and 4 for r yields answer (C).

18. D. Finding the answer requires first finding the distance between the two points given, the center and the second point, which will be our radius. Using the distance formula $d=\sqrt{\left(x_1-x_2\right)^2+\left(y_1-y_2\right)^2}$, we see that $d=5$, which means the radius is five. Going back to the basic equation of a circle, after plugging in 5 for a, 2 for b, and 5 for r, D is the obvious answer. The remaining answers try to confuse us by changing the signs and forgetting to square the radius.

19. E. To find the conventional form of the formula, complete the square. Start by rearranging the equation given into two distinct parts, putting the terms with x next to each other and the terms with y next to each other. $x^2+4x+y^2-2y=4$. To complete the square with the x terms, add a 4 to either side (because $\left(\frac{4}{2}\right)^2=4$), and to complete the square with the y terms, add a 1 to either side (because $\left(\frac{2}{2}\right)^2=1$). $(x^2+4x+4)+(y^2-2y+1)=4+4+1$. Now that we have perfect squares, we can factor those 2 separate parts. $\left(x+2\right)^2+\left(y-1\right)^2=9$. This equation is our answer. See lesson for complete instructions on completing the square.

20. C. If we plug in $y=0$, we maximize the value of x. This value is found to be $x^2+0^2=100\rightarrow x^2=100\rightarrow x=10$.

21. D. The standard equation for an ellipse where $a>b$ is $\frac{\left(x-h\right)^2}{a^2}+\frac{\left(y-k\right)^2}{b^2}=1$, with a center at $\left(h,k\right)$, vertices at $\left(h\pm a,\ k\right)$, and co-vertices at $\left(h,\ k\pm b\right)$. To make this easier, though, we essentially need the center point (which is our (h,k)) and the vertical and horizontal "radii" lengths (our a and b; horizontal radius squared goes under the x and vertical squared under the y). From the graph, the ellipse is shown to be centered $\left(-6,2\right)$, so this is our $\left(h,k\right)$. We need the equation thus to say (x+6) and (y-2). We can eliminate all but D and E. Now we can eyeball the "radii" to be $a=4$, and $b=3$ and the equation must square these values. Choice D squares these values (to get 16 and 9 respectively as denominators) so it is correct.

22. D. Though this question includes the word ellipse, it is more accurately a **"Function as a Model"** problem (see chapter of that name in this book). We are given a formula $\left(p=\frac{\pi}{2}\sqrt{2\left(h^2+w^2\right)}\right)$ and some values $\left(h=5,w=2\right)$. Plug in and solve to get answer (D), $\frac{\pi}{2}\sqrt{58}$.

23. E. A central concept to ellipse is that they contain an infinite number of points, just like circles. Although they have a defined center and vertices and co-vertices, there are infinite values in between on the ellipse, so the answer is (E).

24. A. The graph shows an ellipse centered at $(10,-3)$, with vertices at $(4,-3)$ and $(16,-3)$ and co-vertices at $(10,-1)$ and $(10,-5)$. The standard equation for an ellipse where $a>b$ is $\frac{(x-h)^2}{a^2}+\frac{(y-k)^2}{b^2}=1$, with a center at (h,k), vertices at $(h \pm a, k)$, and co-vertices at $(h, k \pm b)$, so the equation for the ellipse shown must be $\frac{(x-10)^2}{6^2}+\frac{(y-(-3))^2}{2^2}=1$, which simplified becomes $\frac{(x-10)^2}{36}+\frac{(y+3)^2}{4}=1$. However, that version is not a possible answer, so try rewriting the formula. Multiply the entire expression by both denominators to eliminate the fractions. $36\times 4\times\left(\frac{(x-10)^2}{36}+\frac{(y+3)^2}{4}\right)=1\times 36\times 4$ $\rightarrow 4(x-10)^2+36(y+3)^2=144$, answer (A).

25. C. The ellipse is centered at $(2,0)$, extends 2 units in either x-direction, and extends 5 units in either y -direction, implying that $h=2, k=0, a=2$ and $b=5$. Plugging those values into the formula gives us $\frac{(x-2)^2}{4}+\frac{y^2}{25}=1$. Simplify that by multiplying both side by the least common denominator to get $25(x-2)^2+4y^2=100$, answer (C).

26. A. The center of the circle is at $(0,7)$ and the radius is 7 because the circle is tangent to the x -axis, meaning that one point on the circle is exactly 7 units away from the center, which is essentially the definition of a radius. Consequently, the circle equation with these values inputted is $x^2+(y-7)^2=49$.

27. B. The center is at $(-3,4)$ and the distance between the center and the given point on the circle (found visually or with the distance equation) is 9, which means the radius is 9. Thus, the circle equation filled in becomes $(x-(-3))^2+(y-4)^2=81$ $\rightarrow (x+3)^2+(y-4)^2=81$.

28. C. We can find the center of the circle by taking the midpoint of $(3,8)$ and $(-3,4)$. We find that the midpoint is $(0,6)$. Now plug into the standard circle equation form to get $(x-0)^2+(y-6)^2=r^2$. The answer is $(-3,8)$ because if $(3,8)$ works in the above equation, we know changing the sign on the x value won't change the answer at all as the negative sign will go away once we square it. Thus this point, too, must be on the circle. If you want, you could solve for the radius in the equation by plugging in one of the points ($(3)^2+(8-6)^2=13$ so the radius is $\sqrt{13}$) and then could verify or reject each answer choice after plugging into the equation ($x^2+(y-6)^2=13$).

29. D. Because the distance between $(-1,-1)$ and $(5,5)$ is 6 units in the x direction and the y direction, the only two points that will be 6 units away from both $(-1,-1)$ and $(5,5)$ (possibilities for the center of the circle) are $(5,-1)$ and $(-1,5)$.

30. C. This is a circle centered at the origin. We know that the farthest point out to the right (i.e. max value of x) will be on a radius that is horizontal, i.e. with a y-value the same as its center. If we plug in $y=0$, we maximize the value of x. This value is found to be $x^2+0^2=100 \rightarrow x^2=100 \rightarrow x=10$.

CHAPTER

22 GRAPH BEHAVIOR

SKILLS TO KNOW

- Finding domain and range
- Recognizing when a function is undefined
- Recognizing horizontal and vertical asymptotes
- Recognize basic parent graphs/functions and understand transformations
- How coefficients and constants affect functions
- How to identify and execute compressions
- Less common but still relevant topics:
  - Trigonometric functions and graphs (these have their own chapter in Book 2 and are not directly covered in this chapter, except for rules that apply to all graphs)
  - Even, odd, and one-to-one functions
- Understand translations and reflections, along with associated vocabulary

NOTE: I cover some graph behavior topics in **Chapter 16: Quadratics and Polynomials**. I recommend doing that chapter first.

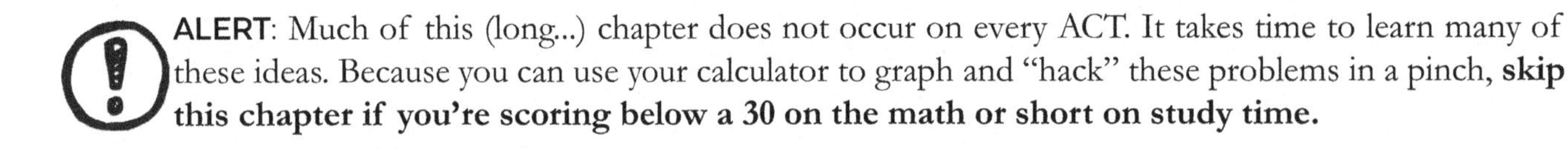

ALERT: Much of this (long...) chapter does not occur on every ACT. It takes time to learn many of these ideas. Because you can use your calculator to graph and "hack" these problems in a pinch, **skip this chapter if you're scoring below a 30 on the math or short on study time.**

DOMAIN AND RANGE

Two terms you'll need to know to interpret graphs on the ACT® are domain and range.

> **DOMAIN:** all the possible values of x in a function or relation. If it's a possible value of x, it is part of the domain. You can find the domain of a function if you have its equation, its full set of values, or its graph.

Domains can be continuous. The graph below represents a function in which all values of x are greater than zero, so the domain of the inequality is $x > 0$.

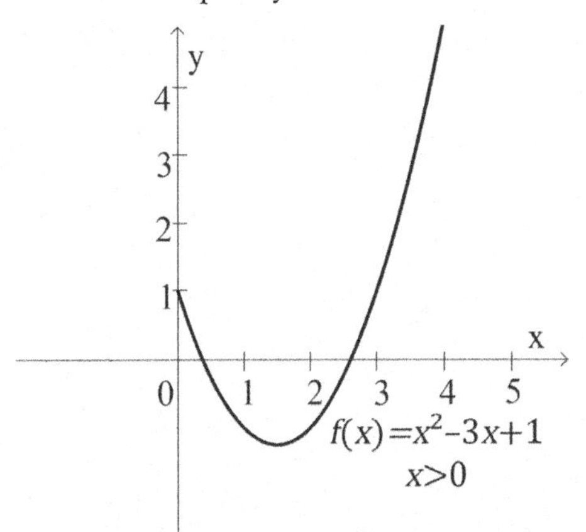

Domains can also be piece-wise, meaning made up of a collection of different intervals of values or individual values, or a set of specific numbers. In such cases, the domain is only the specific x -values for which the function is defined, not the values' entire range.

The chart below shows points that represent values of the given function. What is the domain of the function?

x	y
1	3
2	4
3	5
4	8
5	2

The only possible x -values are all real integers between zero and 6, i.e. 1, 2, 3, 4, 5. So the domain of this function is $\{1,2,3,4,5\}$.

One of the following sets is the domain for the function graphed below. Which set is it?

A. $\{0,1,2,3,4\}$

B. $\{1,2,3,4,5\}$

C. $\{x\,|\,0\le x\le 5\}$

D. $\{x\,|\,0\le x<4\}$

E. $\{x\,|\,0\le x<5\}$

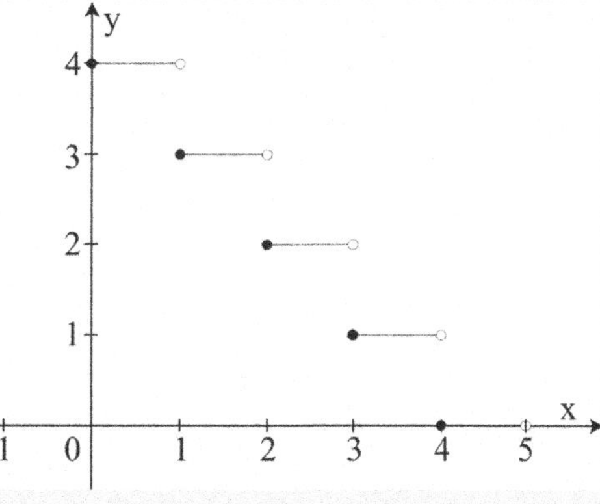

We can eliminate choices A and B because they indicate that the function is only defined at integer values, which we can clearly see is not true because of the line segments that extend from each dot on the graph.

Although this piecewise function is composed of several separate parts, if you choose any point for $0\le x<5$, you will always find a unique y value to match it to, so therefore the function is defined for all $x\,|\,0\le x<5$. Though you may be tempted by answer choice C, this choice erroneously includes $x=5$ in the domain. This point should not be included because there is an open circle at $x=5$, which indicates that 5 is not a possible value of x.

Answer: **E.**

RANGE: the complete set of y –values (or $f(x)$ values) where a function or relation is defined.

TIP: Remember that the word **RANGE** means <u>**three different things**</u> in math:

1. **Range** can simply mean a **span of values**. For example **a word problem** might say "At a drugstore, the cost of a box of tissues **ranges** from $2.00 to $5.00, inclusive, that means that boxes of tissues can be as inexpensive as $2.00, as expensive as $5.00, or anything in between.
2. When you are dealing with **data sets** and measures of central tendency (i.e. **averages, medians, modes, etc.**) the **range** is the **largest number in a set minus the smallest number** in the set. In these cases "range" will be a single number.
3. When you are dealing with **a function** (or relation, i.e. anything that can be written as an equation, denoted by a graph in the standard coordinate plane, or written as a set of paired values), as we are in this chapter, the **range** is **all the possible values of y or $f(x)$**. (Remember $f(x)$ and y are interchangeable ideas).

Because the ACT® is a multiple-choice test, if you don't know which "range" you are being asked for, you can usually figure out which meaning of range is at play by looking at the answer choices.

What is the range of the function $\frac{3x^2-2x+4}{x+2}$ on the interval $(2<x<5)$?

A. $y \neq -2$ **B.** $(3<y<\infty)$ **C.** $\left(3<y<\frac{69}{7}\right)$ **D.** $\left(-\infty<y<\frac{69}{7}\right)$ **E.** $(-\infty<y<\infty)$

This question is simply asking what y -values the function takes on the given interval. First, we plug the function into a graphing calculator to make sure that there isn't a local maximum within the interval. After discovering that the function is roughly linear on that interval, we can plug in the endpoints to get our range. Plugging in $x=2$ gives $f(2)=3$ and $x=5$ gives $f(5)=\frac{69}{7}$. Therefore our range is from 3 to $\frac{69}{7}$, non-inclusive.

Answer: **C.**

GENERAL RULES FOR POLYNOMIALS

Domain & Range

The **domain** of the following types of functions is **all real numbers**:

- Linear Equations, ex: $2x+1$
- Polynomials, ex: x^7+2x
- Power Functions, ex: $y=x^9$

In other words, you can plug ANY VALUE of x into these types of equations and expect to get an answer.

Range of polynomial functions:

- The range of **odd degree polynomial** functions (Example: $y = x^3$) is **all real numbers**.

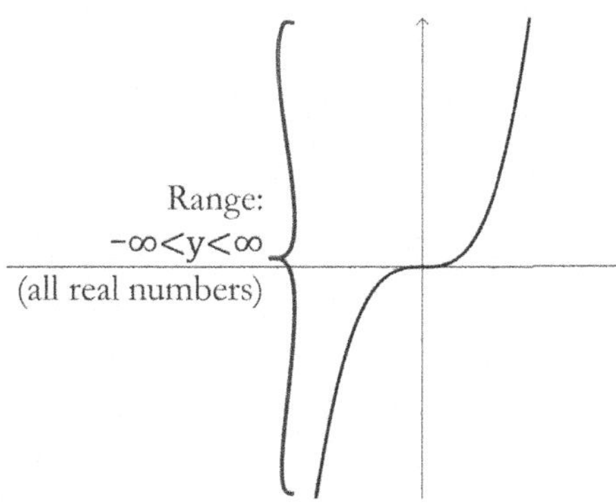

- The range of **even degree polynomial functions** (Example: $y = x^4 + x^3 + 7$) is **all numbers greater than the minimum value or less than the maximum value**.

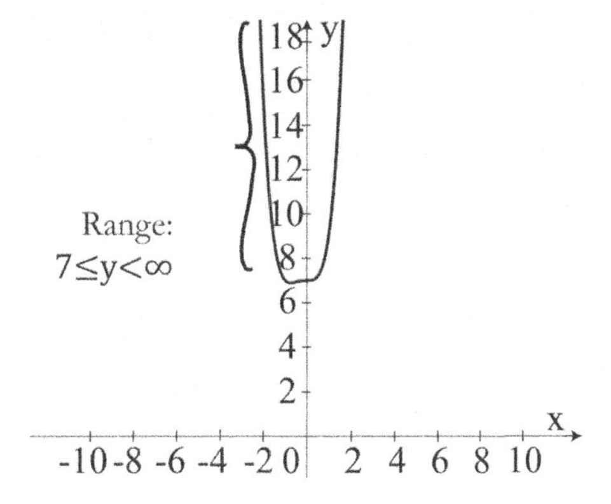

In other words, there will be some curve's vertex, such as the vertex of a parabola, that is a minimum or maximum y-value. The range is all the values to one side of that vertex such that the function is defined.

Undefined Values in Domain

For rational functions, the domain excludes points when the denominator equals zero. Whatever values of x make that denominator equal to zero, thus, are excluded from the domain.

There are two cases possible when a denominator equals zero: **holes** and **vertical asymptotes**.

- A **"hole,"** or removable discontinuity in the graph, occurs when the denominator is a factor of the numerator, and when simplified, the factored element in the numerator cancels with an expression in the denominator.
    - For example, $\frac{x-2}{x^2-4} = \frac{x-2}{(x-2)(x+2)}$.

  Because $(x-2)$ occurs in the numerator and denominator, when this element is equal to zero, the graph will have a "hole" at that point, i.e. when $x = 2$. We can reduce this by canceling these repeating elements and adding in a stipulation, that $x \neq 2$: $\frac{1}{x+2}$.

- In all other cases, the graph has a **vertical asymptote** at any x-values that make the denominator equal to zero.
  - For example, $\frac{x-2}{x-4}$ cannot be factored or reduced in any way. As a result, $x=4$, which makes the denominator equal to zero, defines a vertical asymptote.

TIP: To find when a function is undefined, simply set the denominator to zero and solve for x!

When is the function $y=\frac{x-4}{x^3+x^2-6x}$ undefined?

To find where this function is undefined, let's set the denominator equal to zero and solve. Factoring out an x we get that $x\left(x^2+x-6\right)=0$, so the function is undefined at $x=0$. Further simplifying we get that $(x)(x+3)(x-2)=0$, so the function is also undefined at $x=-3$ and 2. Notice that **the numerator has no part in determining where the function is undefined.** The numerator does indicate, however, in conjunction with the denominator, whether this undefined point is a hole or a vertical asymptote. For this problem, we don't need that information, though.

Answer: The function is undefined at $x=-3$, 0, and 2.

The equation $\frac{\left(x^2-1\right)}{x^2+4x-5}$ is graphed on the standard (x,y) coordinate plane below.

At what values of x is the function undefined?

A. –1 and 1
B. 1 and –5
C. –1 and 5
D. 1 only
E. –5 only

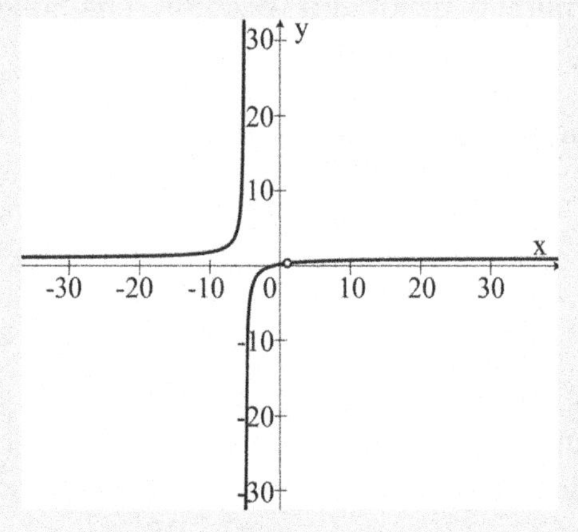

This question is tricky because of the graph. Looking at it, it appears that there is only one place where the function is undefined because of the presence of only one vertical asymptote. However, if we take a closer look at the function itself, factoring the numerator and denominator, we get:

$$\frac{(x+1)(x-1)}{(x+5)(x-1)}$$

We can clearly see that the function is undefined at both $x=-5$, which is clear on the graph, and $x=1$, which is much less obvious on the graph.

The reason for this is that the term $(x-1)$ exists both in the numerator and denominator, basically multiplying the expression $\frac{(x+1)}{(x+5)}$ by $\frac{(x-1)}{(x-1)}$ or 1. This adds a **hole**, or a removable discontinuity, to the graph of $\frac{(x+1)}{(x+5)}$, not altering the shape of the graph in any way but making the function undefined when x is 1. Therefore, our function is undefined at $x=-5$ and 1.

Answer: **B.**

Horizontal and Slant Asymptotes

Occasionally, the ACT® may ask you to calculate a **horizontal or slant asymptote** given a rational equation. Please note these questions are rare on the ACT®, and students scoring below a 33 should focus on other skills first. Students also taking the SAT® Math Level II Exam may find these skills helpful, also. In any case, we'll keep this brief.

TIP: remember that you always have your graphing calculator. If memorizing the ideas below is too much for you, you can plug the equation for an asymptote question into your calculator, let it work while you move on, and come back to it to trace and estimate the answer.

To calculate whether a rational function has a horizontal or slant asymptote, we first need to know the highest degree term in the numerator and the highest degree term in the denominator. These are the only two terms you're going to pay attention to in this process. Let's set the highest degree term in the numerator equal to ax^m and the highest degree term in the numerator equal to bx^n. Whatever isn't the highest degree we won't worry about. Given the function:

$$f(x)=\frac{ax^m+rest\ of\ polynomial\ that\ isn't\ the\ highest\ degree\ part}{bx^n+rest\ of\ polynomial\ that\ isn't\ the\ highest\ degree\ part}$$

We know that if:

- $m>n$ and the difference is more than 1, there is no horizontal asymptote. The graph will start at $\pm\infty$ and end up at $\pm\infty$.
- $n>m$ then the horizontal asymptote is $y=0$. As x approaches $\pm\infty$, y approaches 0.
- $m=n$ then the horizontal asymptote is the ratio of the leading coefficients, $\frac{a}{b}$.
- $m>n$ but only by 1, there is potentially a slant (or oblique) asymptote (this one is less important to know than the others).

We only really care about the largest terms, because as we approach infinity, the rest of the expression doesn't contribute much to the overall number on the top or bottom. The biggest number makes the biggest difference. We can also use the logic of this to solve if we have trouble memorizing the above ideas. I.e. to check a horizontal asymptote, try plugging in something HUGE.

In the standard (x,y) coordinate plane, when $w \neq 0$ and $z \neq 0$, the graph of $f(x)=\frac{2x-4x^2-4}{x^2+3x+2}$ has a *horizontal* asymptote at:

A. $y=2$ **B.** $y=-2$ **C.** $y=4$ **D.** $y=-4$

E. There is no horizontal asymptote.

In order to determine whether or not there is a horizontal asymptote we need to look at the terms with the highest degrees of the two polynomials. Be careful! The terms are not listed in order from greatest degree to least!

The numerator has a highest degree of 2 in the form of $-4x^2$. The denominator also has a highest degree of 2 in the form of x^2. Based on the rules stated earlier, this means that there is a horizontal asymptote at the ratio of the coefficients of the two terms. $-4x^2$ has a coefficient of -4 and x^2 has a coefficient of 1, so our horizontal asymptote is at $y=\frac{-4}{1}=-4$.

Answer: **D.**

BASIC PARENT FUNCTIONS AND GRAPHS

MEMORIZE THESE! It is useful to know the basic shapes of linear, parabolic, n- degree polynomial, exponential, logarithmic, and absolute value functions. That being said, you always have your calculator! If you forget the shape of a graph, you can always try graphing the function. Still, memorizing the basic shapes below can help you save precious time and move quickly.

NOTE: Also see our chapters on **Conics (this book)**, **Trig Graphs (Book 2)**, and **Coordinate Geometry (this book)** for more examples of graphs and their standard forms.

Graph of an exponential function:

Standard Form: $f(x)=a^x$

Example: $y=2^x$

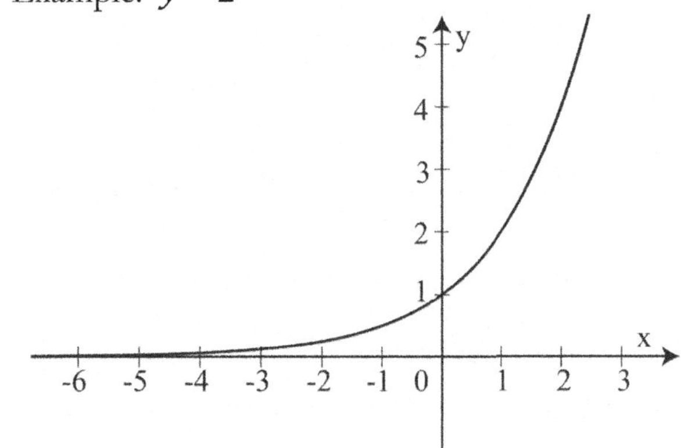

Graph of a square root:

Standard Form: $a\sqrt{x}$

Example (Parent Graph): $y=\sqrt{x}$

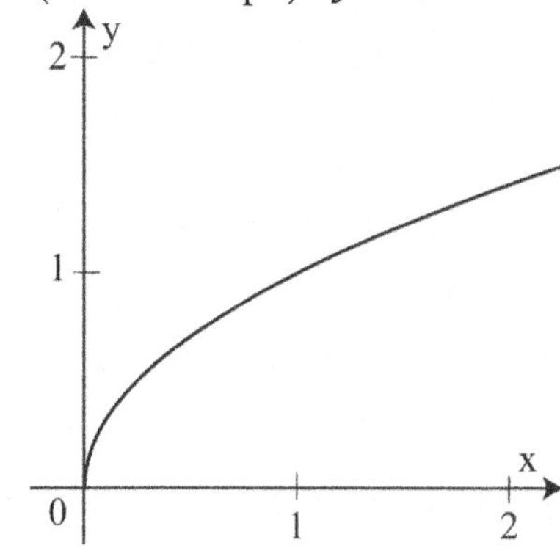

Graph of a cubic function:

Vertex Form: $y = a(x-h)^3 + k$

Example (Parent Graph): $y = x^3$

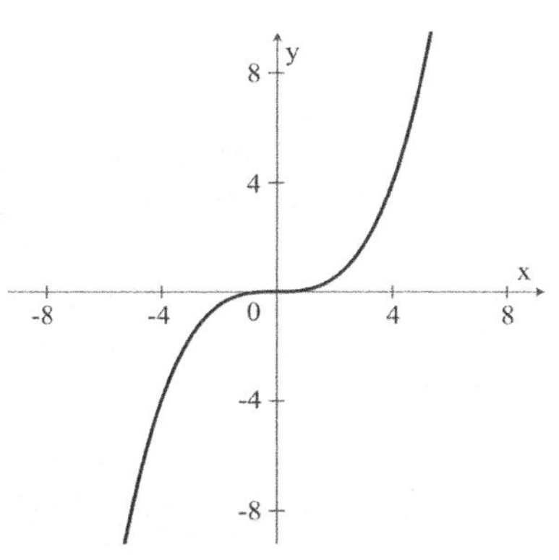

Graph of a parabola (quadratic function):

Vertex Form: $f(x) = a(x-h)^2 + k$

Example (Parent Graph): $y = x^2$

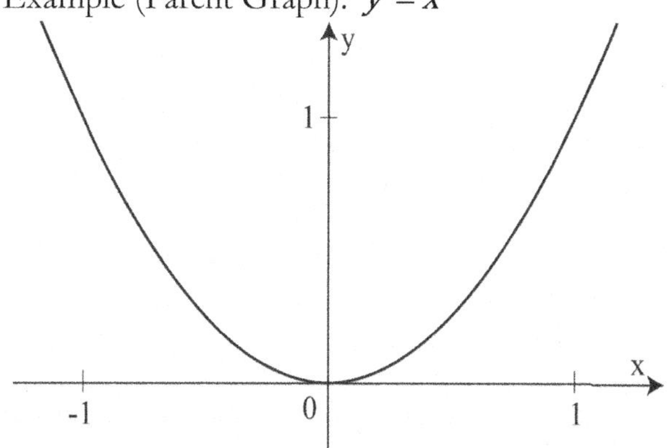

Graph of a linear equation:

Slope-Intercept Form: $y = mx + b$

Example (Parent form): $y = x$

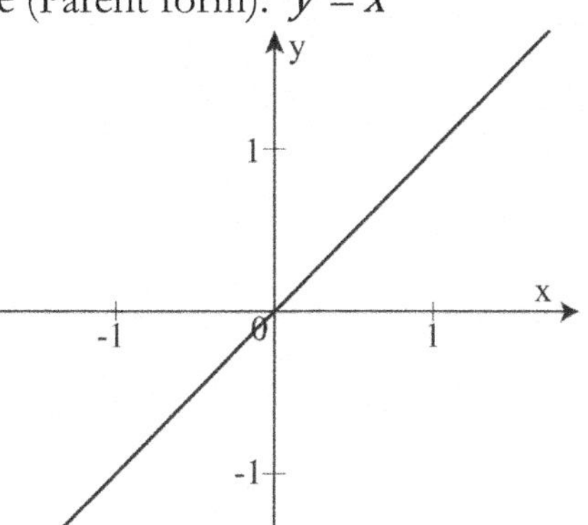

Graph of a logarithmic function:

Standard Form: $y = \log_a x$

Example: $y = \log_{10} x$

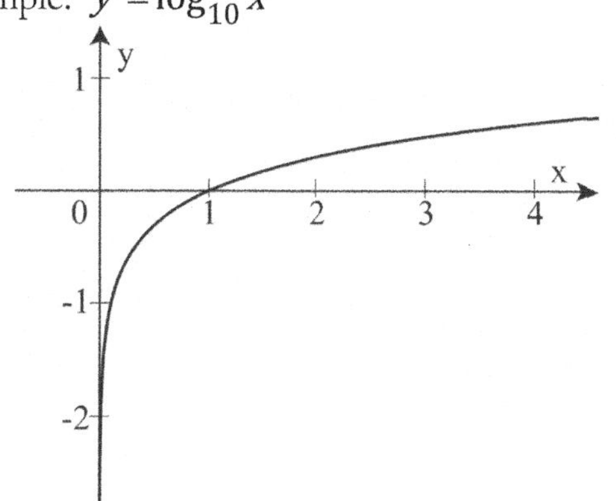

Graph of an absolute value function:

Vertex Form: $y = a|x-h| + k$

Example (Parent Graph): $y = |x|$

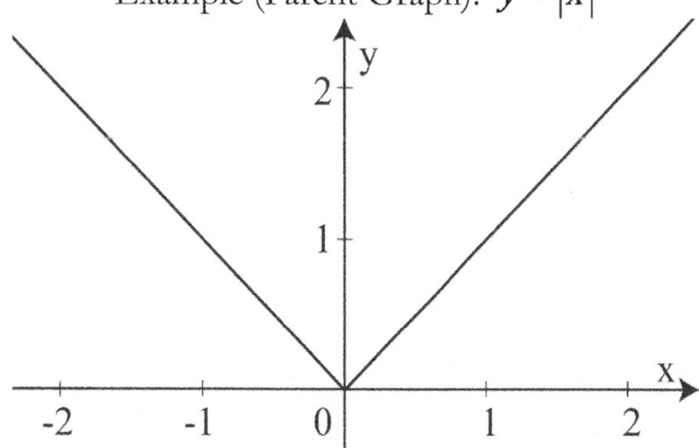

Horizontal and Vertical Shifts

Additionally, you'll need to know **how to "move"** the graph of a parent function. Though the constants and coefficients that "move" these parent functions sometimes vary across different equation types, two constants are universal to all graph translations: h **and** k.

Horizontal Shift $(x-h)$:

For ALL functions, if you replace **all instances of** x in a function with "$x-h$," you'll find that the graph moves "h" **units to the right.**

Vertical Shift $(y-k)$:
If you replace **all instances of** y in a function with "$y-k$," you'll find that the graph moves "k" **units upward.**

What is great about h and k is that they do the same "shift" no matter what function you apply this rule to. For instance, the parent graph of a parabola is: $y=x^2$. You might know the vertex form of a parabola is $y=a(x-h)^2+k$.

But what you might not realize, is that the x and y shift follow the rule above. Just move the k to the other side of the equation, and you'll see that the original y has been replaced with $y-k$, and the original x has been replaced by $x-h$: $y-k=a(x-h)^2$

Again this trick for vertical and horizontal shift works for EVERY type of function. Even trig functions!
$F(x)=\sin x$, for example, can move h units to the right simply by writing $F(x)=\sin(x-h)$.

The graph of $f(x)=|x|$ is shown below. One of the following graphs is the graph of $y=f(x-3)+2$. Which one?

A.

B.

C.

D.

E.

For this problem, we simply apply our h and k rule. When we move to the new equation, we have replaced every instance of x with $x-3$. When we do that, we move the graph 3 units to the right. That means only A and C remain as viable choices.

Now let's think about the y value. If I take the plus two and move it next to y, I get $y-2=f(x-3)$.

Let's think about this for a second. $f(x-3)$ was the original equation, but we replaced x with $x-3$. Still, that whole f(something) is the value of the entire equation that equals the y value. Now in this version, we replace $f(x)$ (what y equals) with $y-2$. That means we move 2 "upwards" on the graph. Clearly choice A is correct. It moves the vertex of our "dart" shaped graph up 2 and 3 to the right.

Again the idea is to **replace every x with $x-h$**, and **replace every y with $y-k$**, and you've **moved h to the right and k up,** no matter the graph or equation!

Answer: **A.**

In the standard (x,y) coordinate plane, the graph of $y=x^3$ is moved four units to the right and three units down. Which of the following equations represents this new function?

A. $f(x)=(x-4)^3+3$ **B.** $f(x)=(x+4)^3-3$ **C.** $f(x)=(x-4)^3-3$

D. $f(x)=(x-3)^3+4$ **E.** $f(x)=(x+3)^3-2$

Here we simply apply our h, k rule. H is the units to the right, or 4. K is the units up, or negative 3 (because we move "down" instead of up, we change the sign to negative). We replace the x in $y=x^3$ with $(x-h)$ or $(x-4)$ and we replace the y value with $(y-k)$ or $(y--3)$ or $(y+3)$.

$$y+3=(x-4)^3$$
$$y=(x-4)^3-3$$

Answer: **C.**

The graph of function $y=f(x)$ is graphed in the standard (x,y) coordinate plane below.

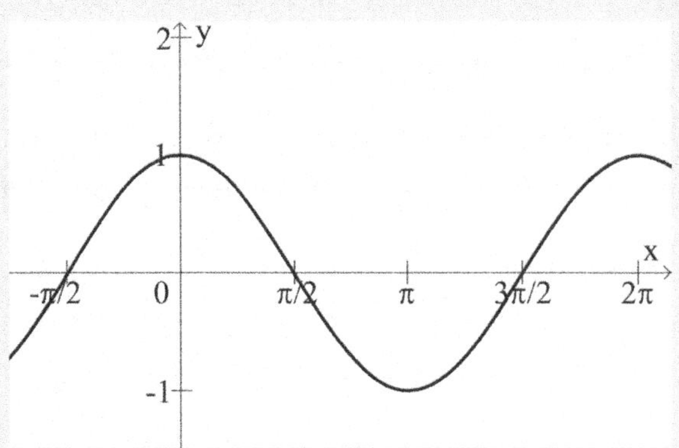

The points on the graph of the function $y=4+f(x+2)$ can be obtained from the points of $y=f(x)$ by a shift of:

A. 4 units to the right and 2 units up
B. 4 units to the left and 2 units down
C. 2 units to the right and 4 units up
D. 2 units to the left and 4 units up
E. 2 units to the left and 4 units down

To solve, we can simply think about our h and k shifts. The "h" value always "hugs" the x. That means $(x+2)=(x-h)$.

Because $2=-h$, we know that $h=-2$. In other words, we aren't moving 2 to the right, but 2 to the left. When we have x plus a value, we move that distance to the left, not to the right. Now we can eliminate choice (C), (A), and (B), as none of these are 2 units to the left.

For the k movement, we can move the k to the side of the equation where y is, and get: $y-4=f(x+2)$.

Our "$y-k$" is now "$y-4$". Thus we are moving 4 units UPWARD. So only answer choice (D) remains.

Remember, if we replace y with $y-k$ and we replace x with $x-h$, we shift the graph h units to the right and k units up. If the signs in front of k or h are different (i.e. positive), then we are moving left or down respectively.

True, this is a trigonometric graph, but you don't need to know ANYTHING about trig to get this right. If you know the h, k rule, you will be set.

Answer: **D.**

WHAT "a" DOES: DIRECTION, SLOPE, AND STEEPNESS OF GRAPHS

Direction

Often, standard or vertex forms of polynomials will include a coefficient or number a, typically the coefficient of the largest power term in a polynomial or linear function. The value by which an entire function is multiplied can also be a version of "a" or create similar effects (flipping a graph, widening it, or stretching it vertically).

For example, notice where "a" is in the vertex form of a parabola $f(x)=a(x-h)^2+k$

For most functions with a vertex form or standard form, the sign (positive or negative) of "a" (typically the coefficient of the largest polynomial term) dictates whether the function is "right side up" or "upside down" as well as the slope or "steepness" of the graph. Exactly what effect a has will vary depending on the specific type of graph, and the exact placement of this variable (or another variable in a similar position) but this general principle is good to know. **The letter "a" in these equations is kind of like a "slope" for the graph.**

This value can also tell us if we need to do a vertical "flip" of our graph. If "a" is negative, and we're dealing with a polynomial or linear equation, or we can clearly see that an entire function has been multiplied by -1 (for example, the graph of $-f(x)$), we flip the shape of the parent graph vertically "upside down" or mirror it across the x-axis before applying other shifts.

Take for example a parabola in the form $f(x)=a(x-h)^2+k$

Thus the graph of: $f(x)=4(x-3)^2+2$...

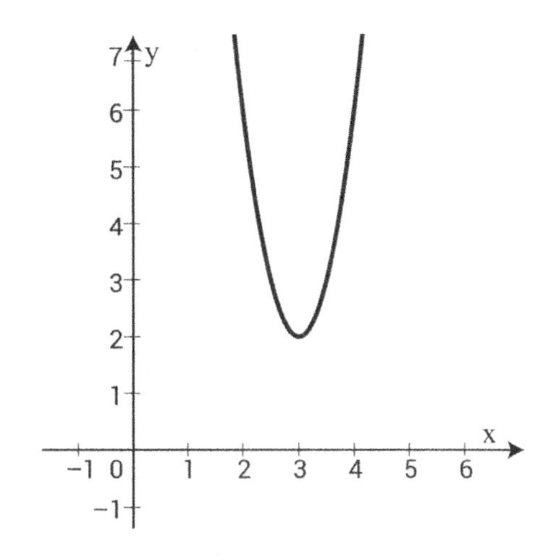

...Would be flipped upside down if the value for a (4) were made negative: $f(x)=-4(x-3)^2+2$

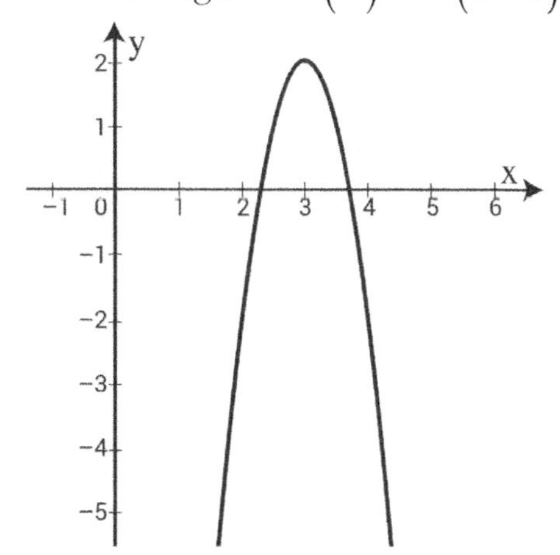

Steepness or Slope

This value also dictates the steepness of the graph, and it functions similarly to slope. The greater the absolute value of a slope, the steeper the line. The same is true for other styles of graphs: if you multiply the largest term or all the terms by a number, if that number's absolute value is greater than one, you'll be making the graph steeper or taller in some way, and if that number's absolute value is fractional, you'll be making that graph flatter or wider in some way.

For example, compare the graph below of $f(x)=(x-3)^2+2$ with a value of $a=1$ with the two graphs above. The below graph is less "steep" than the graphs in which $a=4$ or -4 above.

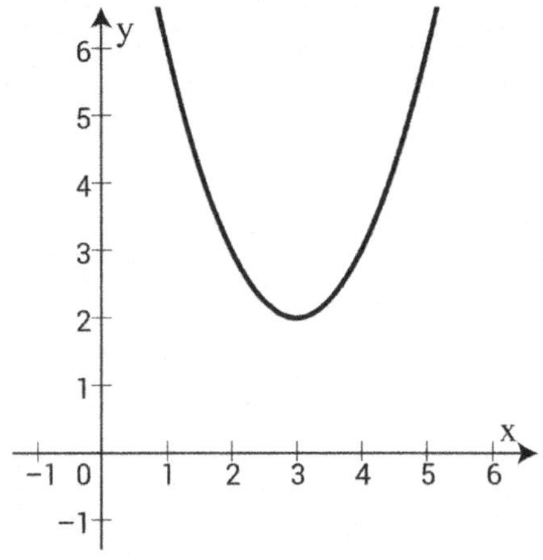

Fractional values of a, on the other hand, would flatten lines and curves, making them less "steep." See $f(x)=\frac{1}{2}(x-3)^2+2$ below:

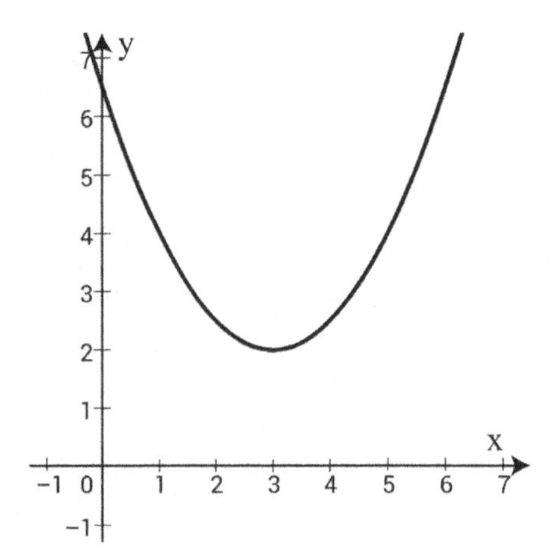

This principle holds for polynomials and absolute value equations, among others.

TIP: If you're ever unsure of what a value in an equation is doing on a graph, make up some numbers and test your theory using your graphing calculator. It is a huge resource on this style of question.

The graphs of $f(x)$ and $g(x)$ are shown in the standard (x,y) coordinate planes below. One of the following expresses $g(x)$ in terms of $f(x)$. Which one?

$y=f(x)$ $y=g(x)$

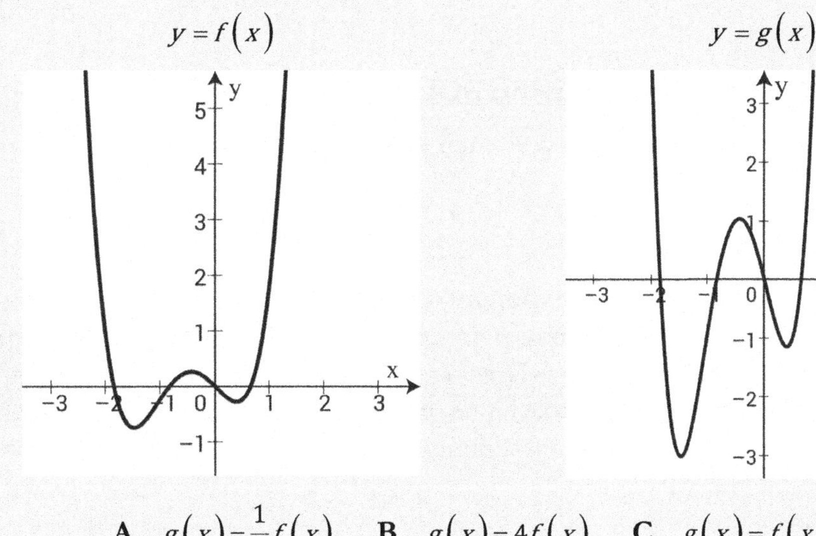

A. $g(x)=\frac{1}{4}f(x)$ **B.** $g(x)=4f(x)$ **C.** $g(x)=f(x)-4$

D. $g(x)=f(x+4)$ **E.** $g(x)=f(x-4)+4$

At first glance, you might think this is a nearly impossible question. We have two polynomials and have literally no idea what polynomial either is! But don't freak out just yet. Look down and analyze what we have to work with. We know these functions are related, and that they have basic shape parameters that are the same. We also know this is NOT a simple h or k shift. If it were, the shape in each function would be EXACTLY THE SAME. The shapes are not. So that eliminates choices C through E, which are all h and k only shifts. We need an "a" of some sort, or essentially different coefficients or different monomial elements within the polynomial to create a different shape. Therefore, we are down to A and B.

Choice A has an "a" value of $\frac{1}{4}$. That means $g(x)$ would be "flatter" and "wider" than $f(x)$. In fact, the opposite is true; $g(x)$ looks taller and skinnier.

Choice B has an "a" value of 4. That means $g(x)$ would be "taller" and "skinner" than $f(x)$, with "steeper" sloping curves.

That aligns with the visual difference in the pictures. Therefore, only choice B makes sense and it is correct.

If in doubt, we could also pluck some points and see if this pattern works. At $(0,0)$, multiplying by 4 shouldn't move that point. Both graphs appear to cross this point at about the same curve/place on the graph, so that works. I then see the graph's minimum point around $(-1.5,-.75)$. These are ballpark estimates, but I can use this point and multiply the y-value by 4 to get -3. Indeed, $g(x)$ appears to be around -3 when $x=-1.5$. By checking individual, plucked points in this manner, I can couple my instincts with mathematical evidence. True, I sometimes don't have time to do all this on the ACT®, but it's a decent strategy to know.

Answering this question without multiple-choice answers would be nearly impossible, but given the choices and what we know about graph behavior, we can deduce the answer fairly easily if we know our rules.

Answer: **B**.

END BEHAVIOR: EVEN VS. ODD DEGREE POLYNOMIAL FUNCTIONS

End Behavior is the behavior of a graph as x approaches the far left or far right of the graph, i.e. or (negative or positive infinity).

Even Degree Polynomial Functions

Functions whose highest degree is an **even exponent** (i.e. the highest degree term in the polynomial is x^2, x^4, x^6, etc.) start and end with the end points of the function **both pointing upward** or **both pointing downward**. As the x-value gets larger over time or smaller over time, as you move to the right or left on the graph, $f(x)$ will approach one of $\pm\infty$, depending on whether the coefficient of the leading term (i.e. the term with the highest numbered exponent) is positive or negative. See a summary of this concept below.

Degree: even
Leading Coefficient: positive
End Behavior: $f(x)$ approaches $+\infty$ at both ends of the graph (upward facing)
Domain: all reals
Range: all reals $\geq$ maximum
Example: $y = x^2$

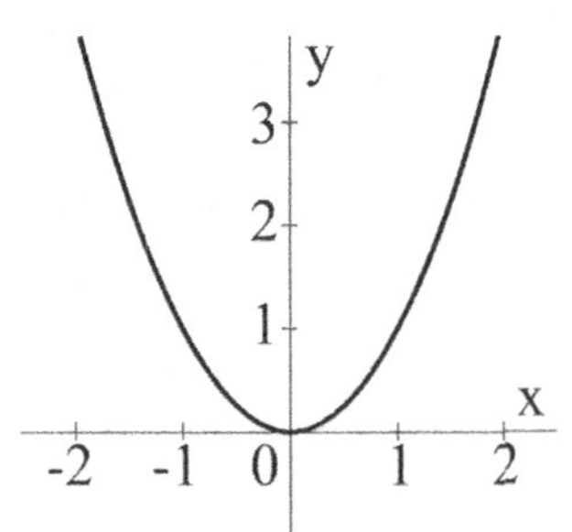

Degree: even
Leading Coefficient: negative
End Behavior: $f(x)$ approaches $-\infty$ at both ends of the graph (downward facing)
Domain: all reals
Range: all reals $\leq$ maximum
Example: $y = -2x^6 + 3x^5 + 4x^4$

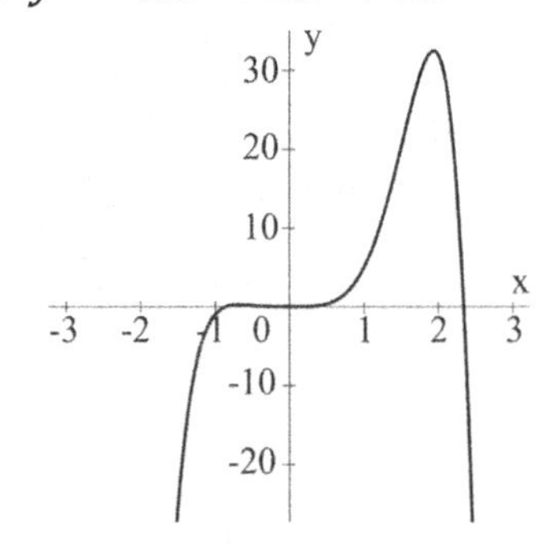

When the **coefficient of the leading term** (i.e. the highest degree polynomial term in the function) is positive, the ends of an even degree polynomial point **upwards**. I remember this as it looks like a "happy face" and happy faces are **positive**. Because the coefficient of the leading (and only) term of $y = x^2$ is positive one, the parabola above on the left is upward facing.

In contrast, the polynomial graphed above at the right, $y = -2x^6 + 3x^5 + 4x^4$, has ends that both point downward. As a result, it looks like a funky sad face, because the **negative** coefficient (-2) in front of the leading term make the graph look **"sad."**

Though **even degree polynomials** of greater degree may not form an exact "u" shape, and may be more complex in the middle, the end behavior will still hold the same pattern, both shooting upward or downward.

Odd Degree Polynomial Functions

Odd degree polynomial functions (functions whose largest degree term includes an odd exponent, such as x^3, x^5, etc.) with a positive coefficient in front of the greatest polynomial term (leading term) start low and end high; in other words, they start at $f(x) = -\infty$ and end at $f(x) = +\infty$. If the leading term has a negative coefficient, these functions start high and end low; in other words, they start at $f(x) = +\infty$ and end at $f(x) = -\infty$.

TIP: To remember this idea, think about the end behavior of a line: it is essentially an odd degree polynomial. Positive sloped lines have a positive coefficient going "uphill" from left to right. Negative sloped lines have a negative coefficient and go "downhill" from left to right. More complex polynomials are less simple, but their end behavior is the same, sloping uphill or downhill.

Degree: odd
Leading Coefficient: positive
End Behavior: At graph left, $f(x) \to -\infty$. At graph right, $f(x) \to +\infty$ (upward sloping)
Domain: all reals
Range: all reals
Example: $f(x) = x^3$

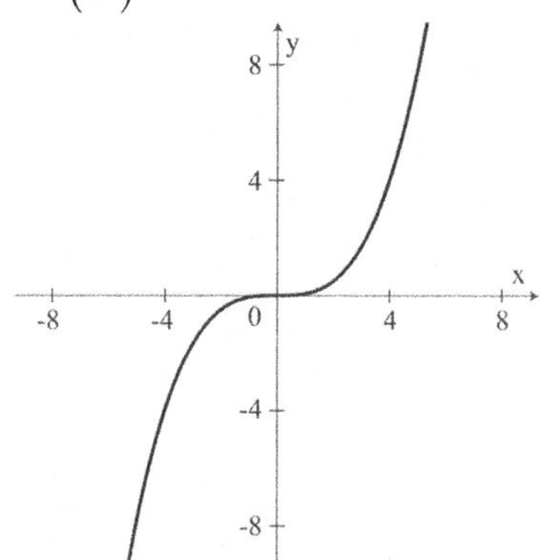

Degree: odd
Leading Coefficient: negative
End Behavior: At graph left, $f(x) \to +\infty$. At graph right, $f(x) \to -\infty$ (downward sloping)
Domain: all reals
Range: all reals
Example: $f(x) = -3x^5 - 2x^4$

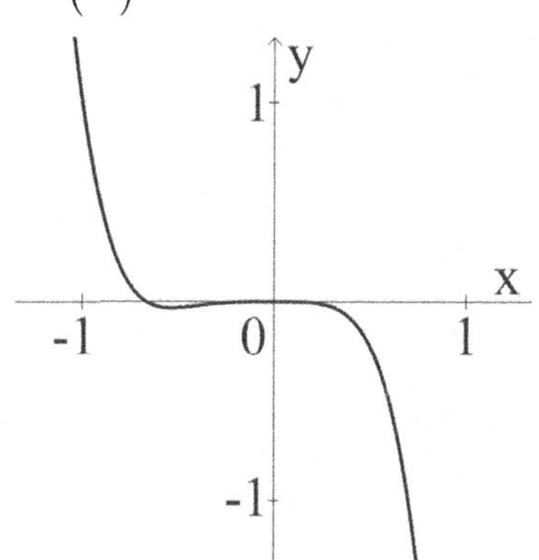

For example, the graph of $f(x) = -3x^5 - 2x^4$ (above right) shows a function that begins at positive infinity and ends at negative infinity.

Because the function has a negative coefficient in front of its leading term (i.e. the three in front of x^5 is negative), the overall endpoint behavior will start high and end low. If the coefficient were positive, the graph would start low and end high, as is the case above to the left, $f(x) = x^3$.

"Doing the Disco": How To Remember End Behavior Trends

I remember end behavior by thinking of the iconic dance move in the 1970's, disco dancing. In traditional disco dancing, you might see a dancer strike a pose something like the fellow in this picture. As you can see, one arm is **up** and the other **down**. In the same way that **"7" in 1970's is an odd number**, I remember that **odd exponents** create end points such that one goes **"up"** and the other **"down."**

BOUNCES, INFLECTIONS, AND SLICES

The behavior of a graph along the x-axis is another element that may be tested. When roots occur once in a polynomial, the graph "**slices**" through the x-axis. **Inflection points** occur with multiple roots that occur an odd number of times. "**Bounces**" occur with multiple roots that occur an even number of times.

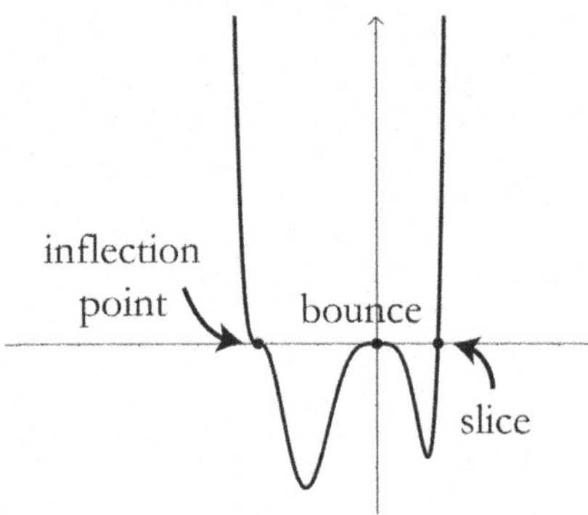

$x^4(x-1)(x+2)^3$		
x^4	$(x-1)$	$(x+2)^3$
4th degree (even)	1st degree (single)	3rd degree (odd)
Bounce ("kisses" the x-axis, reverses direction)	**Slice** (cuts through in a near straight manner)	**Inflection** (curves in different directions—concave vs convex—but does not "change" overall direction)

EVEN, ODD, AND ONE-TO-ONE FUNCTIONS

These are three classifications/properties of functions that are tested infrequently but still useful to know for those overachievers out there (seeking 34+ scores).

Even Functions

Even functions have graphs that are symmetric about the y-axis, such that for all x, $f(x)=f(-x)$. For example, $f(x)=x^2$ is an even function.

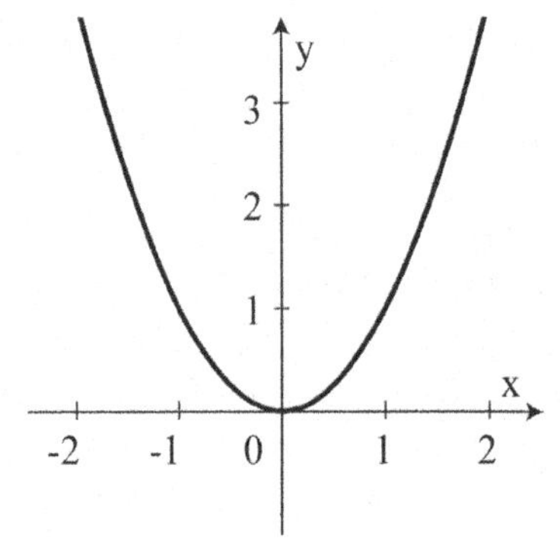

Odd Functions

These are the exact opposite of even function in that for every x, $f(x)=f(-x)$. In other words, the function is symmetrical about the origin. $f(x)=x^3$ is an odd function.

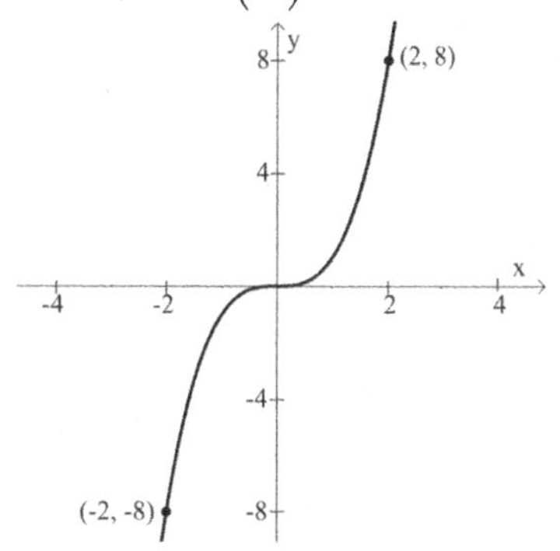

For example, in this graph, when $x=2$, $y=8$ and when $x=-2$, $y=-8$. Every ordered pair that is positive corresponds to an ordered pair with two negative values that are otherwise numerically the same.

One-To-One Functions

One-to-one functions are functions that have only one x-value for every y-value. In other words, not only must the function pass the required **vertical line test**, but it must also pass the **horizontal line test** to be considered a one-to-one function. Functions that reverse directions, with multiple peaks and valleys, are never one-to-one functions. Relatively linear functions are often one-to-one functions.

Example of a one-to-one function:

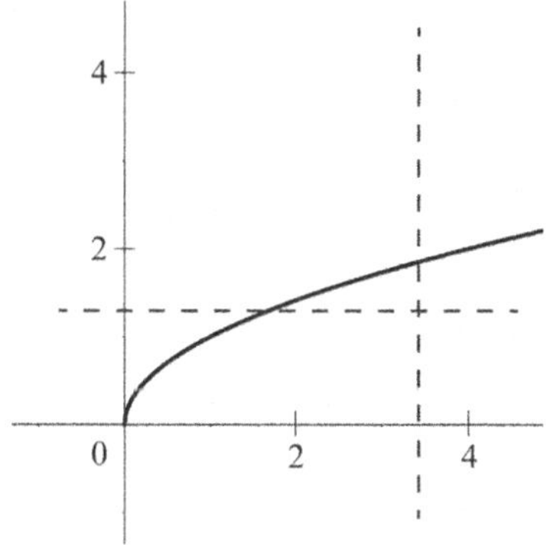

passes the vertical and horizontal line tests

Example of a function that is NOT one-to-one:

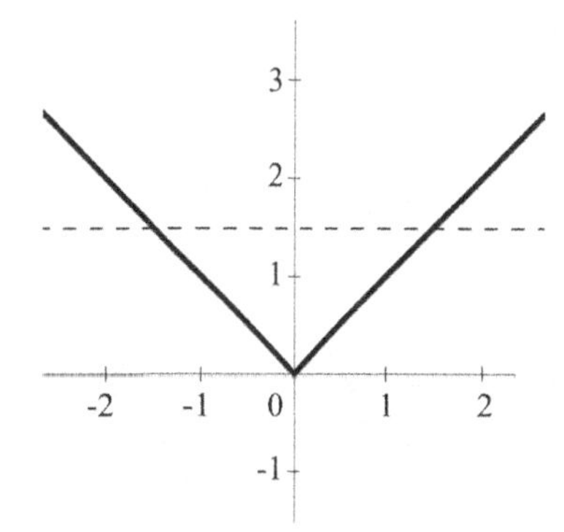

fails the horizontal line test

A function if a *one-to-one* function if and only if each x in the domain of $f(x)$ corresponds to a unique $f(x)$ and each $f(x)$ corresponds to a unique x. Which of the following graphs is a *one-to-one* function?

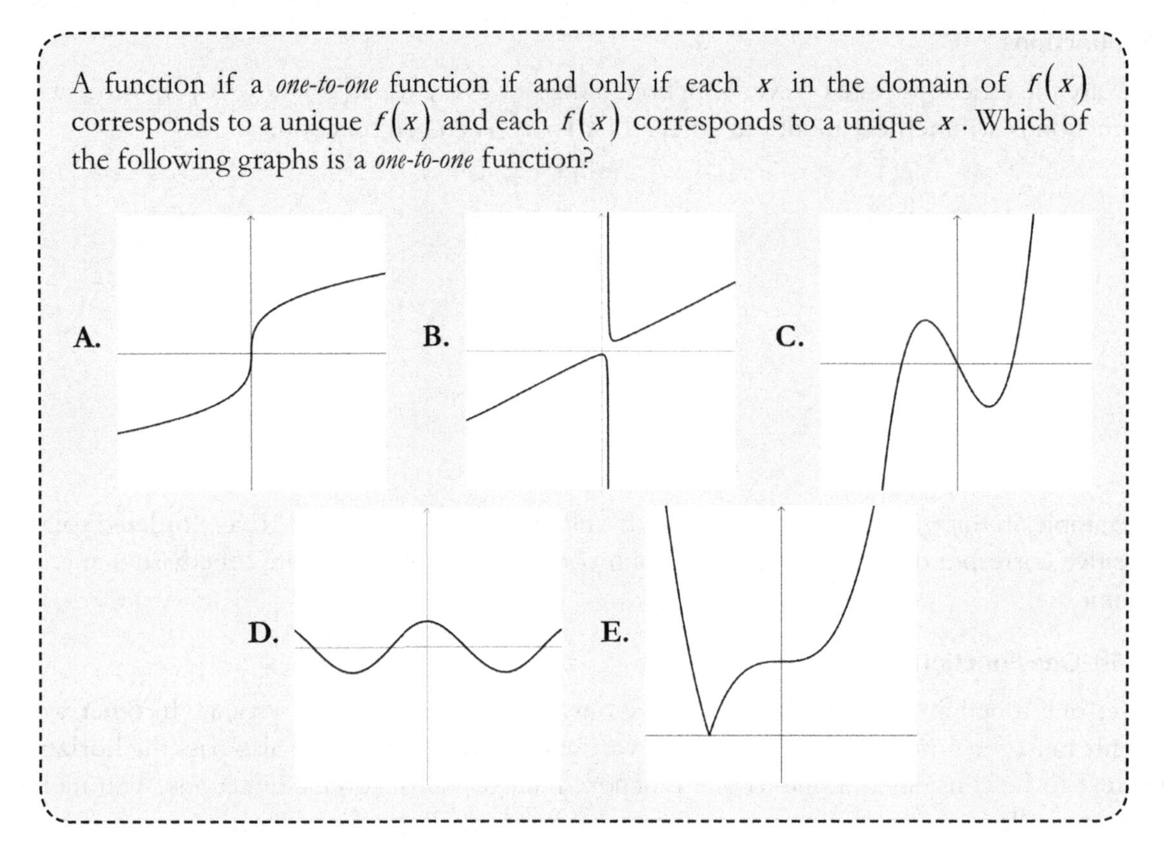

Remember that when it comes to one-to-one functions we have to use the two different line tests. All of these graphs depict functions, and therefore satisfy the vertical line test. However, only one of the functions satisfies the horizontal line test, where every y-value has its own singular x-value, and that is choice A.

Answer: **A.**

1. The domain of the function $f(a)=\frac{1}{3-\frac{1}{|a|}}$ includes all real values of a except:

   A. $\frac{1}{3}$
   B. $-\frac{1}{3}$
   C. 3,0
   D. −3,0
   E. $0, \frac{1}{3}$, and $-\frac{1}{3}$

2. The y-values of $g(x)$ vary directly with the square of $(x-3)$ for all real numbers. The graph of $y=g(x)$ in the standard Cartesian plane is which of the following?

   A. A line
   B. A hyperbola
   C. An ellipse
   D. A parabola
   E. A circle

3. Which of the following lists the real values of x that make the expression $\frac{5x-1}{x^3-6x^2-55x}$ undefined?

   A. 0 only
   B. −5 only
   C. −5,11 only
   D. −11,0,5 only
   E. −5,0,11 only

4. In the standard (x,y) coordinate plane, when $w \neq 0$ and $z \neq 0$, the graph of $f(x)=\frac{5x^2-w}{x^2+z}$ has a *horizontal* asymptote at:

   A. $y=5$
   B. $y=\frac{-w}{z}$
   C. $y=0$
   D. $y=z$
   E. There is no horizontal asymptote.

5. The domain of a function is the set of all values of x for which $f(x)$ is defined. One of the following sets is the domain for the function graphed below. Which set is it?

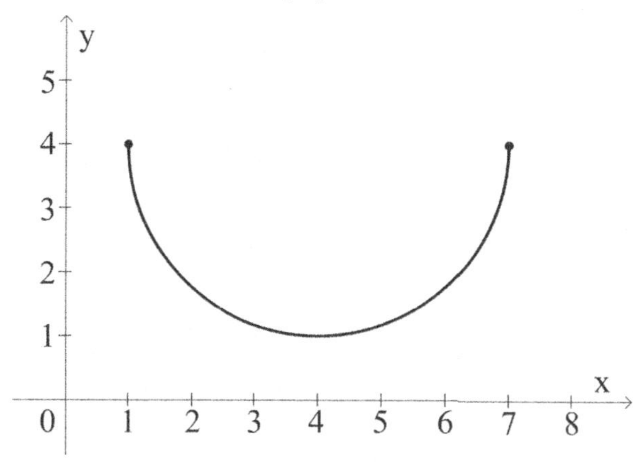

   A. $\{1,2,3,4,5,6,7\}$
   B. $\{1,2,3,4\}$
   C. $\{x:1 \leq x \leq 7\}$
   D. $\{x:1 \leq x \leq 4\}$
   E. $\{x:1 < x < 7\}$

6. The equation $\frac{(x-1)}{x^2-2x-3}$ is graphed on the standard (x,y) coordinate plane below. No point on the graph has which of the following x-coordinate?

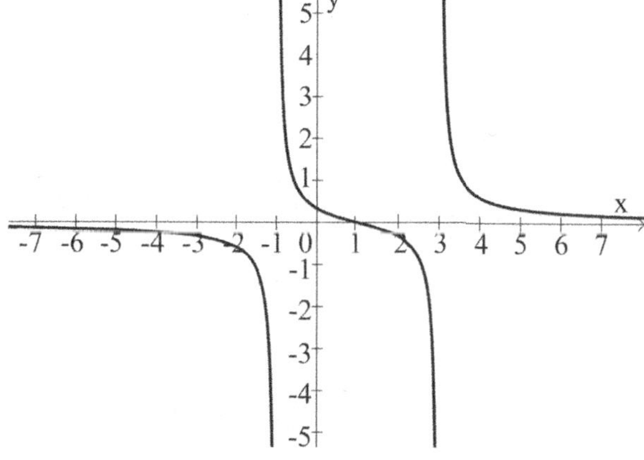

   A. −2
   B. −1
   C. 0
   D. 1
   E. 2

7. Consider the functions $f(x)=|x|+2$ and $g(x)=x^3-1$. Which of the following graphs is the graph of $y=f(g(x))$ in the standard coordinate plane?

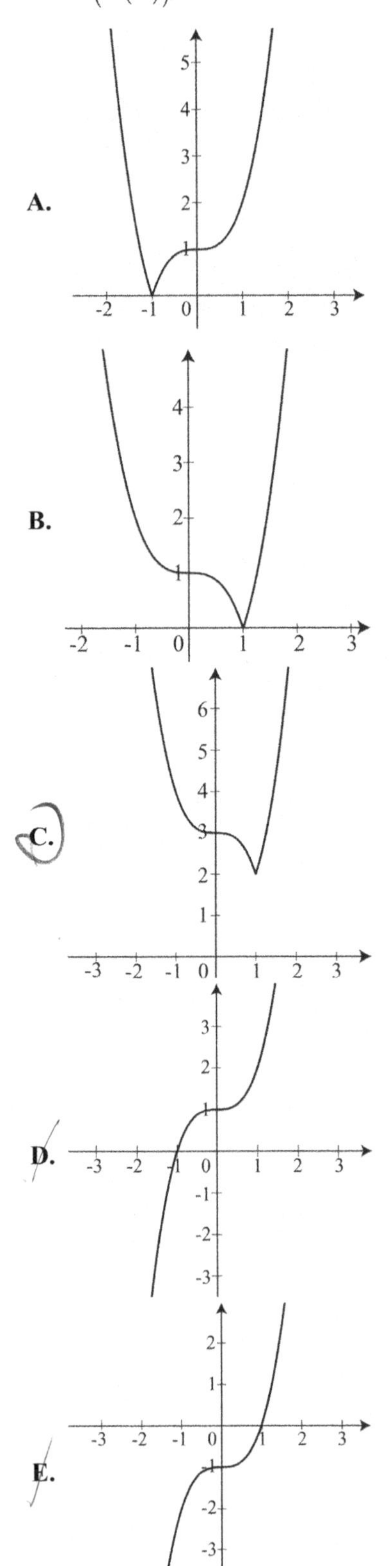

8. A person pedaling a bicycle is pedaling at a constant speed. Let t represent the time that has elapsed since the person started pedaling and let h represent the height above ground of one of the pedals. Below are graphs of one full cycle of the pedals. The pedal is at its maximum height at $t=a$, and is at its minimum height at $t=b$. Which of the following graphs represents the relationship between t and h during this rotation?

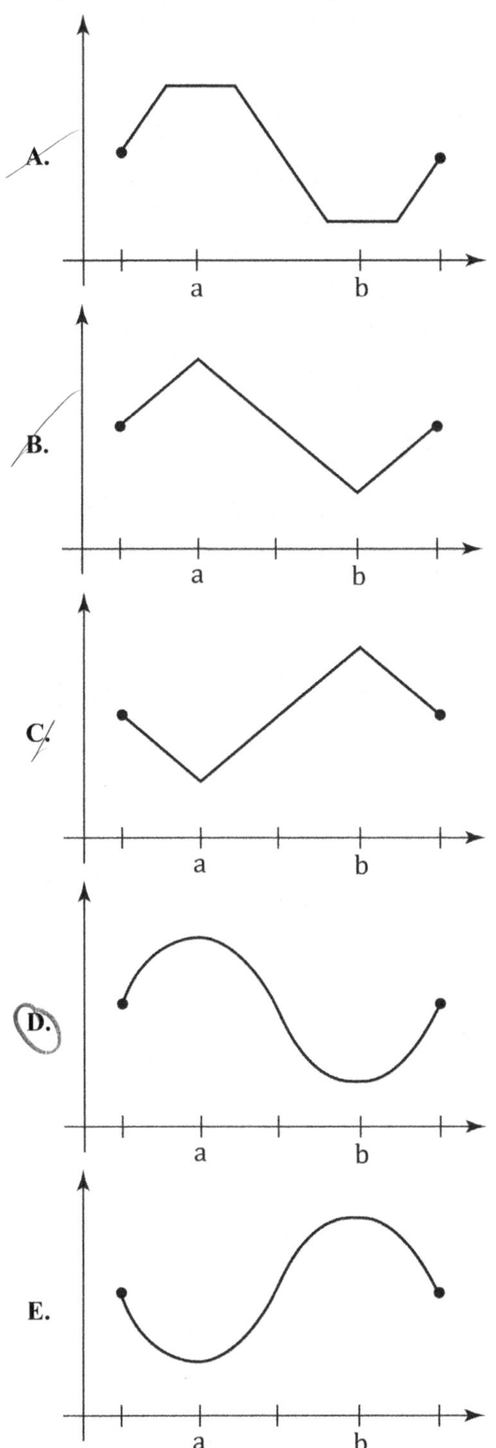

9. A function is a *one-to-one* function if and only if each x in the domain of $f(x)$ corresponds to a unique $f(x)$ and each $f(x)$ corresponds to a unique x. Which of the following graphs is *not* a *one-to-one* function?

A.
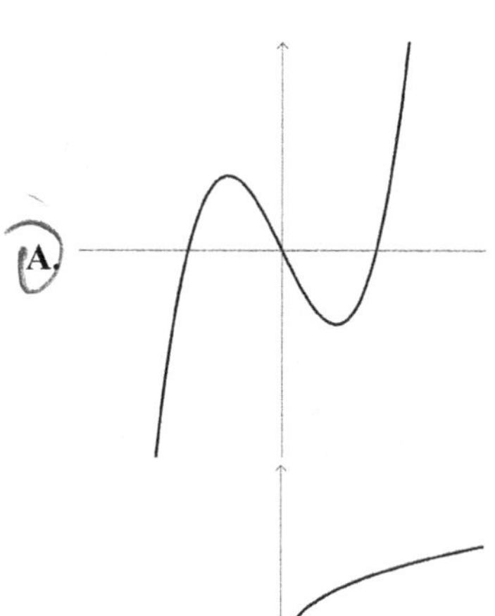

B.
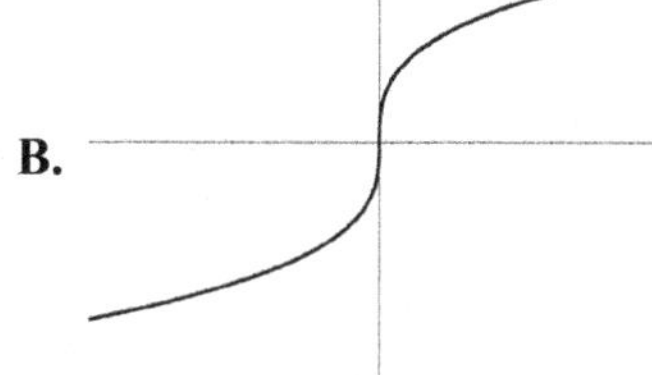

C.
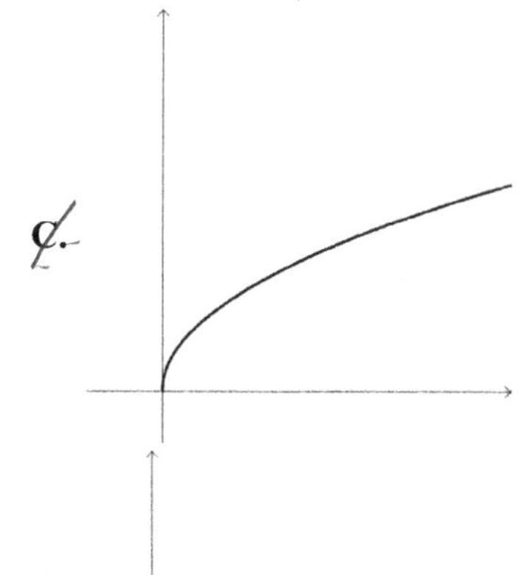

D.
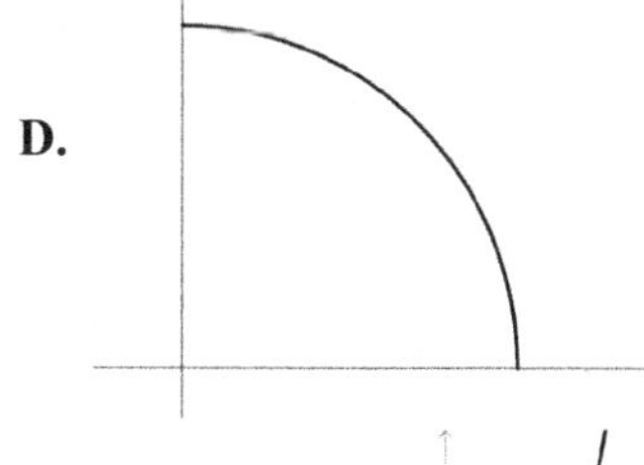

E.
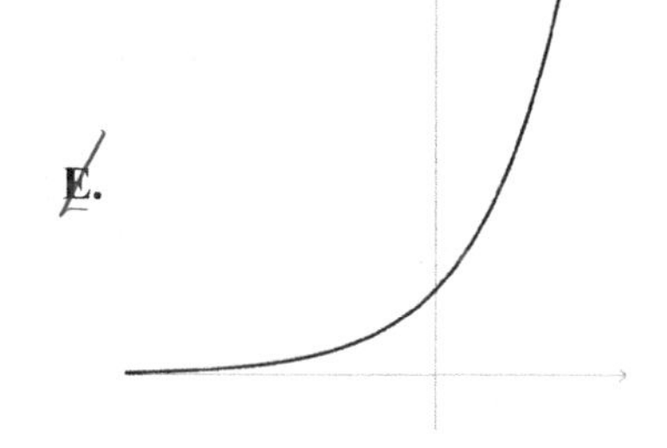

10. One of the graphs below is that of $y = -Kx^3$, where K is a positive constant. Which one?

A.
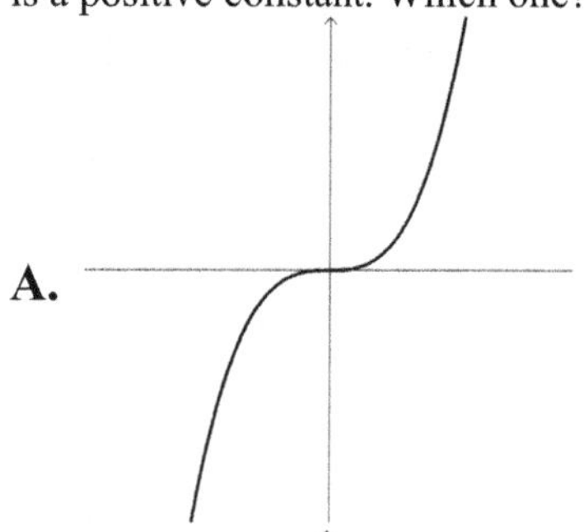

B.
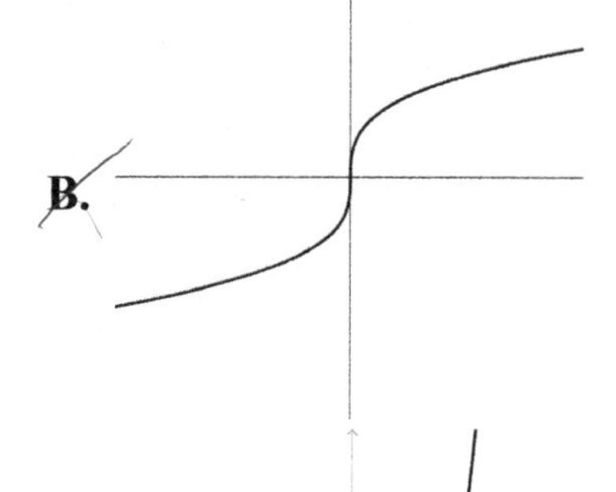

C.
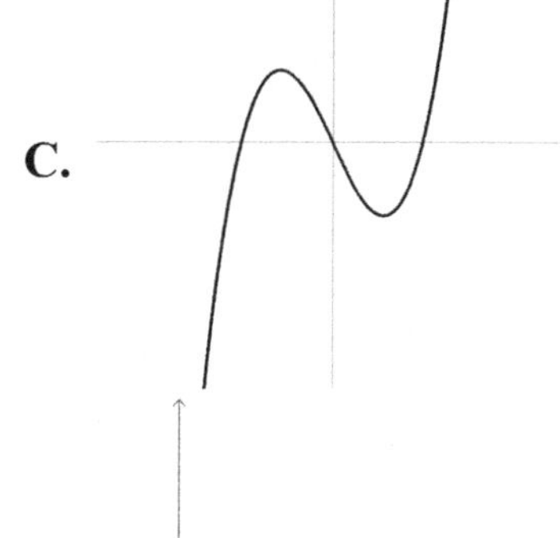

D.
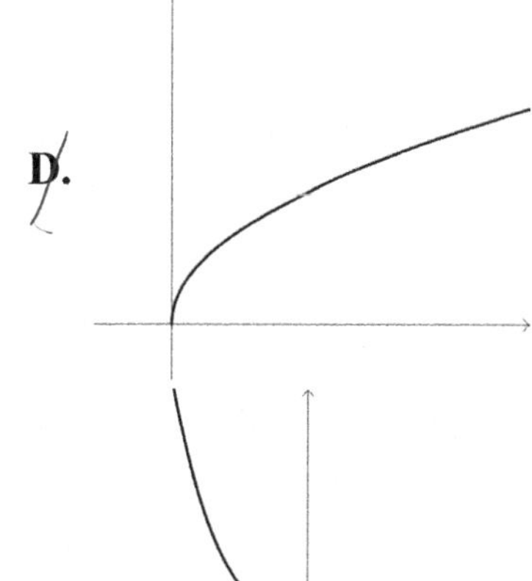

E.
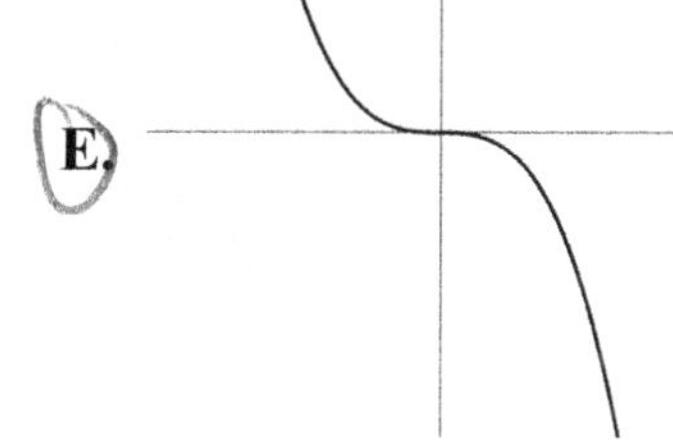

11. A function f is an even function if and only if $f(-x)=f(x)$ for every value of x in the domain of f. One of the functions graphed in the standard (x,y) coordinate plane below is an even function. Which one?

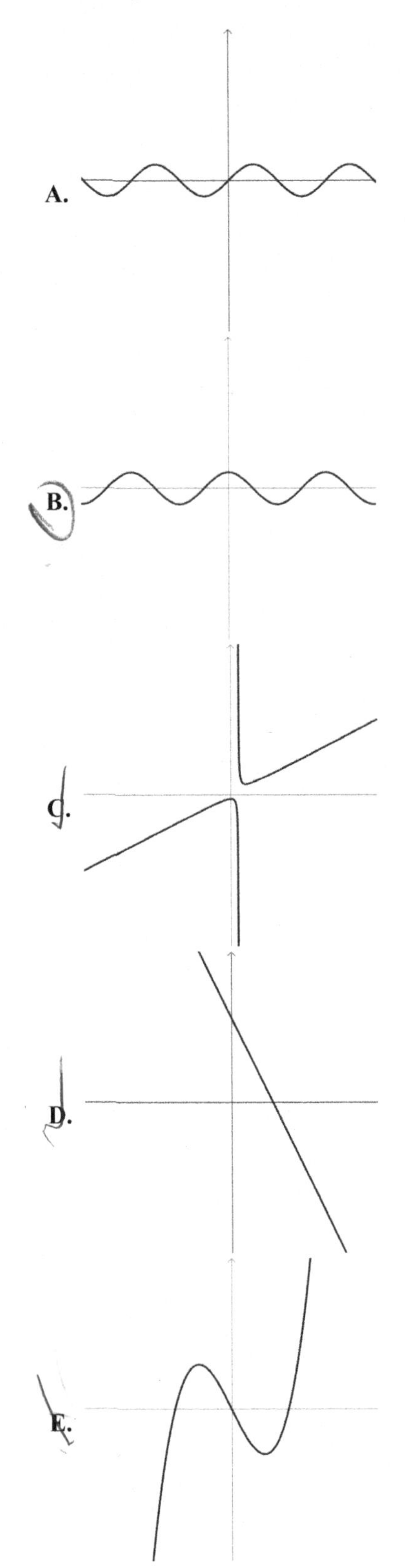

12. The graph of the function $f(x)=\dfrac{-x^3-3x^2-2x}{x^2-1}$ is shown in the standard (x,y) coordinate plane below. Which of the following, if any, is a list of each of the vertical asymptotes of $f(x)$?

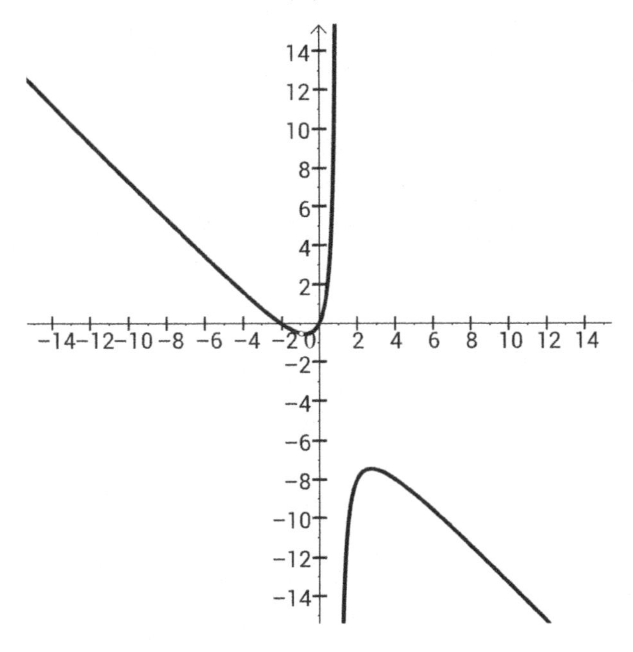

A. $x=-1$ and $x=-2$
B. $y=-x-3$
C. $x=-1$ and $x=1$
D. $x=1$
E. This function has no vertical asymptote.

13. Which of the following is the graph of the function $f(x)$ defined below?

$$f(x)=\begin{cases}-x+2 & x\le -1\\ x^2-1 & -1<x\le 3\\ 2x-3 & x>3\end{cases}$$

A.

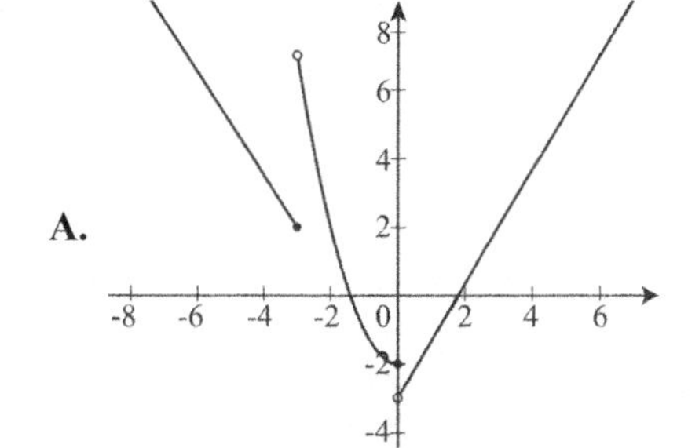

B.

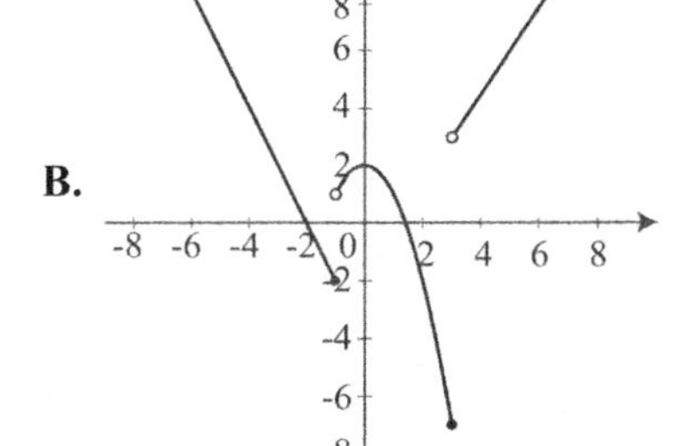

C.

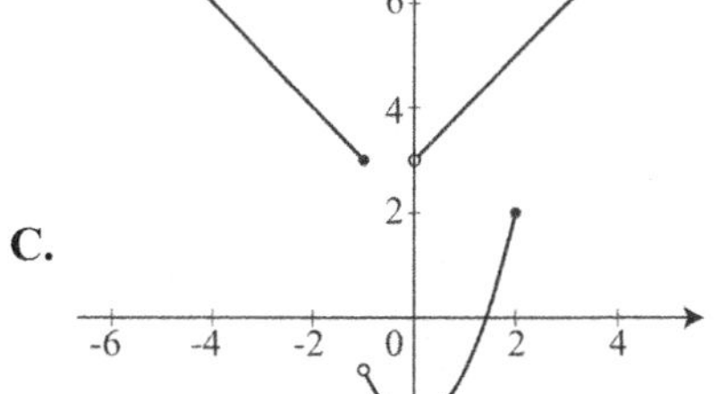

D.

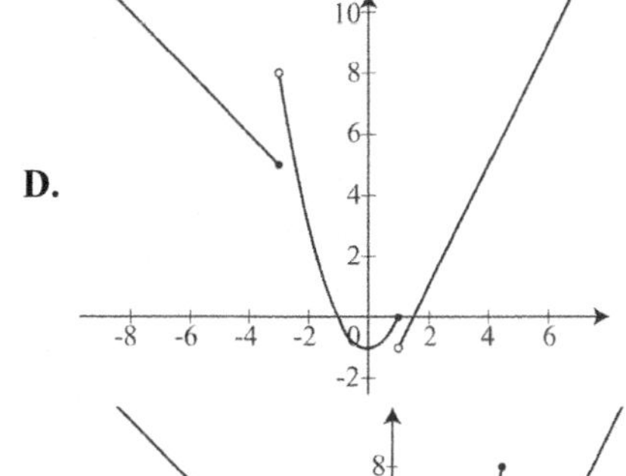

E.

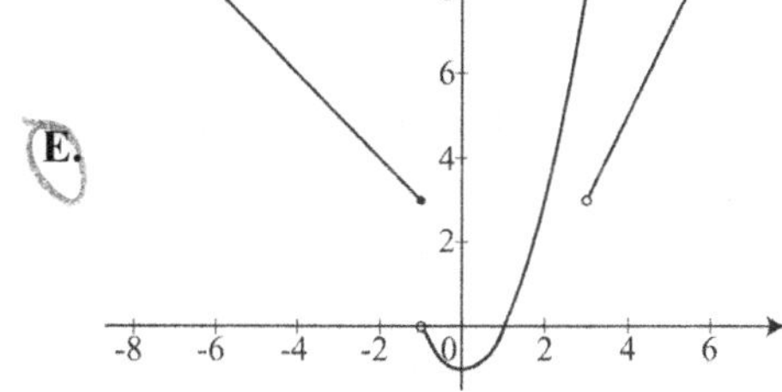

14. At what value(s) of x is $\dfrac{(x+2)(x-1)^2}{x(x+1)}$ undefined?

A. 0 and −1
B. 0 and 1
C. −2 and 1
D. −1 only
E. 0 only

15. One of the following graphs in the standard (x,y) coordinate plane represents the equation $5y-20=-2x$ for $x\ge 0$. Which one is it?

A.

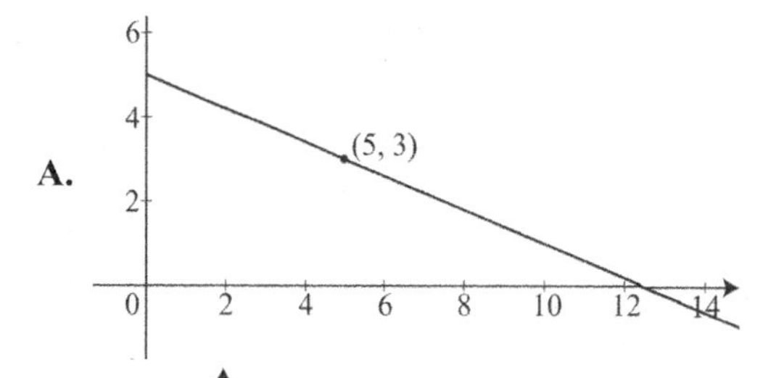

B.

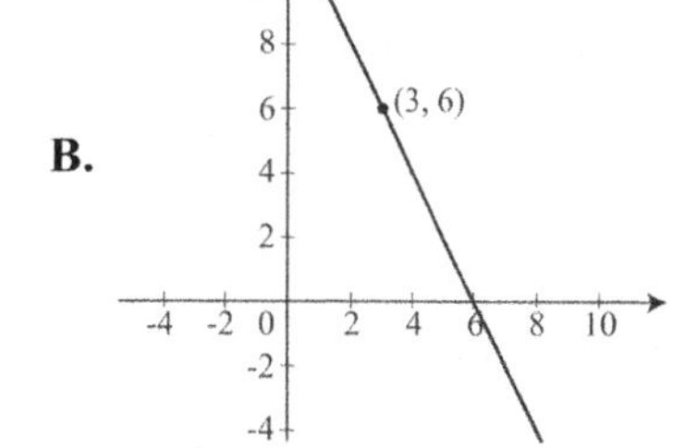

C.

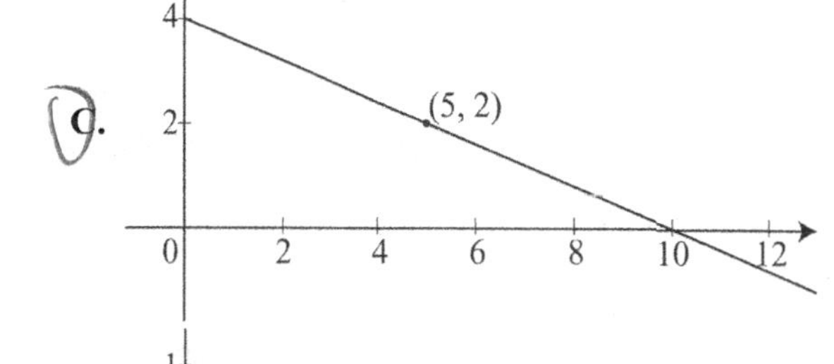

D.

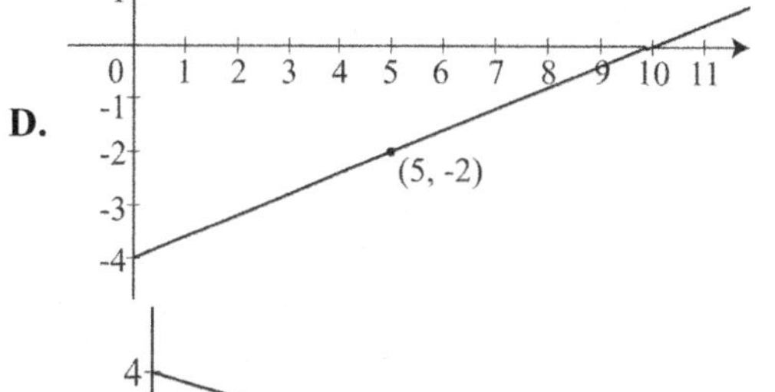

E.

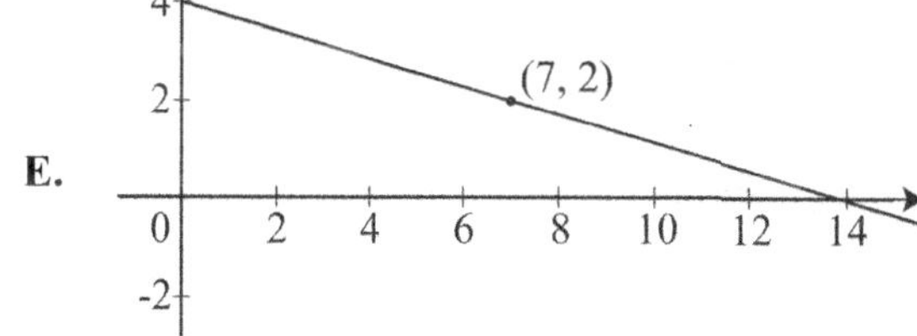

16. The entire graph of $y = f(x)$ in shown in the standard (x,y) coordinate plane below. One of the following sets is the domain of f. Which set?

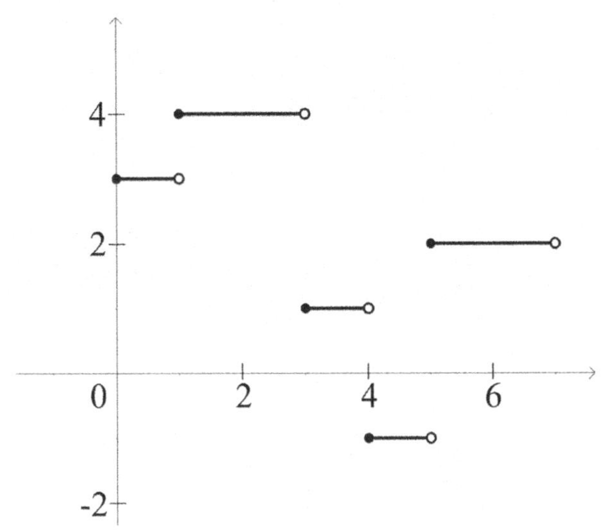

A. $\{-1,1,2,3,4\}$

B. $\{0,1,2,3,4,5,6,7\}$

C. $\{-1,0,1,2,3,4,\}$

D. $\{x \mid 0 \le x < 7\}$

E. $\{x \mid -1 \le x < 4\}$

17. Shown in the standard (x,y) coordinate plane below, the graph $x = (y-1)^2 + 2$ is restricted by one of the following conditions. Which one?

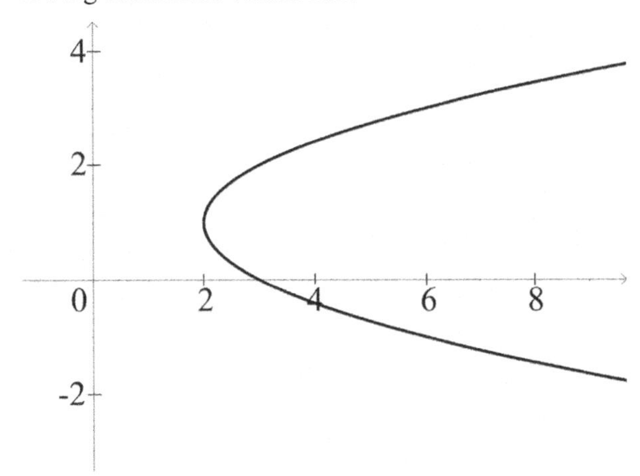

A. $x \le 2$
B. $x \le 1$
C. $x \ge 2$
D. $y \ge 1$
E. $y \ge 2$

18. If $f(x) = \dfrac{5}{x+2}$ and $g(x) = x^2 - 3$, which of the following number lines shows the domain of $f(g(x))$?

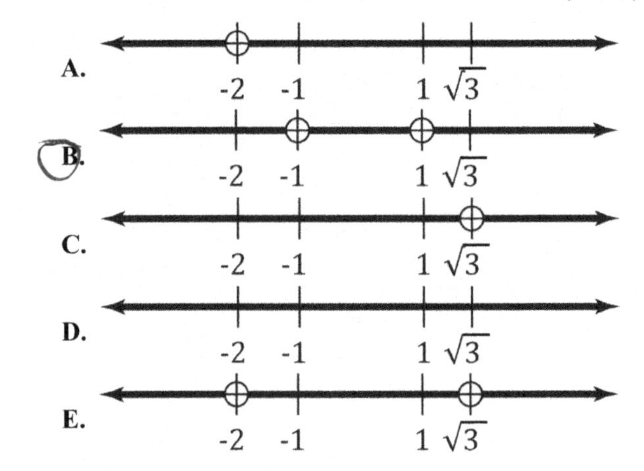

19. For all $0 < n < 1$, which of the following statements describes the function f defined by $f(x) = n^x$?

A. f is constant for all x
B. f is increasing for all $x \ge 0$
C. f is decreasing for all $x \ge 0$
D. f is increasing for $0 \le x < 1$ and decreasing for $x \ge 1$
E. f is decreasing for $0 \le x < 1$ and increasing for $x \ge 1$

For questions 20-22:

The graph of $y = g(x)$ is shown in the standard (x,y) coordinate plane below with two points labeled.

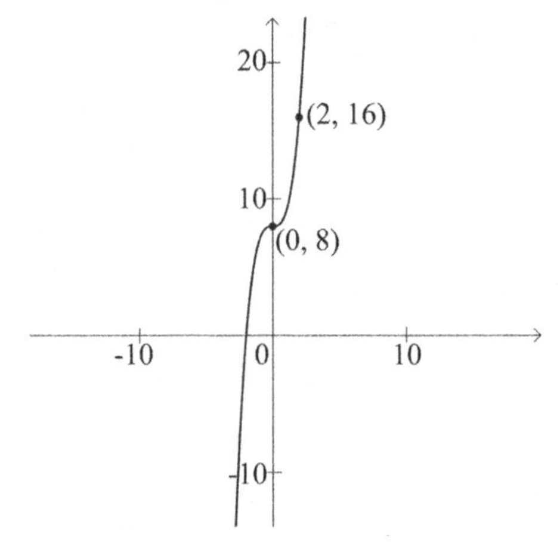

20. What is the x-intercept of the graph of $y = g(x)$?

A. -8
B. -2
C. 0
D. 2
E. 8

21. The function $y = g(x)$ can be classified as one of the following types of functions. Which one is it?

A. Constant

B. Linear

C. Quadratic

D. Cubic

E. Trigonometric

22. Which of the following equations corresponds to the reflection of $y = g(x)$ across the line $y = x$?

A. $y = x^3 - 8$

B. $y = (x-8)^3$

C. $y = \sqrt[3]{x+8}$

D. $y = \pm\sqrt[3]{x-8}$

E. $y = \sqrt[3]{x-8}$

23. One of the graphs below is that of $y = -e^X + K$, where K is a positive constant. Which one?

A.

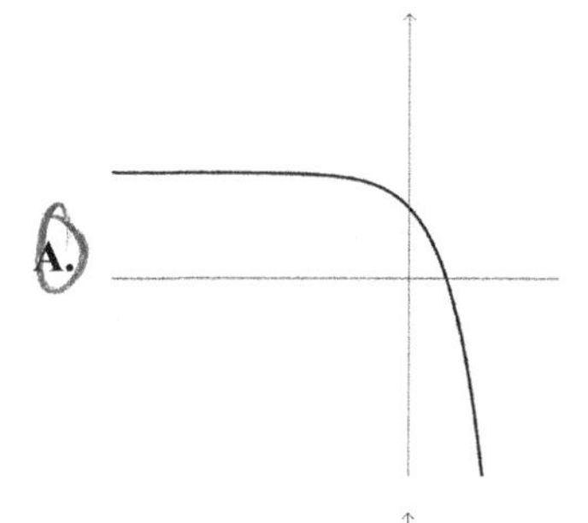

B.

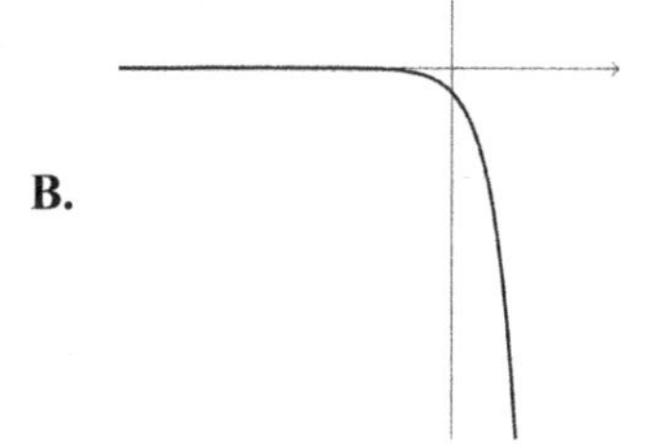

C.

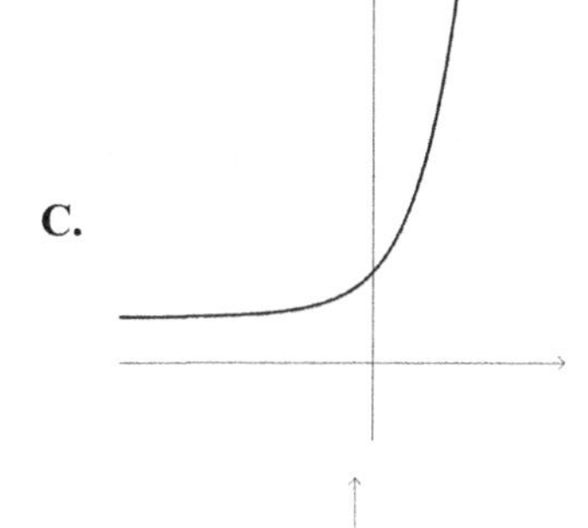

D.

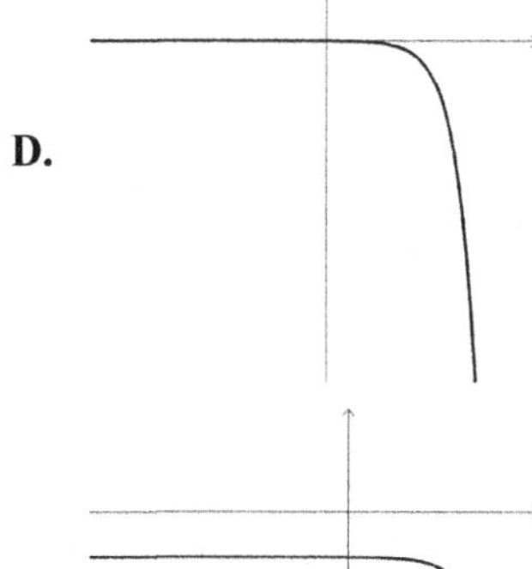

E.

24. A function f is an even function if and only if $f(-x)=f(x)$ for every value of x in the domain of f. Which of the functions graphed in the standard (x,y) coordinate plane below is not even function?

A.

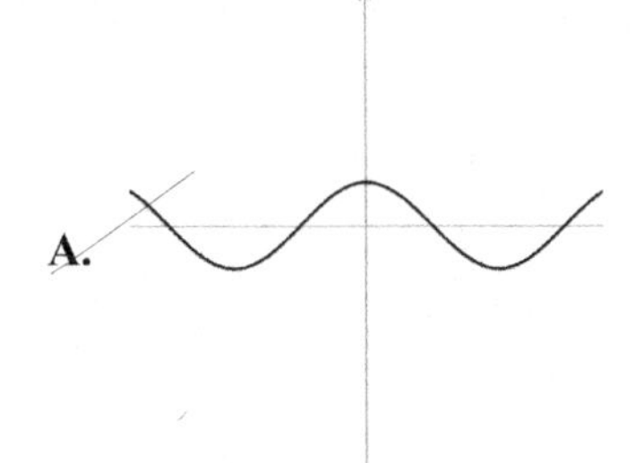

B.

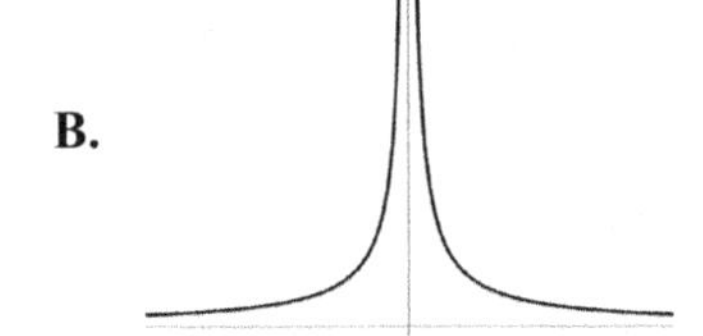

C.

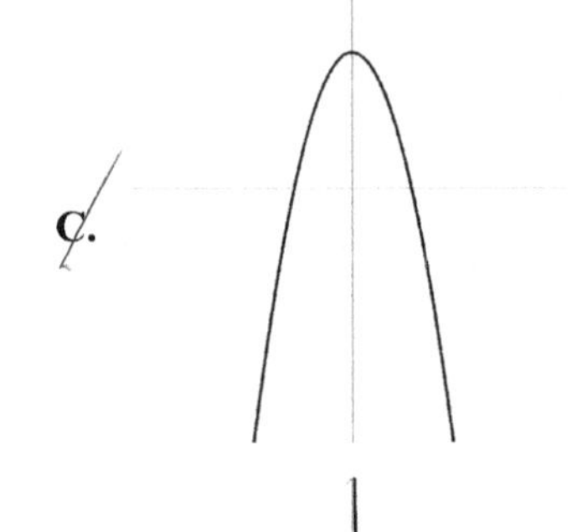

D.

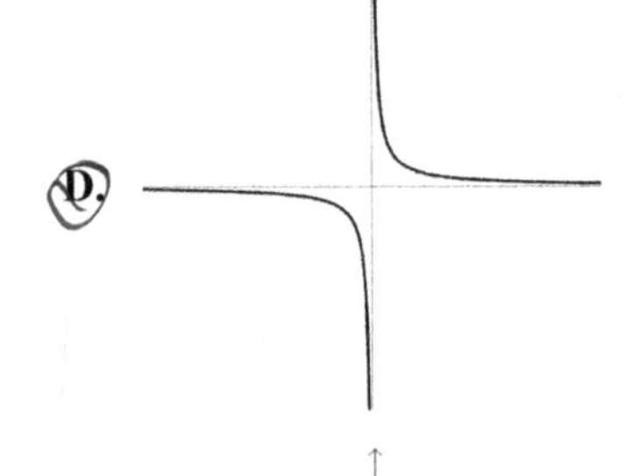

E.

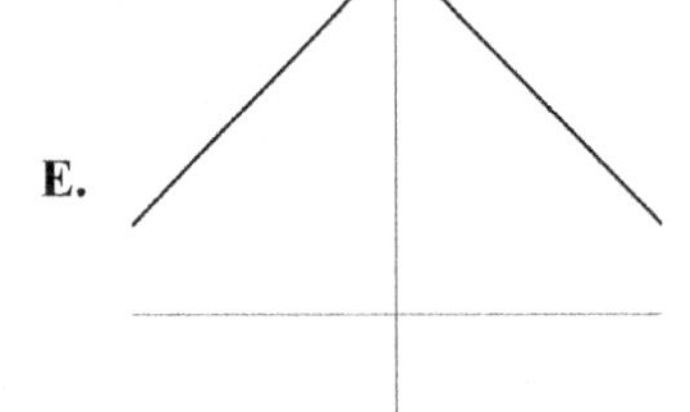

ANSWER KEY

1. E	**2. D**	**3. E**	**4. A**	**5. C**	**6. B**	**7. C**	**8. D**	**9. A**	**10. E**	**11. B**	**12. D**	**13. E**	**14. A**
15. C	**16. D**	**17. C**	**18. B**	**19. C**	**20. B**	**21. D**	**22. E**	**23. A**	**24. D**				

ANSWER EXPLANATIONS

1. **E.** The domain of the function $f(a)=\frac{1}{3-\left(\frac{1}{|a|}\right)}$ does not include the values of a that make $3-\left(\frac{1}{|a|}\right)=0$ or $|a|=0$ because if the denominator of a fraction is 0, the fraction is undefined. Solving this equation for the values of a, we get $3=\frac{1}{|a|} \rightarrow 3|a|=1 \rightarrow |a|=\frac{1}{3}$. This means $a=\pm\frac{1}{3}$. Solving $|a|=0$, we get $a=0$. So, the values of a that are not in the domain of the function are 0 and $\pm\frac{1}{3}$.

2. **D.** Since $g(x)$ varies directly as the square of $(x-3)$, we know that $g(x)=k(x-3)^2$ for some constant k. This is the expression for a parabola with center at $(3,0)$. For more on direct variation, see Chapter 12 in this book.

3. **E.** The expression is undefined if the denominator equals zero. So, setting $x^3-6x^2-55x=0$, we wish to find the values of x that make the expression undefined. Factoring out an x, we get $x\left(x^2-6x-55\right)=0$. Since we factored out an x, we know that the value $x=0$ makes the expression undefined. We solve for the rest of the values by using the quadratic equation (Use a calculator program if you like. See supertutortv.com/BookOwners). $x=\frac{-b\pm\sqrt{b^2-4ac}}{2a}=\frac{-(-6)\pm\sqrt{(-6)^2-4(1)(-55)}}{2}=\frac{6\pm\sqrt{36+220}}{2}=\frac{6\pm\sqrt{256}}{2}=\frac{6\pm16}{2}\rightarrow x=-5,11$. So the values of x that make the denominator equal to zero and the expression undefined are $x=0,-5,11$.

4. **A.** To find the horizontal asymptote, we can use the parameters earlier in the chapter or evaluate x at a very large value and see where the expression approaches. For very large x, the constants w and z become negligible, and so the expression approaches $f(x)=\frac{5x^2}{x^2}=5$. So, the horizontal asymptote is at $y=5$.

5. **C.** Looking at the graph, we can see that the curve starts where $x=1$, and only extends to the point where $x=7$. Thus, the answer is $\{x:1\le x\le 7\}$. We need $\le$ and $\ge$ not $<$ and $>$ because $x=7$ is included in the set (we know this as the dots are solid); choice E is thus incorrect. See inequalities chapters for a review of inequalities skills.

6. **B.** Looking at the graph there appears to be asymptotes at $x=-1$ and $x=3.3$. The second of these is not an option or close to an option, but -1, answer (B), is. We can also simplify the equation to confirm. $\frac{(x-1)}{x^2-2x-3}=\frac{(x-1)}{(x+1)(x-3)}$. None of the terms cancel. Since when the denominator equals zero, there is a vertical asymptote, the equation tells us that there are asymptotes at $x=-1$ and $x=3$.

7. **C.** First plug the entire g(x) into the f(x) function: $y=f(g(x))=\left|x^3-1\right|+2$. The graph of x^3 in translated one unit down, then when the absolute value is taken, only the parts of the graph that have negative y values are reflected over the x-axis. Finally, this form is translated an additional 2 units up. We thus know only an answer with a range of 2 and above can be correct. C is the only graph with this range of y-values. You can also graph this in your calculator if unsure. See chapter 16 to review composite functions.

8. **D.** This question is more common sense than graphing skills. Imagine the pedal of a bicycle as someone pedals with a constant speed. The pedal follows a fluid, circular path as it rises and falls, so an abrupt linear point in the shape doesn't make a lot of sense. That eliminates answers (A), (B), and (C). The question says that the pedal reached a maximum at $t = a$, and of the two left, only (D) has a maximum at $t = a$, so the answer is (D).

9. **A.** The graph is of a cubic function but because it spikes upward, we can see that for a couple values of y, there are multiple potential x's. To prove this, take a pencil, parallel with the x-axis and slide it upward. We see that along one horizontal line, the graph intersects multiple times. The rest of the answers are all *one-to-one* functions because they pass the 'pencil' rule both horizontally and vertically (aka the horizontal and vertical line test).

10. **E.** One way you could solve this would be by graphing on your graphing calculator and making up a value of K. You can also solve with logical reasoning. The equation $y = -Kx^3$ is a basic cubic function. Because the constant in the equation $y = -Kx^3$ is positive, the negative sign indicates the parent graph will be reflected over the x-axis. Only choice E matches this description: a vertical "flip" of the parent graph of $y = x^3$.

11. **B.** Even functions are symmetrical across the y-axis. The only answer given that complies is answer (B), a cosine function.

12. **D.** From the graph we can see that the function appears to approach an asymptote at $x = 1$, but we can analyze the equation as well to confirm. Simplify the function by factoring $f(x) = \frac{-x^3 - 3x^2 - 2x}{x^2 - 1} = \frac{(-x)(x+1)(x+2)}{(x+1)(x-1)} = \frac{(-x)(x+1)(x+2)}{(x+1)(x-1)}$.
Vertical asymptotes occur when the denominator is equal to zero (and the factors don't cancel), which in this equation is when $x = 1$. At $x = -1$, because the terms cancel, there is only a hole, but not an asymptote.

13. **E.** Answer (E) is the only one that has the jumps at the right places and the correct functions for each segment. For example, answer (B) jumps between parts at $x = -1$ and $x = 3$, but the middle part, the parabola piece, is the wrong function (it is facing downward instead of up). We know the parabola faces up because its x^2 coefficient is positive.

14. **A.** The function is undefined when the denominator is equal to zero, which is only possible when $x = 0$ or $x = -1$.

15. **C.** Re-arrange the equation given to a more recognizable form. $5y - 20 = -2x \rightarrow y = -\frac{2}{5}x + 4$. The new equation tells us that the graph has a y-intercept at the point $(0,4)$ and a slope of $-\frac{2}{5}$. Plug in 0 for y and solve to find our x-intercept, which is $(10,0)$. The only graph that contains these two points is answer (C).

16. **D.** This function has a range of integers only, as its pieces are horizontal lines at integer y-value, but the domain concerns horizontal progression, which in this graph is actually continuous from its beginning, where $x = 0$, to its end point, where x approaches but does not equal (implied by the open circle) 7. Every time a segment ends, the next segment begins at the same x value.

17. **C.** Looking at the equation alone we know that the term that includes y (because it has been squared) must at least be 0 if not greater, so when the 2 is added we know that the entire expression (x) must be equal to 2 or greater. Looking at the graph tells us the same thing: it's very clear that only values of x greater than about 2 are part of the graph, because it is a sideways facing parabola that opens to the right and from the graph we can see that the vertex is at 2. Thus by either method we determine that x must be greater than 2, choice C.

18. **B.** The domain of a composite function is limited by the domain of the inner function, as well as any domain restriction of the new, composite function. The inner function, $g(x) = x^2 - 3$, has no domain restrictions, but the composite function does. $f(g(x)) = \frac{5}{(x^2 - 3) + 2} = \frac{5}{x^2 - 1}$, so the domain is undefined when $x = \pm 1$, which is what answer B's graph shows.

For more on composite functions, see Chapter 16.

19. C. One strategy for solving this is to make up a number n that will work given the parameter and then graph this function on your graphing calculator. Then simply evaluate the answers based on what you see. That is the EASIEST method. But we can also think through this logically. We know that n will be a fraction less than 1, such as $\frac{1}{4}$ or $\frac{5}{32}$. These fractions are the same as a greater-than-one number to the power of negative one. For example, $\frac{1}{4}=4^{-1}$ and $\frac{5}{32}=\left(\frac{32}{5}\right)^{-1}$. Thus, the function f is equal to a greater-than-one number to the power of negative one to the power of x or, simplified, to the power of negative x. A negative sign in front of x flips the graph across the y-axis, so this graph will look like a greater-than-one number to the power of x flipped across the y-axis. Sketching this, we see that the function approaches zero asymptotically in the positive direction, and grows exponentially in the negative direction. This graph is always decreasing as we move to the right. One last method is to create a series of test points using a hypothetical equation that fits the parameters. Let's make up $n=\frac{1}{4}$ so $y=\left(\frac{1}{4}\right)^{x}$. Now plug in several x values to see what happens to y.

$$\left(\frac{1}{4}\right)^{-1}=4,\ \left(\frac{1}{4}\right)^{0}=1,\ \left(\frac{1}{4}\right)^{\frac{1}{2}}\rightarrow\sqrt{\frac{1}{4}}=\frac{1}{2},\ \frac{1}{4}^{1}=\frac{1}{4},\ \left(\frac{1}{4}\right)^{2}=\frac{1}{16}$$

We can see over time the function is decreasing because as x increases y decreases.

20. B. The graph shown is a cubic function. We can determine its exact formula by starting with our basic cubic function $y=x^3$. We then move it up 8 in order to match the graph's y-intercept. We now have the equation $y=x^3+8$. This equation matches the graph perfectly, since at $x=2$, we get $y=16$. The x-intercept of this graph can be found by finding where the y-value of the equation is 0: $0=x^3+8$. This becomes $x^3=-8$. Simplify this to yield $x=-2$. In addition, because $y=x^3$ has rotational symmetry, the point $(2,16)$ can be rotated about $(0,8)$ by 180° to reveal that the point $(-2,0)$ is also on the function. Think: this given point is 8 up and 2 right from the center point. Go 8 down and 2 left to find the x-intercept.

21. D. The function looks like $y=x^3$, so we can tell it is a cubic function with just a visual inspection. Use your graphing calculator to graph if you're uncertain.

22. E. A reflection across the line $y=x$ is the inverse function, where the x and corresponding y-values of a function are switched. Thus, from our original equation $y=x^3+8$ we get $x=y^3+8$. Solving this equation for y gives us $x-8=y^3\rightarrow\sqrt[3]{x-8}=\sqrt[3]{y^3}\rightarrow y=\sqrt[3]{x-8}$.

23. A. The graph of $y=-e^x+K$ is merely a modification of the graph of $y=e^x$. Remember, e is a constant number around 2.73. The graph of $y=e^x$ represents exponential growth of a number larger than one, which means that it increases sharply in the positive x-direction and flat-lines at a constant value in the negative x-direction. However, in our equation, e^x has a negative coefficient; this means that our graph represents negative exponential growth: it decreases sharply in the positive x-direction, and still flat-lines in the negative x-direction. We can also think of this as flipping a traditional exponential growth graph over the x-axis. Our equation contains the positive constant K. This means that our equation will have a y-intercept of $1+K$, because when $x=0$, $y=e^0+K\rightarrow y=1+K$, and since K is positive, our equation will have a positive y-intercept. The only answer choice that has a positive y-intercept and decreases sharply in the positive x-direction is choice A. Alternatively, you could graph the equation in your graphing calculator and use some arbitrary positive value for K, such as 3, and compare your graph with the graphs given.

24. D. An easy way to visualize this problem is that each X value and its corresponding negative value, $-X$, will have the same y value. If this is the case, all choices except D have a single y value for the both the positive and negative X value.

CHAPTER

23 TRANSLATIONS AND REFLECTIONS

SKILLS TO KNOW

- Basic reflectional and rotational symmetry
- How to identify the translation/reflection of a given graph or function
- How to translate/reflect a given function and/or select correct graph, equation, or name new points

NOTE: This chapter is a continuation of **Chapter 22: Graph Behavior.** Again, unless you're aiming for a score of 30+, you can likely skip or skim this chapter. Note these questions are rare.

Translating an equation can be split into 3 categories: **shifting**, **compressing**, and **reflecting/ rotating**.

Two of these sections we covered already in **Graph Behavior, shifting** (vertical and horizontal shifts via constants such as "h" and "k") and **compressing** (altering a graph's width or apparent slope with a constant such as "a"). For this chapter, we will focus on problems that involve **reflections** and **rotations**.

REFLECTIONS AND ROTATIONS

Reflectional symmetry: the property a figure has if half of the figure is congruent to the other half over an axis.

Rotational symmetry: also known as radial symmetry—the property a figure has if it is congruent to itself after some rotation less than 360°.

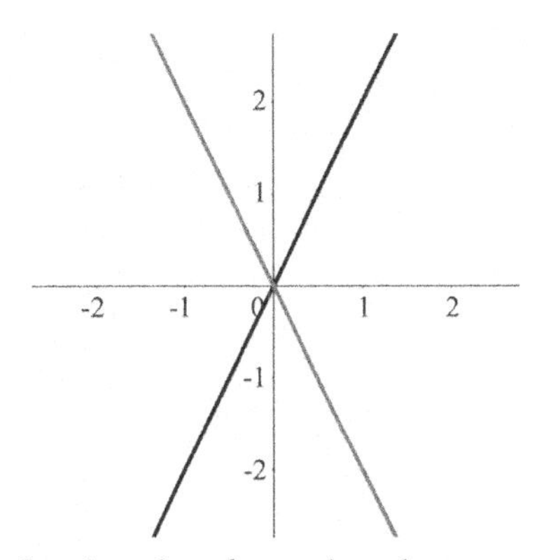

This figure has both reflectional and rotational symmetry over the x or y axis.

How to "reflect" a function algebraically

To reflect over the x-axis: Solve for y or $f(x)$ and distribute a negative sign to the entire other side of the equation. Alternatively, simply place a negative sign in front of every instance of y (no rearrangement necessary).

To reflect over the y-axis: Solve for x and distribute a negative sign to the entire other side of the equation. Alternatively, simply place a negative sign in front of every instance of x (no rearrangement necessary).

To memorize this, remember that **to reflect over one axis, you make the OTHER variable negative.** I.e. make every instance of x negative to reflect across the y-axis and make every instance of y negative to reflect across the x-axis.

As you can see in the picture above, simple reflection could be $y=2x$ reflected to $y=-2x$. This reflection would be over the x-axis because we solved for y and applied a negative sign to the other half of the equation. But you might notice this reflection is also over the y-axis. That's because a direct variation equation (one with no "y" intercept that goes through the origin) reflects over both axis when you add a negative to either side of the equation. We can see this algebraically by solving for x per our rules above (that denote a y-axis reflection requires first solving for x) to get $\frac{1}{2}y=x$ and then applying the negative to get $-\frac{1}{2}y=x$. We could then multiply both sides by negative two to see that this is the same as $y=-2x$.

Symmetry Questions

Oftentimes on the ACT®, you are asked more conceptual questions that require you to understand the notion of reflectional or rotational symmetry.

Which of the following letters of the alphabet does NOT have at least 1 reflectional symmetry and at least 1 rotational symmetry?
(Note: the angles of rotation for the rotational symmetry must be less than 360°.)

A. H **B.** I **C.** O **D.** U **E.** X

Let's take this problem one half at a time. First up is reflectional symmetry. H, I, O, and X are symmetrical about a centered horizontal axis, and U is symmetrical about a centered vertical axis. So no eliminations here.

Now, for rotational symmetry, H, I, O, and X are symmetrical at 180°. However, U is not symmetrical at any rotation less than 360° meaning that it does not have any rotational symmetry.

Answer: **D.**

TRANSLATIONS

Translation problems we first covered in Graph Behavior, but we'll look at more complex examples here that blend the idea of reflections and or rotations with traditional horizontal/vertical shifts and compression.

Now, let's take everything we learned and do full translations.

1. Determine vertical shift
2. Determine horizontal shift
3. Determine compression
4. Determine any reflections across the axis or rotations about the origin.

ABSOLUTE VALUE

For absolute values, most of the rules still apply, but reflections over the y-axis are more complex, as it's tough to solve for x. As a result, your best move is to apply the negative sign to the **variable x itself.** I.e. to reflect across the y-axis, put a negative sign on every instance of x in the equation. To reflect across the x-axis, put a negative on the y-value.

The function $y=\left|\frac{x}{2}+1\right|+3$ is graphed in the standard (x,y) coordinate plane below.

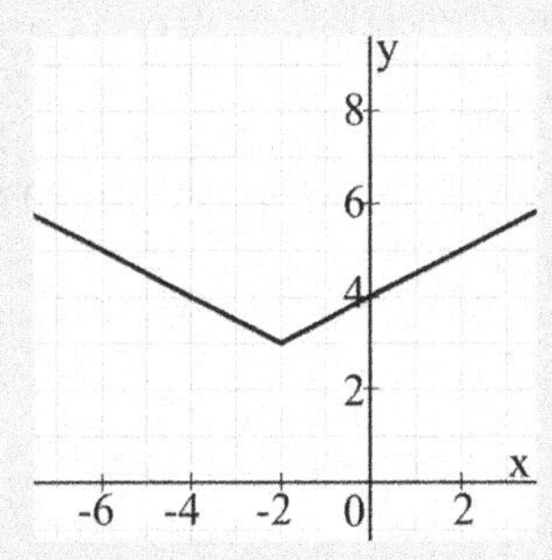

One of the following graphs in the standard (x,y) coordinate plane shows the result of shifting the function 6 units to the right and reflecting it about the y-axis. Which graph is it?

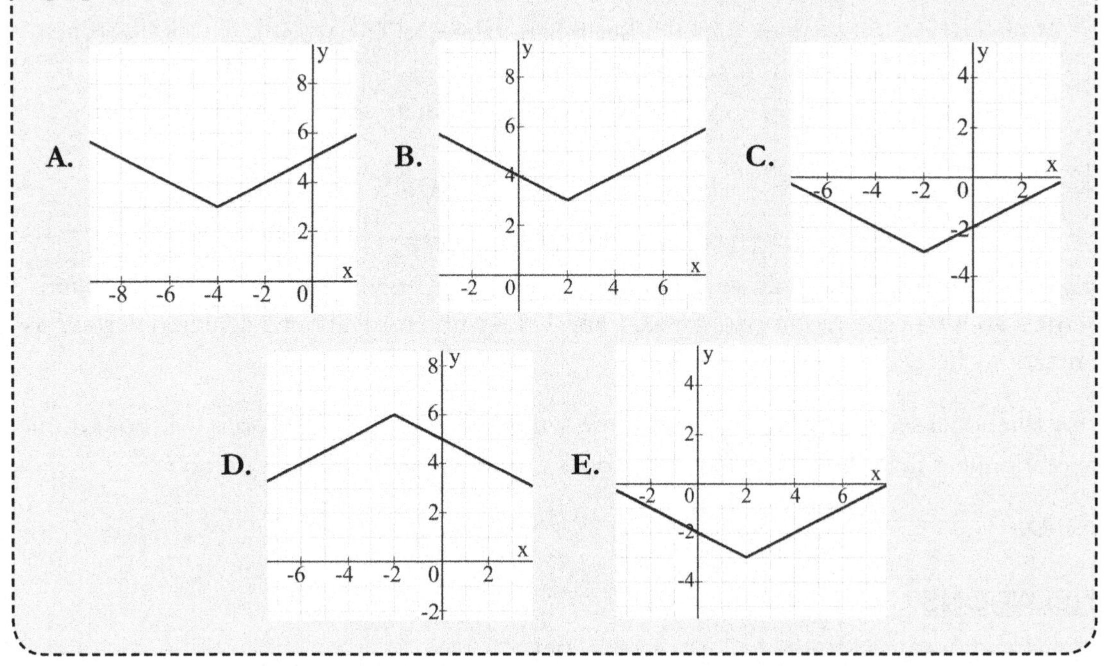

The fastest solution on this question is to "eyeball" your answer by translating an individual point or two. Find the vertex of the original function. Then, per the translation instructions, count six over to the right. Now we're at positive 4 for our x-value. Now reflect this point over the y-axis: we walk this point back to the left keeping the vertical height the same to reach the y-axis. Once we reach the y-axis, we note the distance we travelled: 4 units. Now we go another 4 units to the left to create our reflected point. Thus we know the vertex is at negative four. Only choice A has a vertex with an x-value at -4, so it is the answer.

We can also solve algebraically (though this is total overkill unless you want a 35+):

First, we need to factor out the coefficient of the x term from both values inside the absolute value. Remember, the "a" in any equation should be isolated from the "$x-h$" term in order to see our horizontal shift properly. We don't want $ax-h$, we want $x-h$ to be its own little piece of the equation. Here the coefficient is $\frac{1}{2}$, because x is divided by 2, so I factor out $\frac{1}{2}$ from both $\frac{x}{2}$ and 1 to form this equivalent equation:

$$y=\left|\frac{1}{2}(x+2)\right|+3$$

Now I can move my 3 if I'd like to the other side to see the effect on y (when I look for vertical shifts, I want $y-k$ to replace any instance of y in the root version of the equation).

$$(y-3)=\left|\frac{1}{2}(x+2)\right|$$

Now I'm ready to make another transformation, applying that transformation to the $x-h$ and $y-k$ terms:

To shift the equation to the right 6 units, we need to subtract 6 from the $x+2$ term. Remember, from the last chapter, we replace every instance of x with $(x-h)$ where h is the number of units to the right—but here we already have an $(x+2)$ term indicating. Still, we can replace x with our new "$x-h$" in position and get what we need.

$$(y-3)=\left|\frac{1}{2}((x-6)+2)\right|$$

Which simplifies to:

$$(y-3)=\left|\frac{1}{2}(x-4)\right|$$

Finally, we reflect it over the y-axis by placing a negative sign in front of the x:

$$(y-3)=\left|\frac{1}{2}(-x-4)\right|$$

Now, I need to factor out that negative sign in front of the x so I get a clean "$x-h$" term. In doing so, I also factor the -1 out of -4:

$$(y-3)=\left|-\frac{1}{2}(x+4)\right|$$

Furthermore, I know the "negative" will disappear given the absolute value, so this is equivalent to:

$$(y-3)=\left|\frac{1}{2}(x+4)\right|$$

Now we can use our knowledge of horizontal and vertical shift to find our graph. First, our vertex of an absolute value equation would be $(0,0)$. We can find the horizontal and vertical shift of this vertex by looking at our new h and k: h is now indicating we move 4 to the left $(+4)$ so our x value of the vertex is -4. k indicates that we move 3 up, so our y value of the vertex is positive 3. We also know the slope of the line should be positive $\frac{1}{2}$ and negative $\frac{1}{2}$ before and after the vertex. Additionally, we can assume this is upward facing given that there is no negative in front of the absolute value expression.

Given these parameters, the answer is A.

Answer: **A.**

Quadrilateral $ABCD$ has vertices $(-1,-1)$, $(-5,-3)$, $(-4,3)$, and $(-7,6)$ in the standard (x,y) coordinate plane. Suppose $ABCD$ is translated 5 units to the right and 4 units up, forming quadrilateral $A'B'C'D'$. Which of the following shows the coordinates of the vertices of $A'B'C'D'$?

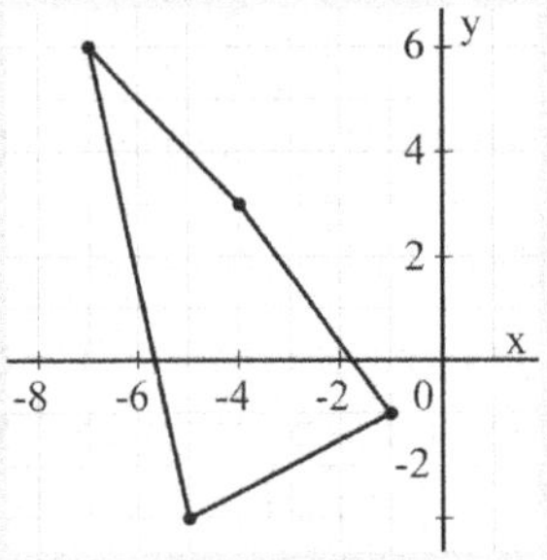

A.

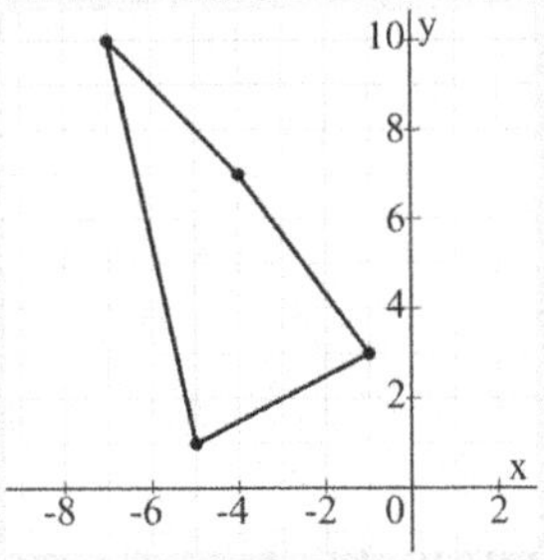

B.

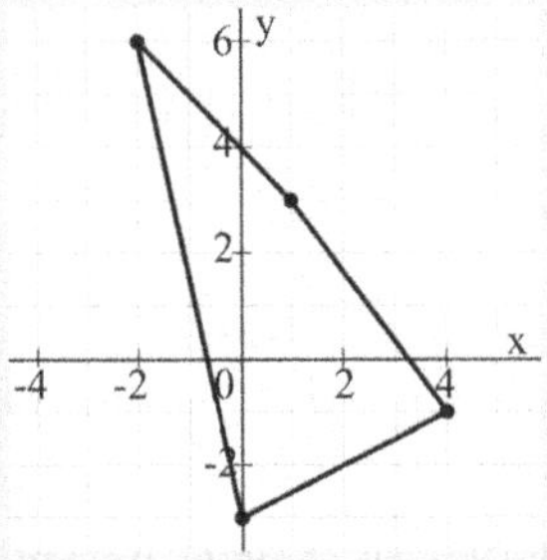

C.

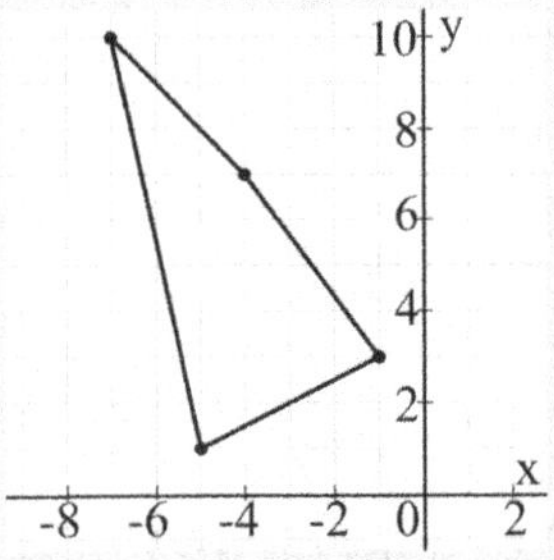

D.

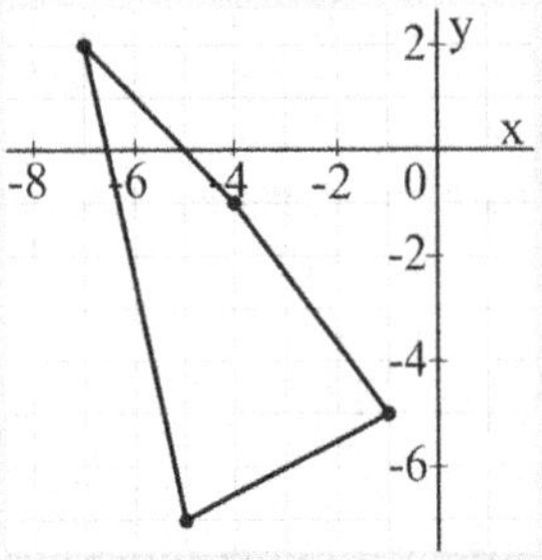

E.

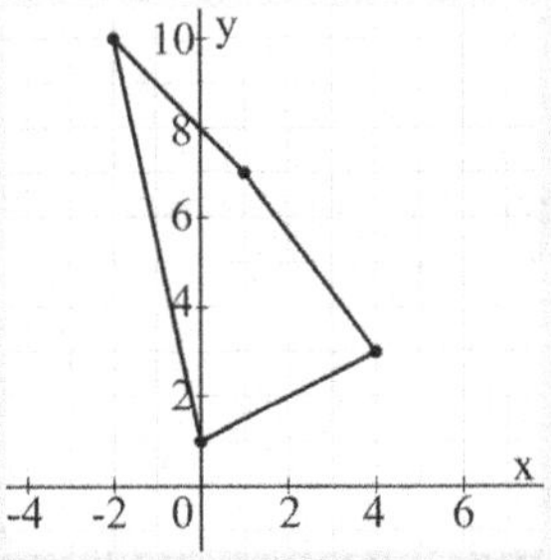

For this question, we can increase every x value by 5 and every y value by 4. We can also just count on the graph for each point and find the transformed quadrilateral. Thus, the new vertices are : $(-1+5,-1+4)$, $(-5+5,-3+4)$, $(-4+5,3+4)$, $(-7+5,6+4)$ = $(4,3)$, $(0,1)$, $(1,7)$, $(-2,10)$.

Answer: **E.**

In the standard (x,y) coordinate plane the graph of $y=\sqrt{x}$ is shifted 3 units up and 4 units to the left, and is then reflected over the y-axis. Which of the following is an equation of the translated graph?

A. $y=\sqrt{(-x+3)}+4$ **B.** $y=\sqrt{(-x+4)}+3$ **C.** $y=\sqrt{-(x+4)}+3$

D. $y=\sqrt{(-x+4)}-3$ **E.** $y=\sqrt{-(x+3)}-4$

For this question, we know that in $y=m(x-h)+k$, the k determines vertical shift, and the h determines horizontal shift. Because the equation shifts up by 3, we know k is 3; similarly, because the equation is moving 4 units to the left, $h=-4$. We can plug this into $y=\sqrt{(x-h)}+k$. This gives us $y=\sqrt{(x+4)}+3$. Now we need to reflect over the y- axis. This can be done by adding a negative sign in front of (and only in front of) x. Thus, we know that the transformed equation is $y=\sqrt{(-x+4)}+3$.

Answer: **B.**

1. The graph $y = x^3$ is shown below. Which of the equations below describe this parabola shifted 4 units to the left and 2 units up?

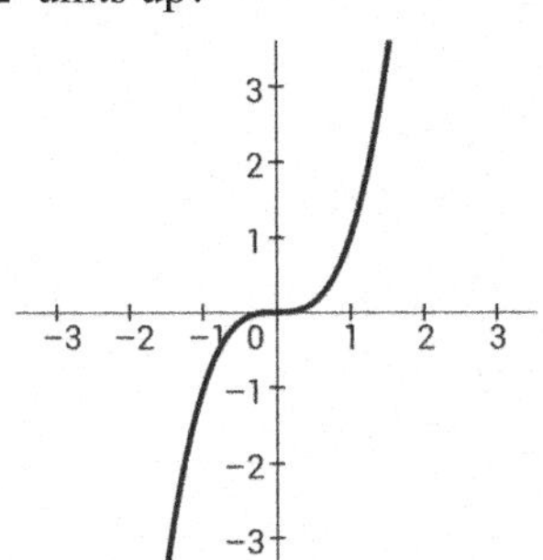

A. $y = (x+2)^2 - 4$
B. $y = (x-2)^2 - 4$
C. $y = (x+4)^2 - 2$
D. $y = (x+4)^2 + 2$
E. $y = (x-4)^2 + 2$

2. In the standard (x, y) coordinate plane shown below, a rectangle has points $(-2,1)$, $(-2,4)$, $(3,1)$, and $(3,4)$. The rectangle is shifted 5 units to the left and 3 units down. Which quadrants does the rectangle exist in?

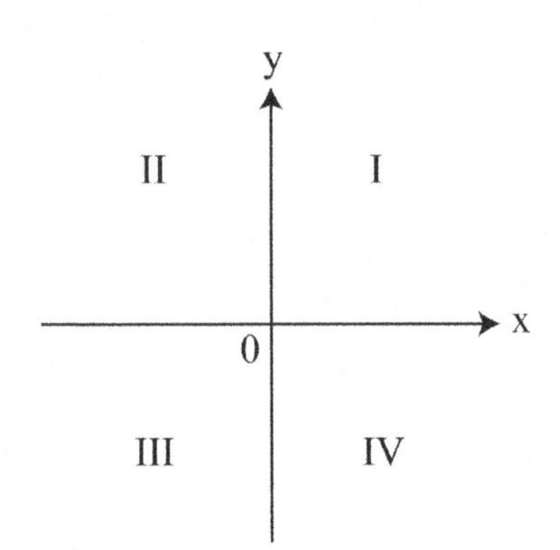

A. I and II
B. II and III
C. I and IV
D. III
E. All quadrants

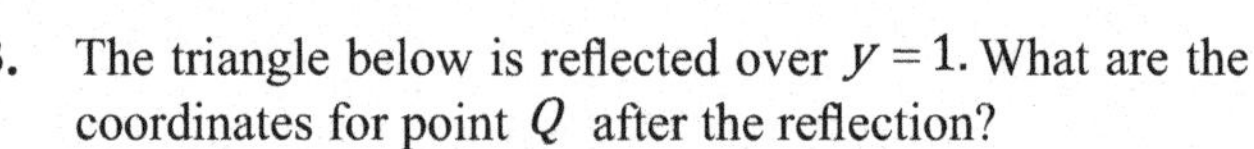

3. The triangle below is reflected over $y = 1$. What are the coordinates for point Q after the reflection?

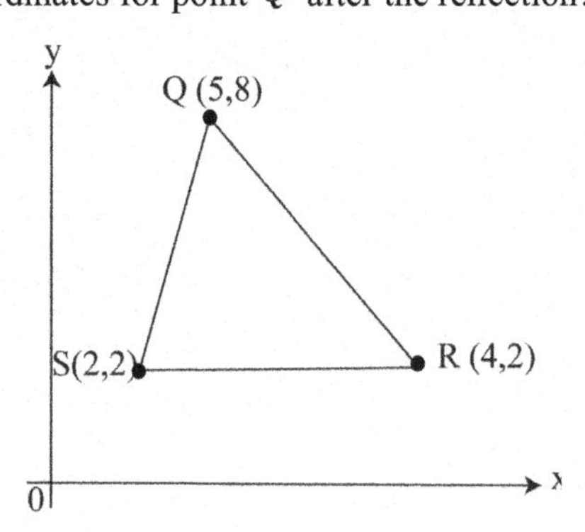

A. $(5,-8)$
B. $(-5,-8)$
C. $(-5,8)$
D. $(5,-6)$
E. $(5,-7)$

4. The graph in the standard (x, y) coordinate plane below is the graph of $y = f(x)$. Which of the following graphs is the graph of $y = f(x+1) - 3$?

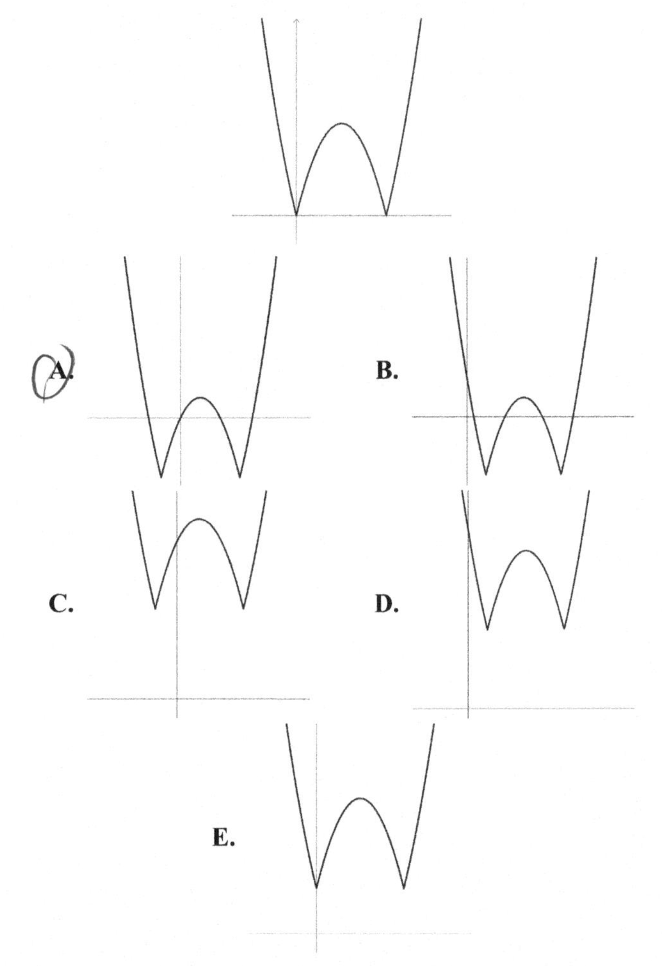

5. Which of the following expressions represents the graph of $h(x)$ in terms of $f(x)$?

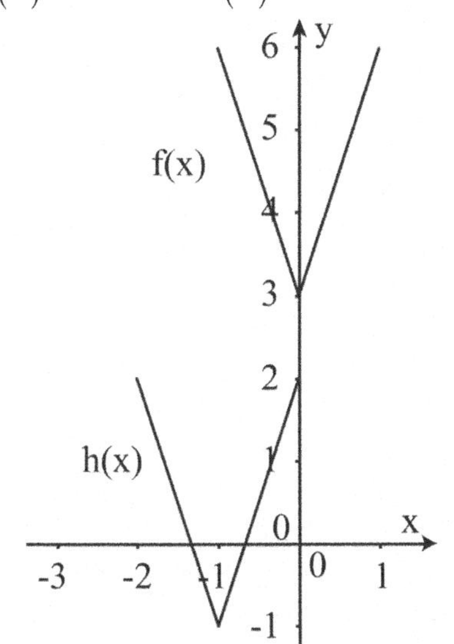

A. $|x+1|-4$
B. $|x-4|+1$
C. $|x-1|+4$
D. $|x+4|-1$
E. $|x-4|+4$

6. The point P is shown below. What are the coordinates of the point P after it is reflected over the line $x=3$?

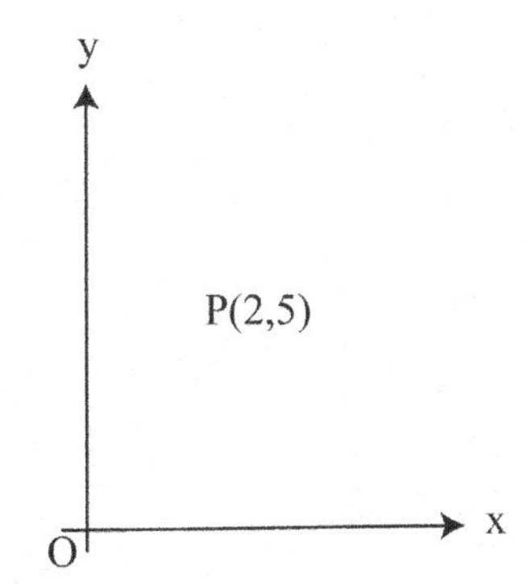

A. $(1,2)$
B. $(2,1)$
C. $(4,2)$
D. $(4,5)$
E. $(5,4)$

7.

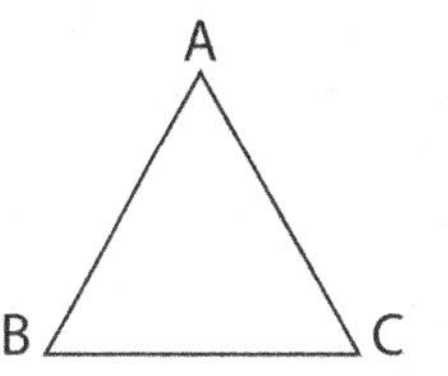

Triangle ABC is pictured above. Which of the following represents the orientation of the triangle after being reflected over AB and then rotated 90° clockwise around point B?

A. 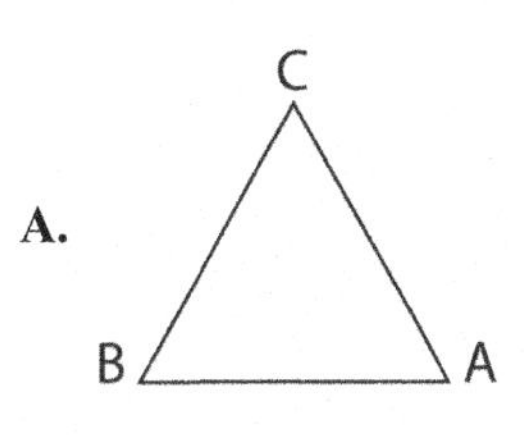

B. 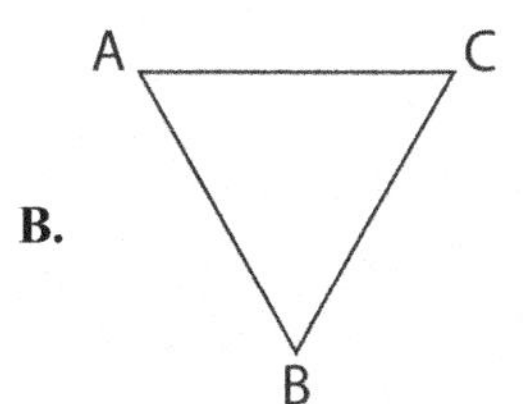

C. 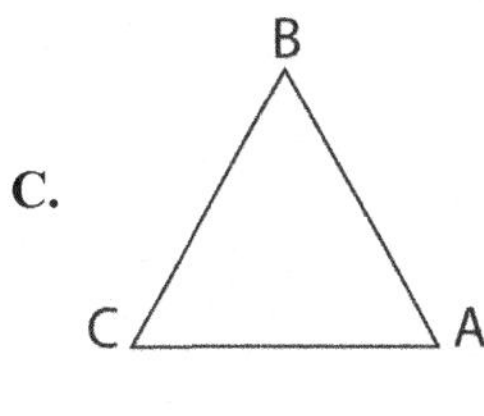

D. 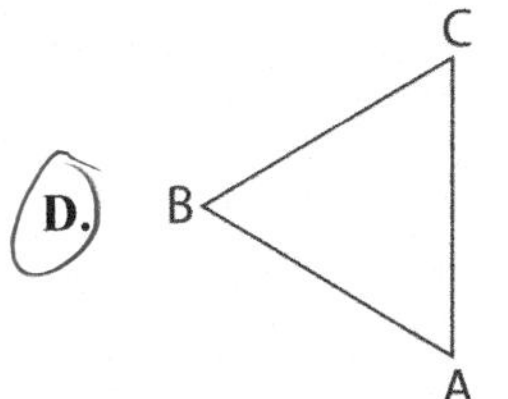

E. 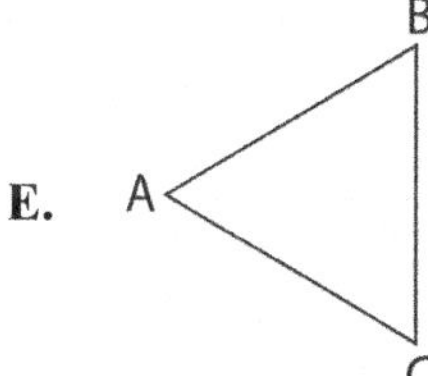

8. In the figure below, $\overline{DE}=7$ and the other sides have lengths of 3, 3, 2, and 3. Let $D'H'G'F'E'$ represent the shape after being reflected over DE. What is the perimeter of the shape $DHGFEF'G'H'$?

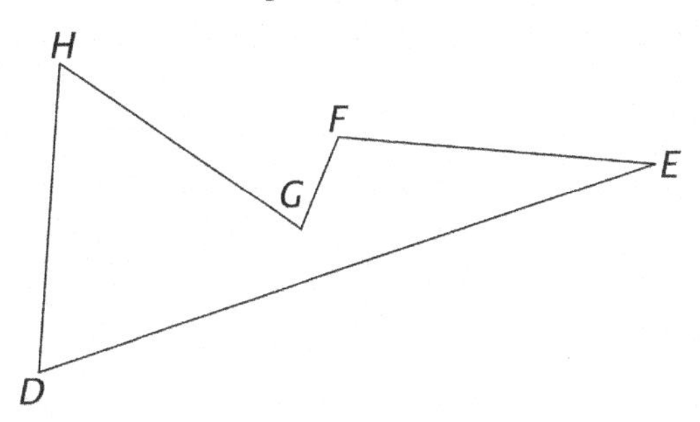

A. 11
B. 18
C. 22
D. 30
E. 36

9. A point at $(-7,3)$ in the standard (x,y) coordinate plane is shifted right 3 units and down 7 units. What are the coordinates of the point?

A. $(-10,10)$
B. $(0,0)$
C. $(-10,-10)$
D. $(-4,10)$
E. $(-4,-4)$

10. The graphs of $f(x)=\sin x$ and $g(x)=\sin\left(x+\frac{\pi}{2}\right)+2$ are shown in the standard (x,y) coordinate plane below. After one of the following pairs of transformations is applied to the graph of $f(x)$, the image of the graph of $f(x)$ is the graph of $g(x)$. Which is it?

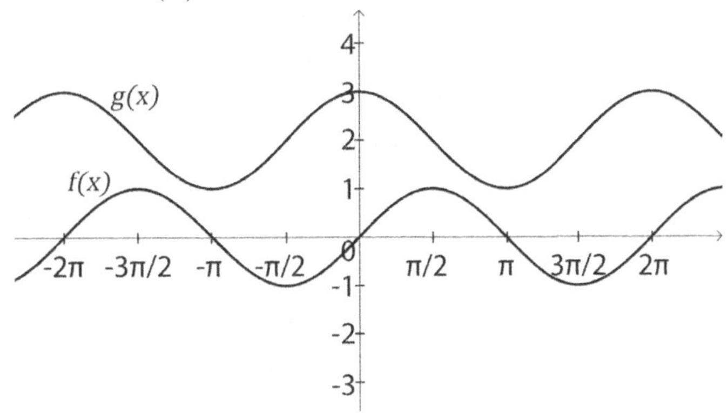

A. Shift $f(x)$ 2 units left and $\frac{\pi}{2}$ units down.

B. Shift $f(x)$ 2 units left and $\frac{\pi}{2}$ units up.

C. Shift $f(x)$ 2 units right and $\frac{\pi}{2}$ units down.

D. Shift $f(x)$ $\frac{\pi}{2}$ units right and 2 units up.

E. Shift $f(x)$ $\frac{\pi}{2}$ units left and 2 units up.

11. Rectangle $ABCD$ has vertices in the standard (x,y) coordinate plane at $A(-2,-4)$, $B(-2,6)$, $C(1,6)$, and $D(1,-4)$. A translation of rectangle $ABCD$ is a second rectangle $A'B'C'D'$ with vertices $A'(5,1)$, $B'(x,y)$, $C'(8,11)$, $D'(8,1)$. What are the coordinates of B'?

A. $(3,10)$
B. $(4,10)$
C. $(5,4)$
D. $(5,0)$
E. $(5,11)$

12. In the standard (x,y) coordinate plane, $P(-5,-3)$ will be reflected over the x-axis. What will be the coordinates of the image of P ?

A. $(-5,3)$
B. $(-3,5)$
C. $(3,-5)$
D. $(3,5)$
E. $(5,-3)$

13. Figure $ABCD$, shown in the standard (x,y) coordinate plane below, has been reflected across a line to figure $A'B'C'D'$. Which of the following lines reflection would best describe this transformation?

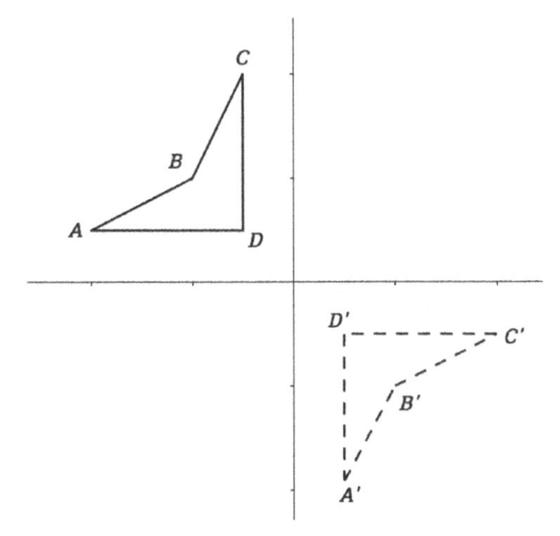

A. $y=0$
B. $x=0$
C. $x=y$
D. $y=\frac{1}{2}$
E. $y=-x$

14. The semicircle $\overset{\frown}{XYZ}$ that is shown below has a diameter of d inches. Let $\overset{\frown}{X'Y'Z}$ be the image of $\overset{\frown}{XYZ}$ reflected across $\overline{XZ}$. Which of the following is an expression for the perimeter, in inches, of the shape formed?

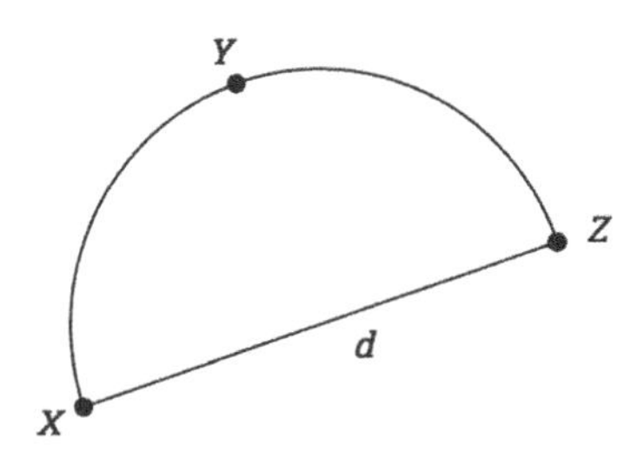

A. $2\left(\overline{XY}+\overline{YZ}\right)$

B. $4\left(\overline{XY}+\overline{YZ}\right)$

C. $\pi\overline{XZ}$

D. $2\pi\overline{XZ}$

E. $4\pi\overline{XZ}$

15. The graph of $f(x)$ and $g(x)$ are shown in the standard (x,y) coordinate planes below. Which one of the following expressions represents $g(x)$ in terms of $f(x)$?

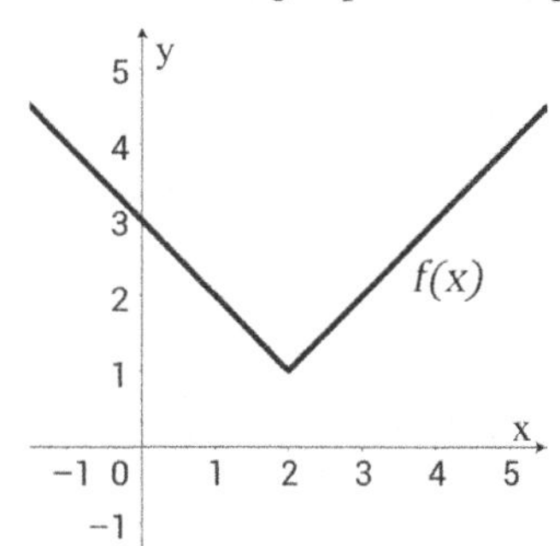

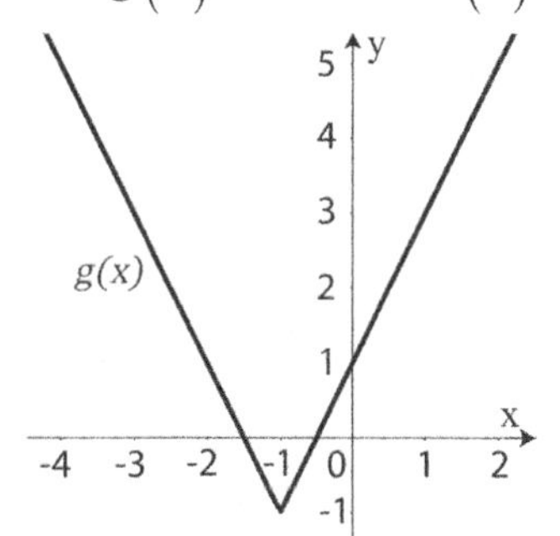

A. $2f(x+3)+2$

B. $2f(x+3)-2$

C. $2f(x-3)+2$

D. $f(x-3)-2$

E. $f(x+3)-2$

16. The point $(7,2)$ and the line $x=4$ are graphed in the standard (x,y) coordinate plane below. After the point has been reflected across the line, what are the coordinates of the point's image?

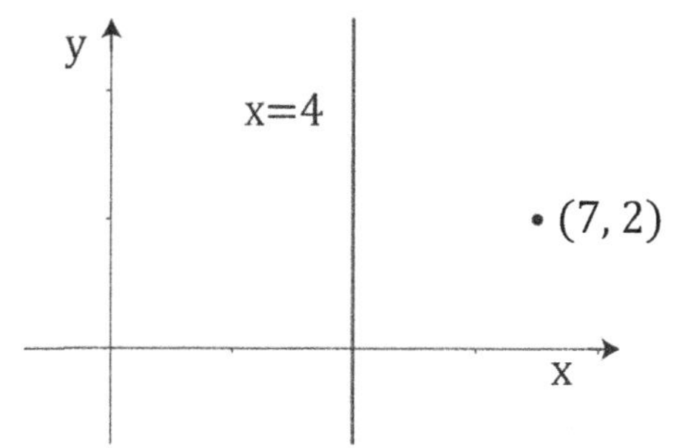

A. $(1,2)$

B. $(-7,2)$

C. $(-1,2)$

D. $(1,7)$

E. $(1,-2)$

17. In the standard (x,y) coordinate plane, a triangle has vertices $(-1,5)$, $(-1,-2)$, and $(2,5)$. What will be the coordinates of the vertices after the triangle is shifted down 3 units?

A. $(-1,2),(-1,1),(2,8)$

B. $(2,5),(-1,1),(2,8)$

C. $(-1,2),(-1,-5),(2,2)$

D. $(-1,8),(1,-1),(5,-2)$

E. $(2,5),(0,-1),(2,2)$

18. A triangle, ΔDEF, is reflected across the x-axis to have the image $\Delta D'E'F'$ in the standard (x,y) coordinate plane; thus, D reflects to D'. The coordinate of point D are $(a,-b)$. What are the coordinates of point D'?

A. (a,b)

B. $(-b,a)$

C. $(-a,b)$

D. $(b,-a)$

E. Cannot be determined from the given information.

19. For an assignment on symmetry, Joey created the shape below. His teacher commented on the symmetry when evaluating this assignment. Which of the following is a true comment about the symmetry of this pattern?

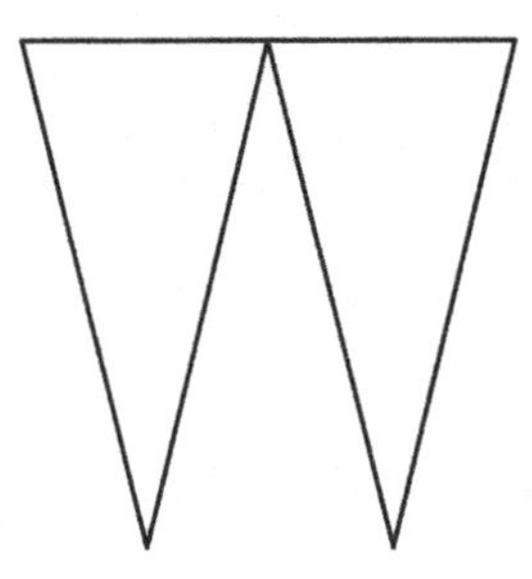

A. The pattern has both a horizontal line and a vertical line of symmetry.
B. The pattern only has a horizontal line of symmetry.
C. The pattern only has a vertical line of symmetry.
D. The pattern has a rotational symmetry of $180°$
E. The pattern has a rotational symmetry of $90°$

20. Which of the following letters of the alphabet has a rotational symmetry of less than $360°$?

A. L
B. W
C. U
D. X
E. N

21. A vector from the origin to the terminal point $(3,-5)$ is shown in the standard (x,y) coordinate plane below. The vector will be rotated $180°$ about the origin, resulting in a new vector. What will be the coordinates of the terminal point of the new vector?

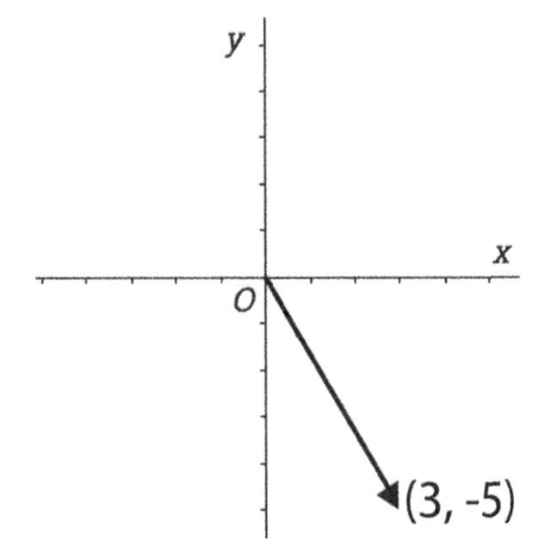

A. $(5,-3)$
B. $(-3,5)$
C. $(-3,-5)$
D. $(3,5)$
E. $(-5,-3)$

22. The graph $\frac{x}{y}=5$ is reflected across the y axis in the standard (x,y) coordinate plane. Which of the following is an equation of the reflection?

A. $xy=5$
B. $\frac{y}{x}=-5$
C. $\frac{x}{|y|}=5$
D. $-x=5y$
E. $\frac{|x|}{y}=5$

23. Grid lines are shown at 1-unit intervals in the standard (x,y) coordinate plane below. Some of the 1 by 1 square are shaded in the grid. Which is the least number of 1 by 1 square that must be unshaded so that the total shaded region will be symmetric about the x-axis?

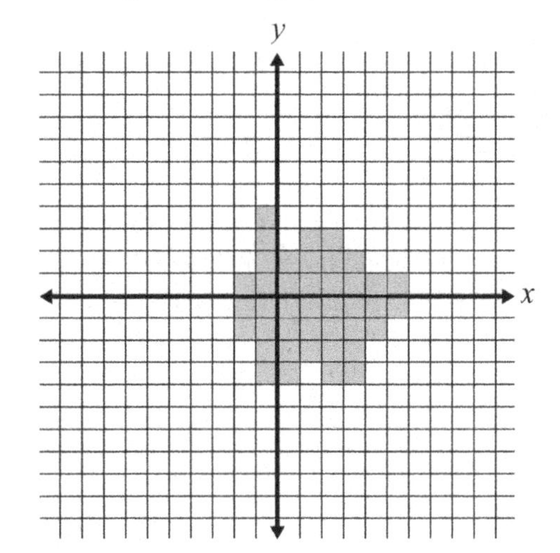

A. 6
B. 7
C. 8
D. 9
E. 10

ANSWER KEY

1. D **2. B** **3. D** **4. A** **5. A** **6. D** **7. D** **8. C** **9. E** **10. E** **11. E** **12. A** **13. C** **14 .C**
15. B **16. A** **17. C** **18. A** **19. C** **20. D** **21. B** **22. D** **23. B**

ANSWER EXPLANATIONS

1. **D.** The formula for a parabola is $y=(x-h)^2+k$. Since h (horizontal shift) undergoes a change of -4, and k (vertical shift) undergoes a change of $+2$, the final equation is $y=(x+4)^2+2$.

2. **B.** The right side of the rectangle lies on the line $x=3$. After the rectangle has been translated 5 units to the left, the right side of the rectangle lies on $x=-2$. Now the rectangle is in quadrant 2. The top side of the rectangle lies on the line $y=4$ and the bottom side on the line $y=1$. After these lines has been translated 3 units down, they line on the lines $y=1$ and $y=-2$, respectively. Thus, the rectangle exists in quadrants II and III.

3. **D.** When the triangle is reflected over the line $y=1$, line SR lies on the horizontal axis. Because the height of the triangle is 6 units, Q lies on the line $y=-6$. The x coordinate of Q does not change because the triangle is reflected over a horizontal line. (5, -6) is thus Q's new coordinate.

4. **A.** Adding to x shifts the graph to the left; for example, compare $y=x^2$ and $y=(x+1)^2$ in your graphing calculator. Subtracting from the entire function by -3 shifts the graph downward, as it represents a negative k and thus downward movement. The only graph that does both is A. If necessary, pluck one point and apply these shifts to that one point.

5. **A.** The formula for the absolute value graph is $y=|x-h|+k$. In the graph of $f(x)$, we can see that the vertex has shifted down 4 and to the left 1. Because h equals movement to the right, and k equals movement upwards, for the graph of $h(x)$, the value of h will be -1 and $k=-4$. Plugging these values into the formula above: $y=|x--1|-4 \rightarrow y=|x+1|-4$ Remember double negatives make a positive! Remember a "plus" next to the x-value means move left!

6. **D.** First, since the point is reflected over a vertical line, the y-axis of point P will not change. Point P lies on the line $x=2$. Because it is one unit to the left of the line of reflection, $x=3$, the coordinates of P after reflection will be one unit away to the right, on line $x=4$.

7. **D.** When the triangle is reflected over $\overline{AB}$, the orientation of line $\overline{AB}$ remains the same, but point C is translated across the line, as in choice C. When the triangle is rotated 90° around point B, line AC rotates from its horizontal position to a vertical line.

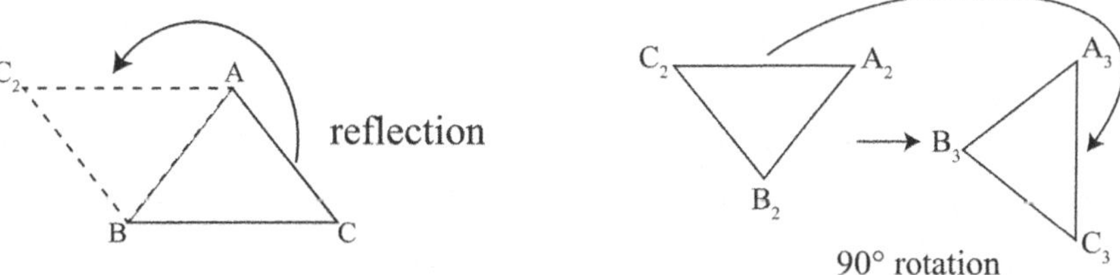

8. **C.** When the shape is reflected over the line $\overline{DE}$, the reflecting line $\overline{DE}=7$ can be ignored for the calculation of the perimeter. Therefore, we can add the other numbers 3,3,2, and 3 to calculate the perimeter of the top part of the shape, and again add the numbers 3,3,2, and 3 to find the perimeter of the bottom part of the shape. The answer is 22.

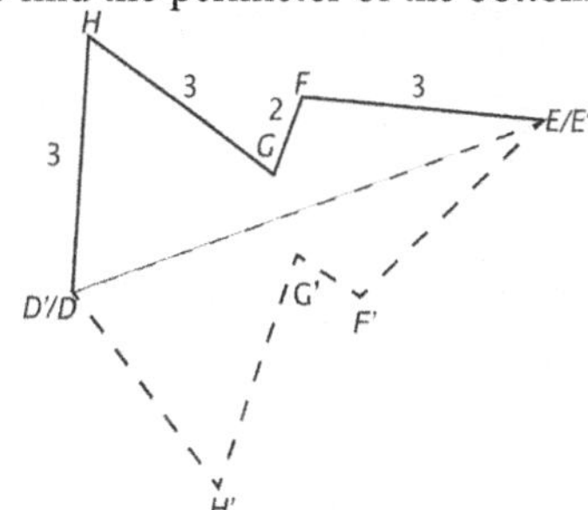

9. **E.** The point lies on the line $x=-7$. Because the point is shifted 3 units right, the point now lies on the line $x=-4$. The point also lies on the line $y=3$. When this point is shifted 7 units down, it will lie on the line $y=-4$. Therefore, the new point lies on $(-4,-4)$. Do these one step at a time, and draw a picture to help!

10. **E.** The formula for this equation, where positive h pertains to horizontal shift to the right and a negative h horizontal

movement to the left, and k denotes vertical shift upward, is $y=\sin(x-h)+k$. We match up this formula to our equation $\sin(x+\frac{\pi}{2})+2$ or rewritten $\sin(x-(-\frac{\pi}{2}))+2$ to see that h is $-\frac{\pi}{2}$, so the graph of $f(x)$ has shifted to the left $\frac{\pi}{2}$ units. We also see k for this equation is $+2$, so the graph of $f(x)$ has shifted up 2 units.

11. E. Notice in the first set how each of the two possible x values repeats twice and each of the possible y values repeats twice among the coordinates? We can sketch the second set of coordinates out and anticipate that the missing coordinate will simply be at the x-coordinate and y-coordinate that don't already "repeat" in the given three coordinates. Thus the x value will be 5 and the y-value 11, as these occur only once among the known three coordinates. Draw to be sure!

12. A. Because the point P is reflected over the x-axis, only the y coordinate of point P will change. P lies on the horizontal line $y=-3$, which is 3 units below the x-axis. Therefore, when the point is reflected, it will be 3 units above the x-axis. The point is at $(-5,3)$. When in doubt, draw a picture and count over to the line of reflection and then count the same distance away from that line of reflection, moving in the same direction.

13. C. Each point and its reflected opposite must be equidistant from the reflecting line. If we look at C and C', we can see they are equidistant from the line $y=x$. We can also imagine folding the paper to get the shapes to overlap. We would fold along the line $y=x$.

14. C. When the semicircle is reflected over $\overline{XZ}$, the new shape is a circle. The formula for the perimeter of any circle is πd.

15. B. Shift $f(x)$ 2 units down and 3 units left; $g(x)$ is our result. Don't confuse the order! Since the equation for the graph is $f(x)=a(x-h)+k$, h must be -3 and k must be -2. So $f(x)=a(x-(-3))+-2$ or $f(x)=a(x+3)-2$ and that leaves choices B and E. We can also see that there is a slope/steepness difference in the two graphs, and that must come from a multiple somewhere, either the ***a*** value in our formula or the 2 in front of the $f(x)$ in the answer choices. Thus B must be correct: $g(x)$ has "steeper" lines, so its "slope" would be multiplied by two compared to that of $f(x)$.

16. A. Because the point is reflected over a vertical line, only the x coordinate will change. The current point and the reflected point must be equidistant to the reflecting line. The point is 3 units right of the reflecting line. Therefore, the reflected point will be 3 units to the left of the reflecting line, at $y=1$.

17. C. Since we are only concerned with vertical shift, the x-values for all numbers will remain the same. Knowing this, we can eliminate all but A and C. Now subtract 3 from the y coordinates of all points: $5-3=2$; $-2=3=-5$; $5-3=2$. Now we replace the respective y-coordinates with these calculations: $(-1, 2)$, $(-1, -5)$, and $(2, 2)$. C is correct.

18. A. Since the point is reflected over a horizontal line, only the y coordinate changes. Thus a remains the same, but the sign of the y coordinate changes, so $(a, -b)$ reflects to become (a, b). Make up numbers and draw a sketch to help.

19. C. The line of symmetry can be viewed as the line where we could fold and see each point overlap. The only such line in this problem is the vertical line of symmetry between the two triangles.

20. D. No letters besides X have rotational symmetry until they have turned 360°. However, the letter X achieves rotational symmetry at 180°. Turn it upside down, rotating about its center, and it looks pretty much the same.

21. B. The vector is rotated 180° about the origin, which mean that the signs of the x and y values are switched. So, the new vector has terminal point at $(-3,5)$. Alternatively, imagine this as extending the line and use slope to get the answer.

22. D. To reflect over one of the axes, make the OTHER value (i.e. not named in the axis name) negative. To reflect across the y axis we make every x term negative. $-\frac{x}{y}=5$ is not given as an answer, but if we multiply both sides by y, we'll get $-x=5y$, which is given.

23. B. To find the least number of squares that must be unshaded, we must go through all the shaded units on one side of the x-axis (above or below the horizontal axis). If the corresponding 1 by 1 unit on the other side of the horizontal x-axis (where a unit is said to correspond if they are mirror images of each other across the x-axis) is not shaded, then the unit that is shaded must be unshaded. After some detailed drawing we discover we must un-shade 7 units: one square at the bottom of the first vertical, two at the bottom of the third vertical, one at the bottom of the fifith vertical, two at the bottom of the 6th vertical and one at the bottom of the 7th vertical bar of shading.

CHAPTER

24 MATRIX ALGEBRA

SKILLS TO KNOW

- Adding and subtracting matrices
- Multiplying a matrix by a scalar
- Solving for variable values in equivalent matrices
- Multiplying matrices
- Finding basic determinants (when given the formula)*

MATRIX BASICS

You've made it to the last chapter! I often like to do matrix algebra with students right before the ACT exam, as multiplying matrices can be tough to remember. While many students are simply rusty when it comes to all this matrix jazz, other students may never have learned it in the first place. Either way, matrix problems present a challenge for many students. These questions do not occur frequently on the test—on any given exam you have perhaps less than a 50% chance of encountering one of them. Nonetheless, you are responsible for knowing the several matrix related tasks above.

First let's familiarize ourselves with the matrix:

m x n Matrix
3 x 3

Columns

Cells

Rows

$$\begin{bmatrix} a & b & c \\ d & e & f \\ g & h & i \end{bmatrix}$$

A matrix is a rectangular array of quantities or values that is usually enclosed by brackets. We usually name matrices with capital letter variables—for example: A, B, C.

Matrices are measured RISE by RUN. RISE counts the number of rows, and RUN the number of columns. The dimensions of a few matrices are listed below:

1×3 Matrix: $\begin{bmatrix} 4 & 3 & 5 \end{bmatrix}$ – ONE row by THREE columns

4×2 Matrix: $\begin{bmatrix} 3 & 4 \\ 6 & 5 \\ 8 & 2 \\ 3 & 4 \end{bmatrix}$ – FOUR rows by TWO columns

Sometimes we refer to the individual positions in a matrix based on the row and column position. For example, see this 2×2 matrix:

$$\begin{bmatrix} \text{Row 1, Column 1} & \text{Row 1, Column 2} \\ \text{Row 2, Column 1} & \text{Row 2, Column 2} \end{bmatrix}$$

For two matrices to be equal, they must not only have the same dimensions, but also have the same values in every position.

ADDING AND SUBTRACTING MATRICES

To be able to ADD or SUBTRACT matrices they must have the same dimensions. Then, you simply add together (or subtract) the values that occupy the corresponding positions.

If $M=\begin{bmatrix}-6 & 3\\ 5 & 2\end{bmatrix}$ and $N=\begin{bmatrix}0 & -3\\ 4 & 1\end{bmatrix}$, find $M+N$ and $M-N$.

1. Find $M+N$:
   Simply add the values in the corresponding positions:
   $$\begin{bmatrix}-6 & 3\\ 5 & 2\end{bmatrix}+\begin{bmatrix}0 & -3\\ 4 & 1\end{bmatrix}=\begin{bmatrix}-6+0 & 3+-3\\ 5+4 & 2+1\end{bmatrix}=\begin{bmatrix}-6 & 0\\ 9 & 3\end{bmatrix}$$
2. Find $M-N$:
   Subtraction is similar. Just make sure you don't reverse the order—in subtraction order matters. Simply subtract the values in the corresponding positions:
   $$\begin{bmatrix}-6 & 3\\ 5 & 2\end{bmatrix}-\begin{bmatrix}0 & -3\\ 4 & 1\end{bmatrix}=\begin{bmatrix}-6-0 & 3--3\\ 5-4 & 2-1\end{bmatrix}=\begin{bmatrix}-6 & 6\\ 1 & 1\end{bmatrix}$$

SPEED TIP: You usually don't need to calculate EVERY position in a final matrix on the ACT. Look at the answer choices and see which positions are different in every answer choice. Then solve down the values of these positions first. For example, A, B, C, D, and E may have different values in the lower right corner, but similar ones in the upper left. If so, start with the lower right corner!

MULTIPLYING MATRICES BY A SCALAR

A scalar is a single value, number, or expression. When multiplying a matrix by a scalar, the scalar is multiplied by every individual value in a matrix. Unlike matrices, which are generally denoted by capital letters, a scalar is a lower case letter such a k or m. It can also be placed to the left of a matrix (just as a number to the left of parenthesis means multiply by each item in the parenthesis).

Here's what a scalar problem looks like:

If $k=7$ and $M=\begin{bmatrix}4 & 5\\ 3 & 2\\ 9 & 1\end{bmatrix}$ find kM.

To solve this problem, we'll multiply each item in M by 7:

$$7\begin{bmatrix}4 & 5\\ 3 & 2\\ 9 & 1\end{bmatrix}=\begin{bmatrix}28 & 35\\ 21 & 14\\ 63 & 7\end{bmatrix}$$

EQUIVALENT MATRICES

When two matrices are equivalent, match up each corresponding position to solve for unknowns and create separate equations.

Solve for a if: $\begin{bmatrix}4 & 2n & a+3\end{bmatrix}=\begin{bmatrix}4 & a+6 & 3n\end{bmatrix}$

Match up the middle terms and set them equal:

$$\begin{bmatrix}4 & 2n & a+3\end{bmatrix}=\begin{bmatrix}4 & a+6 & 3n\end{bmatrix}$$

$$a+6=2n$$

Match up the last terms and set them equal:

$$\begin{bmatrix}4 & 2n & a+3\end{bmatrix}=\begin{bmatrix}4 & a+6 & 3n\end{bmatrix}$$

$$a+3=3n$$

Solve by elimination (you can also solve by substitution):

Don't forget to distribute the negative!

Subtract:

$$\begin{aligned} a+3&=3n \\ -(a+6&=2n) \\ \hline 0-3&=n \\ -3&=n \end{aligned}$$

Substitute:

$$\begin{aligned} a+3&=3(-3) \\ a+3&=-9 \\ a&=-12 \end{aligned}$$

Answer: -12.

MULTIPLYING MATRICES

Though multiplying matrices perhaps has historically appeared on only 10-20% of ACT® exams, you still need to know how to perform this pesky task.

First, understand that matrix multiplication does NOT follow the same, simple rules of matrix addition and subtraction. If you assume it does, you will likely find a nice wrong answer to select. Instead, it's a bit more complicated. Let's break it into steps:

STEP 1: Identify your matrix sizes

Remember matrices are measured RISE by RUN (or ROWS by COLUMNS):

Here we have a 1×2 matrix: $\begin{bmatrix}5 & 1\end{bmatrix}$ and here we have a 2×1 matrix: $\begin{bmatrix}4 \\ 3\end{bmatrix}$

STEP 2: Test to make sure you can multiply

Before multiplying, you need to make sure you *can* multiply the matrices. **Only matrices in which the COLUMN NUMBER of the first matrix matches the ROW NUMBER of the 2nd matrix can be multiplied.** That sounds a bit confusing, but in practice, I just imagine that matrices must "handshake" in order for the matrices to be multiplied.

We can multiply a 1×2 matrix times a 2×1 matrix because the two "2's" in the center match or "handshake." However, just because two matrices $A\times B$ can be multiplied does not mean $B\times A$ can. For example, a 1×4 matrix can be multiplied by a 4×5 matrix (the fours match), but the 4×5 matrix cannot be multiplied by a 1×4 matrix (5 and 1 are not equal).

STEP 3: Set up your destination matrix

When you set up your multiplication problem, you **find the dimensions of the resulting matrix by taking the FIRST and LAST values from the matrix dimensions.** So for a 1×2 by 2×1 we get a 1×1 matrix result:

$$\begin{bmatrix}5 & 1\end{bmatrix}\times\begin{bmatrix}4\\3\end{bmatrix}=\begin{bmatrix}\ldots\end{bmatrix}$$

For a 1×4 by 4×3 we get a 1×3 matrix result:

$$\begin{bmatrix}1 & 3 & 4 & 5\end{bmatrix}\times\begin{bmatrix}1 & 2 & 3\\4 & 5 & 3\\7 & 1 & 2\\4 & 5 & 3\end{bmatrix}=\begin{bmatrix}\ldots & \ldots & \ldots\end{bmatrix}$$

STEP 4: Multiply!

Each element in the final matrix is the sum of the products from corresponding values in a single row and a single column. You are multiplying ROWS times COLUMNS. Whatever number row you are on in the first matrix will correspond to the row you place the answer in in the destination matrix. Whatever number column you are on in the second matrix will also correspond to that column number in the destination matrix. You only take TWO elements at a time to multiply together—those in corresponding order—in order horizontally from the first matrix to those in order vertically in the second matrix—i.e. you take the first item horizontally (from first matrix) times the first item vertically (from 2nd matrix), then add the second item horizontally times the second item vertically, then add to the third item horizontally times the third item vertically. After you finish pairing a row from matrix 1 with the first column in matrix 2, move on and pair that row with the next column for the next sum of products. After you finish pairing a row with each column, move to the next row in the first matrix, and start filling in the 2nd row in the destination matrix. Explaining this is confusing, so pay attention to the matrix multiplication examples below to learn the pattern:

Row 1 item 1 times Column 1 item 1, then Row 1 item 2 times Column 1 item 2:

$$\begin{bmatrix}5 & 1\end{bmatrix}\times\begin{bmatrix}4\\3\end{bmatrix}=\begin{bmatrix}5\times4+1\times3\end{bmatrix}=\begin{bmatrix}20+3\end{bmatrix}=\begin{bmatrix}23\end{bmatrix}$$

Row 1 item 1 times Column 1 item 1, then Row 1 item 2 times Column 1 item 2, then Row 1 item 3 times Column 1 item 3...etc. Add together the items for the 1st row and 1st column into a single blank that bears both features (row 1 column 1 position). Then do Row 1 times Column 2, etc.

$$\begin{bmatrix}a & b & c & d\end{bmatrix}\times\begin{bmatrix}1 & 2 & 3\\4 & 5 & 6\\7 & 8 & 9\\10 & 11 & 12\end{bmatrix}=\begin{bmatrix}1a+4b+7c+10d & 2a+5b+8c+11d & 3a+6b+9c+12d\end{bmatrix}$$

$$\begin{bmatrix}a & b\\c & d\end{bmatrix}\times\begin{bmatrix}1 & 2\\3 & 4\end{bmatrix}=\begin{bmatrix}1a+3b & 2a+4b\\1c+3d & 2c+4d\end{bmatrix}$$

Matrix multiplication is a bit confusing–once you figure out the pattern it's fine, but that pattern can be tough to depict in a book. That's why video teaching is the backbone of Supertutor's teaching elements. **Check out SupertutorTV.com/BookOwners for more information on supporting videos.** Finally, if you don't feel like you "get" it, chances are you just need practice. Matrix algebra is one of those topics that really never makes sense until you just do it over and over again. Don't fret! Just dive in! See SupertutorTV.com/BookOwners for suggestions on more practice in this area.

SPEED TIP: Work smart! **Don't multiply out full matrices on the ACT. Instead, find the elements of the answer choices in the matrices that are different.** For example, if the lower left corner in each answer choice is distinct, solve out that value and not all the values to move more quickly. Use process of elimination to eliminate choices with the wrong dimensions.

FINDING A BASIC DETERMINANT

Finally, you may be asked to find or use the definition of a determinant.

A determinant is usually written as a matrix with "straight" brackets as so: $\begin{vmatrix} a & b \\ c & d \end{vmatrix}$

Or with standard brackets: $\det\begin{bmatrix} a & b \\ c & d \end{bmatrix}$

The great news on the ACT® is that you will never (at least not if the test remains consistent with past tests) be required to memorize the formula for a determinant. Rather, if you have a determinant question, the question will include the formula itself. **In such a case, all you have to do is use that formula! Plug in numbers that are in the same positions as the variables in the given expression.**

The determinant of a matrix $\begin{bmatrix} a & b \\ c & d \end{bmatrix}$ is $ad - bc$.

If the determinant of $\begin{bmatrix} 6 & 4 \\ x & x \end{bmatrix}$ is 24, what is the value of x?

Assume $a = 6$, $b = 4$, $c = x$, $d = x$ (matching items in same position) then plug into the expression $ad - bc$ and set equal to 24:

$$6x - 4x = 24$$

$$2x = 24$$

$$x = 12$$

Answer: 12.

1. What is the matrix product $\begin{bmatrix} 0 & 1 & 2 \end{bmatrix}\begin{bmatrix} x \\ 3x \\ 4x \end{bmatrix}$?

A. $\begin{bmatrix} 11x \end{bmatrix}$
B. $\begin{bmatrix} 12x \end{bmatrix}$
C. $\begin{bmatrix} 6x \end{bmatrix}$
D. $\begin{bmatrix} 32x \end{bmatrix}$
E. $\begin{bmatrix} 8x \end{bmatrix}$

2. For which values of a and b is the following matrix equation true?

$$\begin{bmatrix} 1 & 4 \\ 2b & \frac{1}{3}b \end{bmatrix} - \begin{bmatrix} 3 & -2 \\ 6 & -8 \end{bmatrix} = \begin{bmatrix} -2 & 6 \\ a & a \end{bmatrix}$$

A. $a = \frac{-54}{5}$, $b = \frac{42}{5}$
B. $a = \frac{54}{5}$, $b = \frac{-42}{5}$
C. $a = \frac{54}{5}$, $b = \frac{42}{5}$
D. $a = \frac{-54}{5}$, $b = \frac{-42}{5}$
E. $a = \frac{54}{4}$, $b = \frac{42}{5}$

3. Matrix $P = \begin{bmatrix} -3 & 6 \\ -2 & 11 \end{bmatrix}$ and $Q = \begin{bmatrix} 1 & 4 \\ 7 & -5 \end{bmatrix}$. What is $P - 2Q$?

A. $\begin{bmatrix} -4 & 2 \\ -9 & 16 \end{bmatrix}$
B. $\begin{bmatrix} -2 & 10 \\ 5 & 6 \end{bmatrix}$
C. $\begin{bmatrix} 2 & -10 \\ -5 & -6 \end{bmatrix}$
D. $\begin{bmatrix} -8 & 4 \\ -18 & 16 \end{bmatrix}$
E. $\begin{bmatrix} -5 & -2 \\ -16 & 21 \end{bmatrix}$

4. The determinant of a matrix $\begin{bmatrix} x & y \\ z & w \end{bmatrix}$ is $xw - yz$. If the determinant of $\begin{bmatrix} 2a & 13 \\ a & a \end{bmatrix}$ is 24, which of the following is a value of a?

A. 4
B. 24
C. 8
D. −8
E. $\frac{3}{2}$

5. The 3×3 matrix $\begin{bmatrix} -3 & 6 & -5 \\ 9 & 2 & -2 \\ 4 & 3 & 1 \end{bmatrix}$ is multiplied by a scalar n.

The resulting matrix is $\begin{bmatrix} 9 & -18 & 15 \\ -27 & a & 6 \\ -12 & -9 & -3 \end{bmatrix}$. What is a?

A. 2
B. −6
C. −3
D. 3
E. 6

6. The determinant of any 2×2 matrix $\begin{bmatrix} x & y \\ z & w \end{bmatrix}$ is $xw - yz$. If the determinant of $\begin{bmatrix} (x+4) & 4 \\ 9 & (x-3) \end{bmatrix}$ is 8. What are all the possible values of x?

A. −4 and 3
B. 8 and −7
C. −8 and 7
D. 4 and −3
E. 7 and 8

7. $\begin{bmatrix} a & b \\ c & d \end{bmatrix} \times \begin{bmatrix} b & c \\ d & a \end{bmatrix} = ?$

A. $\begin{bmatrix} ab & bc \\ cd & da \end{bmatrix}$
B. $\begin{bmatrix} ab+bd & ac+ba \\ cb+d^2 & c^2+da \end{bmatrix}$
C. $\begin{bmatrix} a+b & b+c \\ c+d & d+a \end{bmatrix}$
D. $\begin{bmatrix} 2a & 2b \\ 2c & 2d \end{bmatrix}$
E. $\begin{bmatrix} ab+c^2 & b^2+cd \\ ad+ca & bd+da \end{bmatrix}$

8. Which of the following matrices for $\begin{bmatrix} a & b \\ c & d \end{bmatrix}$ makes the following expression true?**

$$\begin{bmatrix} a & b \\ c & d \end{bmatrix} \times \begin{bmatrix} 1 & 3 \\ 5 & 7 \end{bmatrix} = \begin{bmatrix} 7 & 13 \\ 23 & 37 \end{bmatrix}$$

A. $\begin{bmatrix} 7 & \frac{13}{3} \\ \frac{23}{5} & \frac{37}{7} \end{bmatrix}$

B. $\begin{bmatrix} 6 & 10 \\ 18 & 30 \end{bmatrix}$

C. $\begin{bmatrix} 7 & 0 \\ 1 & 4 \end{bmatrix}$

D. $\begin{bmatrix} 0 & 1 \\ 1 & 0 \end{bmatrix}$

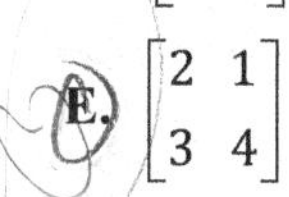

E. $\begin{bmatrix} 2 & 1 \\ 3 & 4 \end{bmatrix}$

*** problem may be more challenging than typical ACT® problems*

9. If $\det\begin{bmatrix} a & b \\ c & d \end{bmatrix} = ad - bc$, then $\det\begin{bmatrix} -c & -a \\ d & b \end{bmatrix} = ?$

A. $-ad - bc$

B. $bc + ad$

C. $ad - cb$

D. $-cb - ad$

E. $-ad + cb$

10. Given the matrix equation shown below, what is $\frac{b}{a-b}$?
(Note: Whenever n is a positive integer, the notation of $n!$ represents the product of the integers from n to 1. For example, $3! = 3 \times 2 \times 1$.)

$$\begin{bmatrix} 4! \\ 2! \end{bmatrix} + \begin{bmatrix} 2! \\ 3! \end{bmatrix} = \begin{bmatrix} a \\ b \end{bmatrix}$$

A. $\frac{4}{9}$

B. $\frac{13}{9}$

C. $\frac{6}{5}$

D. 5

E. $\frac{1}{7}$

11. $\begin{bmatrix} a & a \\ b & b \end{bmatrix} + \begin{bmatrix} a & b \\ c & d \end{bmatrix} + \begin{bmatrix} \frac{1}{a+b} & \frac{1}{a+b} \\ \frac{1}{b+c} & \frac{1}{b+d} \end{bmatrix} =$

A. $\begin{bmatrix} \frac{2a}{a+b} & 1 \\ 1 & 1 \end{bmatrix}$

B. $\begin{bmatrix} 2a + \frac{1}{a+b} & a+b+\frac{1}{a+b} \\ b+c+\frac{1}{b+c} & b+d+\frac{1}{b+d} \end{bmatrix}$

C. $\begin{bmatrix} 2a+ab & 2a+2b \\ 2b+2c & 2b+2d \end{bmatrix}$

D. $\begin{bmatrix} \frac{aa}{a+b} & \frac{ab}{a+b} \\ \frac{bb}{b+c} & \frac{cd}{b+d} \end{bmatrix}$

E. $\begin{bmatrix} \frac{1}{3a+b} & \frac{1}{a+2b} \\ \frac{1}{2b+2c} & \frac{1}{2b+2d} \end{bmatrix}$

12. Emmy owns 2 juice bars (X and Y) and stocks 3 flavors of juice (A, B, and C). The matrices below show the numbers of each flavor of juice in each shop, and the cost for each flavor. Using the values given below, what is the difference between the value of juice inventories for the two shops?

$$\begin{array}{c} \\ X \\ Y \end{array} \begin{array}{c} \begin{matrix} A & B & C \end{matrix} \\ \begin{bmatrix} 25 & 50 & 20 \\ 50 & 100 & 25 \end{bmatrix} \end{array} \qquad \begin{array}{c} \text{Cost} \\ \begin{array}{c} A \\ B \\ C \end{array} \begin{bmatrix} \$5 \\ \$10 \\ \$15 \end{bmatrix} \end{array}$$

A. \$900

B. \$700

C. \$2850

D. \$2550

E. \$1275

13. The number of students who practice an art at a certain conservatory can be shown by the following matrix:

$$\begin{matrix} \text{band} & \text{choir} & \text{painting} & \text{pottery} \end{matrix}$$
$$\begin{bmatrix} 80 & 60 & 60 & 100 \end{bmatrix}$$

The head of the conservatory estimates the ratio of the number of art awards that will be earned to the number of students participating with the following matrix:

$$\begin{matrix} \text{band} \\ \text{choir} \\ \text{painting} \\ \text{pottery} \end{matrix} \begin{bmatrix} .2 \\ .4 \\ .1 \\ .3 \end{bmatrix}$$

Given this, which is the best estimate for the number of art awards that will be earned for the year?

A. 73
B. 74
C. 75
D. 76
E. 77

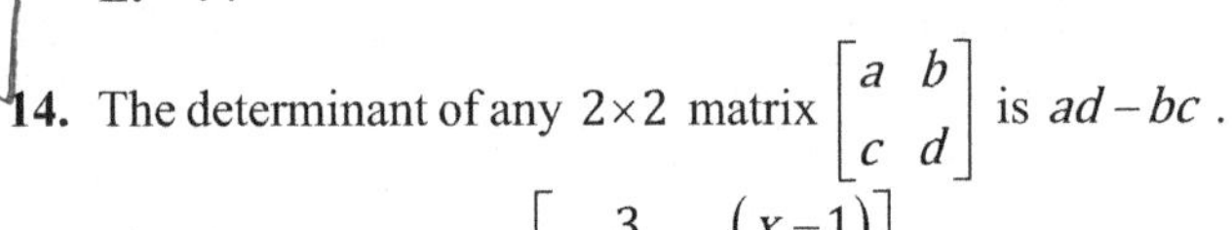

14. The determinant of any 2×2 matrix $\begin{bmatrix} a & b \\ c & d \end{bmatrix}$ is $ad-bc$.

The determinant of $\begin{bmatrix} 3 & (x-1) \\ (x+4) & 2 \end{bmatrix}$ is equal to 0.

What are all possible values of x?

A. $-\frac{\sqrt{17}-3}{3}$ and $\frac{\sqrt{17}+3}{3}$

B. 5 and −2

C. −5 and 2

D. $\frac{\sqrt{17}-3}{3}$ and $-\frac{\sqrt{17}+3}{3}$

E. none of the above

15. What value of x satisfies the matrix equation below? (Assume x is a scalar.)

$$x\begin{bmatrix} 3 & 2 \\ 4 & 1 \end{bmatrix} + \begin{bmatrix} 2 & 2 \\ 5 & 1 \end{bmatrix} = \begin{bmatrix} 11 & 8 \\ 17 & 4 \end{bmatrix}$$

A. $\frac{11}{5}$

B. $\frac{20}{9}$

C. 3

D. $\frac{11}{3}$

E. 4

16. $4\begin{bmatrix} 0 & 1 \\ 4 & -2 \end{bmatrix} - 2\begin{bmatrix} 3 & -3 \\ -1 & 2 \end{bmatrix} = ?$

A. $\begin{bmatrix} 0 & 4 \\ 16 & -8 \end{bmatrix}$

B. $\begin{bmatrix} -3 & 5 \\ 9 & -6 \end{bmatrix}$

C. 19

D. −22

E. $\begin{bmatrix} -6 & 10 \\ 18 & -12 \end{bmatrix}$

17. Graph theory is often used to represent connections between different points. Three satellites in space communicate with one another, as represented by the drawing below.

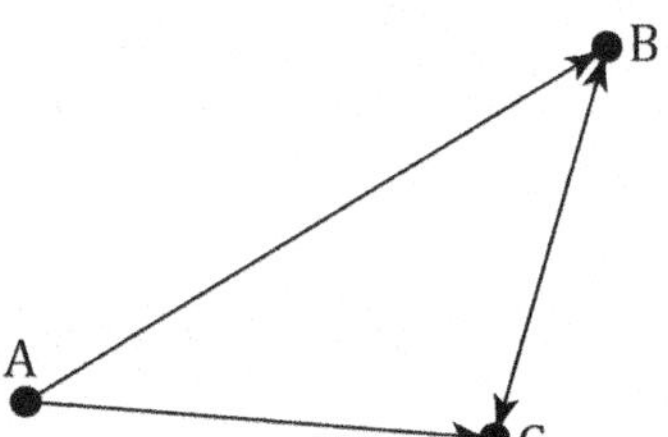

For example, the arrows indicate that satellite A communicates with satellite B and C, but satellite C only communicates with satellite B. The same relationships are demonstrated in the matrix, where, because satellite A communicates with satellite B, there is a 1 in the A row and B column, but because satellite A does not communicate with itself, there is a 0 in A row and A column.

$$\begin{matrix} & A & B & C \\ A & 0 & 1 & 1 \\ B & 0 & 0 & 1 \\ C & ? & ? & ? \end{matrix}$$

Which of the following is the third row of the matrix?

A. 1 1 1
B. 1 1 0
C. 1 0 0
D. 0 1 0
E. 0 0 0

18. For what (x,y) pair is the matrix equation below true?

$$\begin{bmatrix} y & 3x \\ 0 & -1 \end{bmatrix}\begin{bmatrix} -3 & 3 \\ 4 & \frac{x}{2} \end{bmatrix}=\begin{bmatrix} 21 & 9 \\ -4 & -1 \end{bmatrix}$$

A. $(4,24)$
B. $(24,4)$
C. $(1,2)$
D. $(2,1)$
E. $\left(-\frac{3}{4},-7\right)$

19. The 2×4 matrix $\begin{bmatrix} 1 & 2 & 8 & 4 \\ 2 & 5 & 6 & 1 \end{bmatrix}$ represents quadrilateral $ABCD$, with vertices $A(1,2), B(2,5), C(8,6), D(4,1)$ in the standard coordinate plane. After the quadrilateral was reflected over the x-axis, the matrix representing the translated triangle is $\begin{bmatrix} 1 & 2 & 8 & 4 \\ m & -5 & 3m & -1 \end{bmatrix}$. What is the value of m?

A. −2
B. 2
C. 3
D. −1
E. −3

20. $3\begin{bmatrix} 2 & 0 & -1 \\ 3 & 4 & 0 \end{bmatrix}\begin{bmatrix} -2 & 1 \\ 2 & 1 \\ -3 & 8 \end{bmatrix}=?$

A. $\begin{bmatrix} -1 & -6 \\ 2 & 7 \end{bmatrix}$

B. $\begin{bmatrix} -3 & -18 \\ 6 & 21 \end{bmatrix}$

C. $\begin{bmatrix} 0 & 6 & -12 \\ 12 & 15 & 24 \end{bmatrix}$

D. $\begin{bmatrix} 0 & 12 \\ 6 & 15 \\ -12 & 24 \end{bmatrix}$

E. $\begin{bmatrix} -12 & 45 \\ -4 & 84 \end{bmatrix}$

ANSWER KEY

1. A **2. C** **3. E** **4. C** **5. B** **6. C** **7. B** **8. E** **9. C** **10. A** **11. B** **12. B** **13. D** **14. C**
15. C **16. E** **17. D** **18. D** **19. A** **20. B**

ANSWER EXPLANATIONS

1. **A.** Here we have a 1×3 matrix and a 3×1 matrix respectively. To multiply, we must ensure the middle two numbers of the matrix dimensions match or "handshake" (they do: 3 and 3). Our solution will be a matrix made up of the first and last digits of the dimensions (1×1). To multiply matrices, multiply the rows times the columns: Row 1 of the first matrix times column 1 of the second. Add the product of the first item in the row times the first item in the column $(0\times x)$, the product of the second item in the row times the second item in the column $(1\times3x)$ and the product of the third item in the row times the third item in the column $(2\times4x)$. In matrix form, the result is $[(0)(x)+(1)(3x)+(2)(4x)]$ .Simplify this to $0x+3x+8x=11x$. Write the answer in matrix form: $[11x]$.

2. **C.** This problem essentially is four different equations written in one matrix based form. When we add or subtract matrices, we look only at the positions—same position means we add together those elements, i.e. in the matrix problem here $\begin{bmatrix} a & d \\ g & k \end{bmatrix} - \begin{bmatrix} b & e \\ h & l \end{bmatrix} = \begin{bmatrix} c & f \\ j & m \end{bmatrix}$, a, b, and c are all elements in the same position—so they form the following relationship—$a-b=c$. The same is true for all other "same position" letters in the above matrix. Let's now take the matrix at hand and reorganize it into a series of equations based on elements in the same position: $\begin{bmatrix} 1 & 4 \\ 2b & \frac{1}{3}b \end{bmatrix} - \begin{bmatrix} 3 & -2 \\ 6 & -8 \end{bmatrix} = \begin{bmatrix} -2 & 6 \\ a & a \end{bmatrix}$
We get:
$1-3=-2$
$4--2=6$
Those first two only state the obvious—not helping us much—but at least you can use these to verify that you're setting the problem up right. The last two will help more, and with more practice, we can skip ahead to this step to save time:
$$2b-6=a \text{ and } \frac{1}{3}b-(-8)=a$$
Now we have a system of two equations with two unknowns. We can solve by elimination or substitution. Here we will use substitution, substituting for a: First, create a positive from the double negative to get $2b-6=\frac{1}{3}b+8$. Solving for b gives us $b=\frac{42}{5}$. Finally, plug in this fraction using the first (or second) equation to find a . $2\left(\frac{42}{5}\right)-6=a$ so $\frac{84}{5}-\frac{30}{5}=\frac{54}{5}=a$.

3. **E.** To understand the basics of how to subtract two matrices, see explanation at question 2 . $2Q$ implies that we must first use scalar multiplication before performing the subtraction. To do so, multiply the scalar (2) by each item in the matrix. Let's solve for $2Q$:
$$2\begin{bmatrix} 1 & 4 \\ 7 & -5 \end{bmatrix} = \begin{bmatrix} 2\times1 & 2\times4 \\ 2\times7 & 2\times-5 \end{bmatrix} = \begin{bmatrix} 2 & 8 \\ 14 & -10 \end{bmatrix}$$
Now we can set up $P-2Q$, subtracting numbers in the same positions:
$$\begin{bmatrix} -3 & 6 \\ -2 & 11 \end{bmatrix} - \begin{bmatrix} 2 & 8 \\ 14 & -10 \end{bmatrix} = \begin{bmatrix} -3-2 & 6-8 \\ -2-14 & 11--10 \end{bmatrix} = \begin{bmatrix} -5 & -2 \\ -16 & 21 \end{bmatrix}$$

4. **C.** For this problem just apply the formula given:
$$2a\times a-13\times a=24$$
$$2a^2-13a=24$$
$$2a^2-13a-24=0$$
To solve, use the quadratic formula or factoring. I'll use the quadratic formula (you should have this memorized). Use a calculator for the multiplication and roots.

$$\frac{-b\pm\sqrt{b^2-4ac}}{2a}=\frac{+13\pm\sqrt{13^2-4(2)(-24)}}{2(2)}=\frac{13\pm\sqrt{169+192}}{4}=\frac{13\pm\sqrt{361}}{4}=\frac{13\pm19}{4}$$

Now we can split into two solutions. $\frac{13-19}{4}=\frac{-6}{4}\to-1.5$ OR $\frac{13+19}{4}=\frac{32}{4}\to8$

8 is the only answer available, so C is correct. You could also "back-solve" this problem, using the answers, or program your calculator with the quadratic equation for speed (see supertutortv.com/bookowners for more on calculator programs).

5. **B.** A scalar is a single number that is multiplied by each individual value in a matrix. To multiply the given matrix by n:

$$n\begin{bmatrix}-3 & 6 & -5\\ 9 & 2 & -2\\ 4 & 3 & 1\end{bmatrix}=\begin{bmatrix}-3n & 6n & -5n\\ 9n & 2n & -2n\\ 4n & 3n & 1n\end{bmatrix}$$

Now we can set this equal to the result:

$$\begin{bmatrix}-3n & 6n & -5n\\ 9n & 2n & -2n\\ 4n & 3n & 1n\end{bmatrix}=\begin{bmatrix}9 & -18 & 15\\ -27 & a & 6\\ -12 & -9 & -3\end{bmatrix}9$$

Playing "match" we can match up each value in each position and form up to different equations. We actually will only need two, though. ***We need to find "a" <u>not</u> "n"***—but we'll need "n" to find "a". To find n, we have many choices, but let's try for the first row / column value:

$$-3n=9$$
$$n=-3$$

<u>Now don't go looking for that as the answer choice!</u> Plug in that value into another equation—the one that involves $a:2n=a$. Again we found this equation by matching up values in the same positions in the equivalent matrices. We then simplify:

$$2(-3)=a=-6$$

6. **C.** Here apply the formula given, set equal to 8, then FOIL and simplify:

$$(x+4)(x-3)-(4)(9)=8$$
$$x^2+4x-3x-12-36=8$$
$$x^2+x-48=8$$
$$x^2+x-56=0$$

Now apply the quadratic equation or factor. You can program your calculator to do the quadratic equation (see supertutortv.com/bookowners for more). Here I'll factor to: $(x+8)(x-7)=0$, which gives the answers $x=7$ or $x=-8$.

7. **B.** This problem essentially asks the definition of how to multiply matrices. (A) is wrong—you do not multiply in the same way you add and subtract matrices—you cannot just multiply each item in the same position together. Who knows why this isn't true but it's just the definition of matrix multiplication. (B) is correct—it adds the products of rows times columns. (C) adds the two matrices (D) multiplies the first matrix by a scalar of 2 and (E) mixes up rows and columns—it's rows times columns not columns times rows. If you missed this, go back and review the chapter or practice more matrix multiplication.

8. **E.** For this problem you must multiply the matrices. Let's first just focus on the setup to the left of the equals sign and simplify that into a single matrix. If you don't know how to do this, review the example problems earlier in the chapter. Take rows times columns and add the products:

$$\begin{bmatrix}a & b\\ c & d\end{bmatrix}\times\begin{bmatrix}1 & 3\\ 5 & 7\end{bmatrix}=\begin{bmatrix}a(1)+b(5) & a(3)+b(7)\\ c(1)+d(5) & c(3)+d(7)\end{bmatrix}=\begin{bmatrix}a+5b & 3a+7b\\ c+5d & 3c+7d\end{bmatrix}$$

Now let's take what we have and put it with what the original problem was equal to, and match each corresponding position to make four equations:

$$\begin{bmatrix} a+5b & 3a+7b \\ c+5d & 3c+7d \end{bmatrix} = \begin{bmatrix} 7 & 13 \\ 23 & 37 \end{bmatrix}$$

$$a+5b=7$$
$$c+5d=23$$
$$3a+7d=13$$
$$3c+7d=37$$

Now we could do the ridiculous task of solving each of these out—but it's a better idea to backsolve a bit—as most of the matrices are VERY different answers. We'll use the first of these equations ONLY—and check each one. Let's try (A): $a=7$ and $b=\frac{13}{3}$—no way will that give us the integer value 7. It's out. Let's try (B) $a=6$ and $b=10$. $6+50\neq 7$. Let's try (C) $7+5(0)=7$, but we can't assume this is right yet. (D) yields $0+5\neq 7$. (E) works too though—$2+5=7$. Now we've narrowed to C and E. Let's try another of the equations above—this time let's use the last equation. For (E) $c=3$ and $d=4$, so $3(3)+7(4)=9+28=37$—yes. But for (C), $c=1$ and $d=4$, so $3(1)+7(4)=3+28=31$—not 37 E is correct.

**Note that this problem tests skills for which you are responsible on the ACT® but is more difficult than most ACT® problems.*

9. C. This is a very straightforward problem: finding the determinant, made even more simple by giving the definition of a determinant in the question. The trick will be in the format of the answers, as the answers will be rearranged to trick you. If approaching the problem traditionally, $\det\begin{bmatrix} -c & -a \\ d & b \end{bmatrix}=(-c)(b)-(-a)(d)$ simplified becomes $-cb+ad$. While that is not an answer, answer (C), $ad-cb$, is the same by simply rearranging the order of the terms.

10. A. This requires multiple steps. First apply the factorial (the !- factorial is covered in Ch 9 Book 2.) and then sum the matrices until you find the final values for a and b, which are 26 and 8 respectively. $\begin{bmatrix} 4! \\ 2! \end{bmatrix}+\begin{bmatrix} 2! \\ 3! \end{bmatrix}=\begin{bmatrix} a \\ b \end{bmatrix}$ $\rightarrow\begin{bmatrix} 4\times3\times2\times1 \\ 2\times1 \end{bmatrix}+\begin{bmatrix} 2\times1 \\ 3\times2\times1 \end{bmatrix}=\begin{bmatrix} a \\ b \end{bmatrix}\rightarrow\begin{bmatrix} 24 \\ 2 \end{bmatrix}+\begin{bmatrix} 2 \\ 6 \end{bmatrix}=\begin{bmatrix} a \\ b \end{bmatrix}\rightarrow\begin{bmatrix} 24+2 \\ 2+6 \end{bmatrix}=\begin{bmatrix} a \\ b \end{bmatrix}\rightarrow\begin{bmatrix} 26 \\ 8 \end{bmatrix}=\begin{bmatrix} a \\ b \end{bmatrix}$ Finally, plug those values into the expression desired: $\frac{b}{a-b}=\frac{8}{26-8}=\frac{8}{18}=\frac{4}{9}$. $\frac{4}{9}$ is answer (A).

11. B. Remember that to add matrices, simply add each element that is in the same position, and put that sum in that same position in the resultant matrix. Thus the upper left corner is a plus a plus $\frac{1}{a+b}$ which can be rewritten as $2a+\frac{1}{a+b}$. Look down, and see that B matches this. Now double check each other element in answer B to see that it also equals the sum of all elements in its matching position. No need to find a common denominator, as each element in choice B is clearly the sum of all elements in that position without making the answers into single fractions.

12. B. To solve this, we can see that the question is essentially asking us to multiply the matrices and then subtract the two elements in the resultant matrix at the end. Think of the matrices as tables that organize this information. Remember X and Y are just labels as are A, B and C. These are NOT scalars. If you have 25 of Juice A in Shop X, and according to the cost matrix Juice A costs $\$5$, it makes sense that you have $\$125$ worth of Juice A in Shop X: (Shop X flavor) $\times$ (Flavor X value)= $(\$25\times\$5)$. Since you have three juices in each shop, you must sum the totals from each juice. Ultimately, it's as if you were multiplying the matrices, and the resultant matrix tells you the value of the inventory in each shop. To find the total value of inventory for each shop, let's now multiply the matrices. First we check dimensions: 2x3 is multiplied by a $3\text{x}1$. The middle numbers match and the outer numbers form our resultant 2x1 matrix. Now multiply rows times columns:

$$\begin{array}{c} \\ X \\ Y \end{array}\begin{array}{c} A \quad B \quad C \\ \begin{bmatrix} 25 & 50 & 20 \\ 50 & 100 & 25 \end{bmatrix} \end{array} \times \begin{array}{c} \\ A \\ B \\ C \end{array}\begin{array}{c} \textit{Value} \\ \begin{bmatrix} \$5 \\ \$10 \\ \$15 \end{bmatrix} \end{array} = \begin{array}{c} \\ X \\ Y \end{array}\begin{array}{c} \textit{Value} \\ \begin{bmatrix} (25)(\$5)+(50)(\$10)+(20)(\$15) \\ (50)(\$5)+(100)(\$10)+(25)(\$15) \end{bmatrix} \end{array} = \begin{array}{c} \\ X \\ Y \end{array}\begin{array}{c} \textit{Value} \\ \begin{bmatrix} \$925 \\ \$1625 \end{bmatrix} \end{array}$$

Then, $\$1625-\$925=\$700$, which is answer (B).

13. D. For this question, multiply the matrices to find out how many students per department receive an award. Logically this makes sense: we multiply the total kids in band times the rate of awards for kids in band to find the number of kids in band who get awards (16), we multiply the number of kids in choir (60) times the rate of awards in choir (0.4) to get 24 kids with awards in choir, and so forth. To find the total number of students awarded across all the departments, sum the elements in the destination matrix. We have a 1x4 matrix times a 4x1 matrix. The center dimensions match and our destination matrix is the outer paramters: 1x1. We set up the problem as follows:

$$\begin{matrix} \text{band} & \text{choir} & \text{painting} & \text{pottery} \end{matrix} \qquad \begin{matrix} \text{band} \\ \text{choir} \\ \text{painting} \\ \text{pottery} \end{matrix}$$

$$\begin{bmatrix} 80 & 60 & 60 & 100 \end{bmatrix} \times \begin{bmatrix} 0.2 \\ 0.4 \\ 0.1 \\ 0.3 \end{bmatrix} = [80(0.2)+60(0.4)+60(0.1)+100(0.3)] = \left[16+24+6+30\right] = [76]$$

The answer is (D). You can also solve this logically without the matrix setup if that's easier.

14. C. With the formula given, all you need to do is substitute the values given with the corresponding variables:

$$\begin{bmatrix} 3 & (x-1) \\ (x+4) & 2 \end{bmatrix} = 0$$

Following the formula we get $6-(x+4)(x-1)=0$ or $6-\left(x^2+3x-4\right)=0$. After distributing the subtraction sign and simplifying we get $6-x^2-3x-(-4)=6-x^2-3x+4=-x^2-3x+10=0$. After factoring we get $(-x-5)(x-2)=0$ so, $x=-5$ and 2.

15. C. Answer (C) correctly solves for x. The trick with problems like this is that you don't have to solve the entire matrix. Multiply the matrix by x and get the following:

$$\begin{bmatrix} 3x & 2x \\ 4x & 1x \end{bmatrix} + \begin{bmatrix} 2 & 2 \\ 5 & 1 \end{bmatrix} = \begin{bmatrix} 11 & 8 \\ 17 & 4 \end{bmatrix}$$

But since we are only going to focus on one address, let's say the top left, all that matters is: $3x+2=11$. Solve this and $x=3$, which is answer (C).

16. E. Answer (E) is correct. Per the rules of scalar multiplication, we multiply all the elements in the matrix by the scalar, and then subtract the elements with the corresponding ones on the second matrix. To gain speed on problems like this, skim answer choices to find a position in the matrix where the answers are distinct for every lettered answer option and calculate that single position on the matrix. Here every position is distinct across the answers, so I can calculate any position. I'll calculate 4(0)-2(3)=upper left position. Simplifying I get -6, so I find that only in choice E and am done. The complete subtraction is below in case you chose a different position:

$$4\begin{bmatrix} 0 & 1 \\ 4 & -2 \end{bmatrix} - 2\begin{bmatrix} 3 & -3 \\ -1 & 2 \end{bmatrix} = \begin{bmatrix} 4(0) & 4(1) \\ 4(4) & 4(-2) \end{bmatrix} - \begin{bmatrix} 2(3) & 2(-3) \\ 2(-1) & 2(2) \end{bmatrix} = \begin{bmatrix} 0 & 4 \\ 16 & -8 \end{bmatrix} - \begin{bmatrix} 6 & -6 \\ -2 & 4 \end{bmatrix} = \begin{bmatrix} 0-6 & 4--6 \\ 16--2 & -8-4 \end{bmatrix} = \begin{bmatrix} -6 & 10 \\ 18 & -12 \end{bmatrix}.$$

17. D. Satellite C receives messages from both satellite A and B, but it only sends to satellite B, as seen by the double arrow line between C and B, so the correct answer is $0 \ 1 \ 0$. Here a matrix is just a formatting tool. Don't get confused!

18. D. To solve this problem quickly, pick the right place to start. Although convention dictates that we multiply matrices starting with the top left element, and go left to right, we can speed up by first finding the equation that solves for the bottom right element in the final matrix as it will enable us to find x. We take the row with values 0 and -1 and mutiply it by the corresponding values in the column with the 3 and $\frac{x}{2}$ and then set the sum equal to -1, i.e. the circled elements here $\begin{bmatrix} y & 3x \\ 0 & -1 \end{bmatrix} \begin{bmatrix} -3 & 3 \\ 4 & \frac{x}{2} \end{bmatrix} = \begin{bmatrix} 21 & 9 \\ -4 & -1 \end{bmatrix}$: $0(3)+(-1)\left(\frac{x}{2}\right)=-1$, and solve to find $x=2$. (D) is correct since no other answer has 2 for x.

19. A. In this question the matrices merely serve as a way to organize information, in this case coordinate points. Reflecting a polygon over the x axis means that the coordinates stay the same while the y coordinates become negative. Since the top row of the matrix represents the x coordinates, those stay the same, while the bottom row elements, which represents the y coordinates, all become negative. Thus, 2, the left most bottom element, made negative becomes −2. Looking at the translated matrix, m takes the place of where −2 should be, so m is clearly −2.

20. B. This problem combines scalar and matrix multiplication. Scalar multiplication is commutative, but matrix multiplication is not. The scalar factor, 3, can be applied either before or after the two matrices are multiplied correctly (row by column, adding the products). Here I'll apply it last. Again if you're gunning for speed you could complete a portion of this math instead of all of it. First figure out your matrix dimensions taking the "outer" numbers of the dimensions multiplied (a 2x3 by a 3x2 makes a 2x2). Then figure out which position in the 2x2 matrix answer options is unique across each choice. The upper right is! I've circled the elements used to find that upper right number. Solve out these only to save time.

$$3\begin{bmatrix} 2 & 0 & -1 \\ 3 & 4 & 0 \end{bmatrix}\begin{bmatrix} -2 & 1 \\ 2 & 1 \\ -3 & 8 \end{bmatrix} = 3\begin{bmatrix} (2)(-2)+(0)(2)+(-1)(-3) & (2)(1)+(0)(1)+(-1)(8) \\ (3)(-2)+(4)(2)+(0)(3) & (3)(1)+(4)(1)+(0)(8) \end{bmatrix} = 3\begin{bmatrix} -1 & -6 \\ 2 & 7 \end{bmatrix} = \begin{bmatrix} -3 & -18 \\ 6 & 21 \end{bmatrix}$$

About the Author:

As a teenager, Brooke scored perfectly on the PSAT, SAT, and ACT math sections, and won awards at regional and state math competitions. As an adult, she also scored perfectly on the GRE math section and the new SAT math section, and perfectly overall on the SAT and ACT. She has been tutoring for over a decade working with classes, small groups, and individual students to increase their opportunity in the college admissions marketplace, whether that means going to Harvard or surviving community college. She's worked for or contracted with over 10 education companies, and has developed curriculum for three different education firms in everything from test prep to reading comprehension and debate. Her YouTube channel, SupertutorTV, has over 120,000 subscribers and 10,000,000 views. She also has designed a complete video based prep course for the ACT (The Best ACT Prep Course Ever) and for the SAT (The Best SAT Prep Course Ever) available at SupertutorTV.com.

In addition to her education work, Brooke is a filmmaker. Before launching SupertutorTV, Brooke worked for two seasons as a showrunner for a short form series on the Yahoo! Screen platform with Emmy nominated host Cat Deeley (So You Think You Can Dance); co-produced, wrote, and edited a documentary, Dear Albania, for public television; and field produced EPK and digital content for Stand Up to Cancer, featuring top talent from Tom Hanks to Dave Matthews Band.

Brooke graduated with honors from Stanford University with a BA in American Studies and also holds an MFA in Cinematic Arts Production from the University of Southern California. She lives in Santa Monica, California with her husband and energetic toddler.

Made in the USA
Middletown, DE
28 December 2019